Trans-jurisdictional Water Law and Governance

Governance of global water resources presents one of the most confounding challenges in contemporary natural resource governance. With considerable government, citizen and financial donor attention devoted to a range of international, transnational and domestic laws and policies aimed at protecting, managing and sustainably using fresh and coastal marine water resources, this book proposes that sustainable water outcomes require a 'trans-jurisdictional' approach to water governance.

Focusing on the concept of trans-jurisdictional water governance, the book diagnoses barriers and identifies pathways to coherent and coordinated institutional arrangements between and across different bodies of laws at local, national, regional and international levels. It includes case studies from the European Union, Australia, New Zealand, South Africa, the United States and Southeast Asia. Leading specialists offer insights into the pretence and the promise of trans-jurisdictional water governance and provide readers, including students, practitioners, policy makers and academics, with a basis for better analysing, articulating and synthesizing standards of good trans-jurisdictional water governance both in theory and in practice.

Janice Gray is Senior Lecturer in the Faculty of Law and an Affiliate of the Connected Waters Initiative Research Centre at UNSW Australia (The University of New South Wales).

Cameron Holley is Associate Professor in the Faculty of Law and Connected Waters Initiative Research Centre at UNSW Australia.

Rosemary Rayfuse is Professor of International Law in the Faculty of Law at UNSW Australia.

Earthscan Studies in Water Resource Management

For more information and to view forthcoming titles in this series, please visit the Routledge website: www.routledge.com/books/series/ECWRM/

Hydropower Development in the Mekong Region
Political, Socio-economic and Environmental Perspectives
Edited by Nathanial Matthews and Kim Geheb

Governing Trans-boundary Waters
Canada, the United States and Indigenous Communities
By Emma S. Norman

Trans-boundary Water Politics in the Developing World
By Naho Mirumachi

International Water Law and the Quest for Common Security
By Bjørn-Oliver Magsig

Water, Power and Identity
The Cultural Politics of Water in the Andes
By Rutgerd Boelens

Water and Cities in Latin America
Challenges for Sustainable Development
Edited by Ismael Aguilar-Barajas, Jürgen Mahlknecht, Jonathan Kaledin and Marianne Kjellén

Catchment and River Basin Management
Integrating Science and Governance
Edited by Laurence Smith, Keith Porter, Kevin M. Hiscock. Mary Jane Porter and David Benson

Trans-boundary Water Management and the Climate Change Debate
Anton Earle, Ana Elisa Cascão, Stina Hansson, Anders Jägerskog, Ashok Swain and Joakim Öjendal

Rules, Norms and NGO Advocacy Strategies
Hydropower Development on the Mekong River
Yumiko Yasuda

The Mekong: A Socio-legal Approach to River Basin Development
Ben Boer, Philip Hirsch, Fleur Johns, Ben Saul and Natalia Scurrah

Trans-jurisdictional Water Law and Governance
Edited by Janice Gray, Cameron Holley and Rosemary Rayfuse

Trans-jurisdictional Water Law and Governance

Edited by
Janice Gray, Cameron Holley
and Rosemary Rayfuse

Routledge
Taylor & Francis Group
LONDON AND NEW YORK

earthscan
from Routledge

First published 2016
by Routledge

2 Park Square, Milton Park, Abingdon, Oxfordshire OX14 4RN
52 Vanderbilt Avenue, New York, NY 10017

Routledge is an imprint of the Taylor & Francis Group, an informa business

First issued in paperback 2018

British Library Cataloguing-in-Publication Data
A catalogue record for this book is available from the British Library

Library of Congress Cataloging in Publication Data
Names: Gray, Janice, 1956- editor. | Holley, Cameron, editor. | Rayfuse, Rosemary Gail, editor.
Title: Trans-jurisdictional water law and governance / Edited by Janice Gray, Cameron Holley and Rosemary Rayfuse.
Description: New York : Routledge, 2016. | Includes bibliographical references and index.
Identifiers: LCCN 2015048699| ISBN 9781138928275 (hbk) | ISBN 9781315681764 (ebk)
Subjects: LCSH: Water rights. | Water--Law and legislation. | Water resources development--Law and legislation. | Conflict of laws. | Jurisdiction.
Classification: LCC K3496 .T74 2016 | DDC 346.04/691—dc23
LC record available at http://lccn.loc.gov/2015048699

ISBN: 978-1-138-92827-5 (hbk)
ISBN: 978-1-138-36404-2 (pbk)

Typeset in Bembo
by Florence Production Ltd, Stoodleigh, Devon

Contents

Contributors

Lauren Butterly, Associate Lecturer, ANU College of Law, Australian National University, Australia and Associate, Indigenous Law Centre, UNSW Australia

Meg Casey, independent water law consultant, Montana, USA

Cristy Clark, Lecturer, School of Law and Justice, Southern Cross University, Australia

Anita Foerster, Research Fellow, Faculty of Law, University of Tasmania, Australia

Alex Gardner, Professor of Law, School of Law, The University of Western Australia and Adjunct Professor, ANU College of Law, Australian National University, Australia

Lee Godden, Professor of Law, Centre for Resources, Energy and Environmental Law, Melbourne Law School, The University of Melbourne, Australia

Janice Gray, Senior Lecturer, Faculty of Law, UNSW Australia, Australia; Visiting Scholar, Centre for Socio-Legal Studies, University of Oxford, UK; Visiting Scholar, Queen Mary University of London, UK; and Affiliate, Connected Waters Initiative Research Centre, UNSW, Australia

Taylor W. Henshaw, Barrister and Solicitor, Vancouver, BC, Canada and Associate, Global Transboundary International Waters Governance Initiative, The University of British Columbia, Canada

Cameron Holley, Associate Professor, Faculty of Law, and Member of the Connected Waters Initiative Research Centre, UNSW Australia, Australia

Amanda Kennedy, Associate Professor, School of Law and Deputy Director, The Australian Centre for Agriculture and Law, The University of New England, Australia

Michael Kidd, Professor, University of Kwa-Zulu Natal School of Law, South Africa

Paul Martin, Professor, School of Law and Director, The Australian Centre for Agriculture and Law, The University of New England, Australia

Jan McDonald, Professor of Environmental Law, Faculty of Law, and Centre for Marine Socio-ecology, University of Tasmania, Australia

Rebecca Nelson, Senior Lecturer, Melbourne Law School, The University of Melbourne, Australia

Richard Kyle Paisley, Senior Research Associate, Institute of Asian Research, The University of British Columbia, Canada and Director of the Global Transboundary International Waters Governance Initiative

Rosemary Rayfuse, Professor of International Law, Faculty of Law, UNSW Australia, Australia; Conjoint Professor Lund University, Sweden; and Visiting Professor, University of Gothenburg, Sweden

Jacinta Ruru, Professor, Faculty of Law, University of Otago, New Zealand

Poh Ling Tan, IWC Professor for Water Law and Governance, Griffith Law School, Griffith University, Australia

Erika Techera, Professor and Dean of Law, School of Law, The University of Western Australia, Australia

Barton H. "Buzz" Thompson Jr, Professor of Law, Stanford University; Perry L McCarty Director and Senior Fellow, Stanford Woods Institute for the Environment; and Senior Fellow, The Freeman Spogli Institute for International Studies, Stanford University, USA

Marleen van Rijswick, Professor of Law and Director of the Utrecht Centre for Water, Oceans and Sustainability Law, Utrecht University, The Netherlands; Visiting Professor of European Environmental Law, University Panthéon-Assas, France

Marie Waschka, independent water consultant, Brisbane, Australia

Patrick Weiler, Barrister and Solicitor, Vancouver, BC, Canada; former officer on the United Nations Development Programme/Global Environment Facility (GEF) International Waters Learning Exchange and Resource Network (IW:LEARN) and the Good Practices and Portfolio Learning in GEF Trans-boundary Freshwater and Marine Legal and Institutional Frameworks projects

Preface

On 25 September 2015 the United Nations General Assembly adopted Resolution 70/1 entitled 'Transforming our world: the 2030 Agenda for Sustainable Development'. The resolution sets out 17 Sustainable Development Goals (SDGs) intended to 'stimulate action over the next 15 years in areas of critical importance for humanity and the planet'. Specifically, Goal 6 aims to ensure, by 2030, 'the availability and sustainable management of water and sanitation for all'.

The scale of the task is enormous. Only about 2.5% of all water on earth is freshwater, and most of that is locked up in ice-caps, glaciers and permafrost. A mere 0.3% is accessible as surface water. According to the United Nations Development Program, approximately 1.2 billion people, or almost one-fifth of the world's population, live in areas of physical scarcity, and 500 million people are approaching this situation. Another 1.6 billion people, or almost one quarter of the world's population, face economic water shortage, where countries lack the necessary infrastructure to take water from rivers and aquifers. Based on current and projected patterns of consumption, use and abuse, it is estimated that two-thirds of the world's population could be living in water-stressed countries by 2025. In 2015 the World Economic Forum listed human-induced and climate change-exacerbated water-related crises as the top risk globally, in terms of impact. With an ever-increasing global population and decreasing availability of freshwater supplies, it is clear that new and more effective governance approaches are needed to ensure the adequacy, accessibility and quality of this most essential resource.

Designing successful governance approaches is not easy. Despite decades of reforms, sustainable water management continues to be an elusive goal; problems such as pollution of terrestrial and coastal water resources from diffuse and cumulative sources, conflicts over shared water resources, and water scarcity as a result of increased demands from energy, agricultural, urban and cultural uses persist. Each of these problems involves a complex array of causes, conflicts and actors, often operating at different levels (local, state, national, international) whose agendas are either not linked, are at cross-purposes, or are in conflict with one another. Resolving these problems may be assisted by what can best be described as 'trans-jurisdictional' approaches.

The aim of this book is to examine existing and emerging mechanisms for trans-jurisdictional water governance in order to provide readers with a basis for analysing the challenges of, and possible approaches to, trans-jurisdictional water governance, and for better articulating and synthesizing standards of good trans-jurisdictional water governance both in theory and in practice. Leading specialists were asked to reflect on a range of issues, including the reasons for the continuing intractability of effective trans-jurisdictional water governance, the key challenges to its successful implementation, the lessons that might be gleaned from existing trans-jurisdictional water governance practices, and the broader import of the concept of trans-jurisdictionality for the theory and practice of water governance in general. Hard choices were made to confine the book to case studies from a limited number of jurisdictions (Australia, the European Union, the Mekong Basin, New Zealand, South Africa and the United States). However, in doing so it was intended that these case studies would go beyond the usual trans-boundary, international, or comparative approaches to offer a wider scope that considers international trends as well as national and local developments in trans-jurisdictional water governance. While, admittedly, the literature on water law and governance is considerable, the ongoing intransigence of the problems and challenges involved ensures the continuing need for scholarly analysis of new and emerging issues. In this respect it is our hope that this book offers useful and new insights into both the pretence and the promise of trans-jurisdictional water governance.

The papers on which the chapters of this book are based were presented at a workshop held at the Faculty of Law, UNSW Australia in December 2013, at which the authors and other workshop participants discussed the issues. What became clear through the workshop papers and the lively discussions that ensued was that the term 'trans-jurisdictional' meant different things to different people. Nevertheless, everyone could recognize elements of trans-jurisdictional law and governance at different sites and in different contexts. The concept thus served as an essential linking point between the various contributions. Following the workshop, speakers were invited to revise their papers for submission as chapters in this book and a number of other contributions were received from those invitees who had been unable to make the journey to Sydney to attend the workshop.

As in any project of this nature, many thanks are due. We are grateful for the funding received from UNSW Law, which supported the workshop and the editorial work on the book. In particular, we would like to thank Jennifer Wong for her editorial assistance, Antonia Ross and Bonnie Perris for their research assistance and Roisin Dwyer of the Faculty of Law, whose behind-the-scenes contributions in assisting with the workshop organization were invaluable. Cameron Holley also acknowledges funding from the Australian Research Council's Discovery Early Career Researcher Award funding scheme (DE140101216), which has supported his work on this book.

Preparation of this volume has, itself, been a trans-jurisdictional affair, with work being carried out by the editors at their home institution, UNSW

Australia, and at other institutions with which they have visiting appointments. In this respect, Janice Gray acknowledges the intellectual stimulation she has enjoyed with colleagues as a result of her position as visiting scholar at both the Centre for Socio-Legal Studies, University of Oxford and the Faculty of Law, Queen Mary University of London, as well as her appreciation of the assistance and facilities with which she was provided at those institutions. It was in Oxford and London that much of her work associated with this volume was undertaken. Rosemary Rayfuse would similarly like to acknowledge the intellectual companionship and stimulating work environments she enjoys, not only at UNSW Australia, but also at Lund University where she holds a conjoint professorship and at the University of Gothenburg where she holds a visiting professorship. A great deal of her work on the preparation of this volume was carried out at both of her Swedish 'homes'.

Of course, this book would not have been possible without the commitment and hard work of the authors, and we offer our thanks to each of them. In addition, we express our appreciation to the three anonymous reviewers of our book proposal for their thoughtful comments. Finally, particular thanks must go to Routledge Earthscan for its support for this book and for the professionalism, helpfulness and understanding of its staff, particularly Ashley Wright and Tim Hardwick.

Janice Gray, Cameron Holley and Rosemary Rayfuse
Sydney, Oxford/London, Lund/Gothenburg
November 2015

Part I

Introduction

1 The challenge of trans-jurisdictional water law and governance

Janice Gray, Cameron Holley and Rosemary Rayfuse

Introduction

It may be trite, but in a book about water governance it is useful to begin with the observation that water is critically important to human existence. Not only is it necessary to life itself, but as the United Nations World Water Development Report 2015 notes:

> [W]ater is at the core of sustainable development. Water resources, and the range of services they provide, underpin poverty reduction, economic growth and environmental sustainability. From food and energy security to human and environmental health, water contributes to improvements in social well-being and inclusive growth, affecting the livelihoods of billions.
>
> (WWAP, 2015, p. 2)

Yet, as Irina Bokova, Director-General of UNESCO puts it: 'development also places considerable pressure on water resources – agriculture, energy and industry all have impacts on the use and governance of water' (WWAP, 2015, pv).

Recognized since Roman times as a quintessential common resource, effective governance and management of water resources is central to avoiding the tragedy of overuse and depletion exemplified by Hardin's notion of the 'tragedy of the commons' (Hardin, 1968). It can thus come as no surprise that throughout history a good deal of government, citizen and financial donor attention has been devoted to developing a plethora of international and national laws and policies aimed at protecting, managing and ensuring the sustainable use of fresh and coastal marine water resources. Water management approaches that have been adopted include market-based instruments, command and control regulation, integrated water resource planning, integrated coastal zone management, international conventions and agreements, as well as polycentric governance approaches (Pahl-Wostl et.al., 2013).

Nevertheless, despite decades, indeed centuries, of reforms, sustainable water management continues to be an elusive goal (NCCARF, 2012). Problems persist, such as pollution of terrestrial and coastal water resources from diffuse and cumulative sources, conflicts over shared water resources, and water scarcity (or allocation prioritization) as a result of increased demands from energy, agricultural, urban and cultural uses. Each of these problems involves a complex array of causes, conflicts and actors, often operating at different levels (local, state, national, regional, international), whose agendas are either not linked, are at cross-purposes, or are in conflict with one another (Folke, 2007). In short, the intractability of sustainable water governance arises largely from its 'trans-jurisdictional' nature, traversing as it does, physical, administrative, discipline, social and political boundaries at all levels. Achieving sustainable water governance thus requires a 'trans-jurisdictional' approach.

In this chapter we outline the concept of trans-jurisdictionality and its application in the context of water law and governance. We begin with a discussion of the concept of trans-jurisdictionality and then turn to a discussion of the challenges of such an approach in the context of sustainable water governance. The purpose of this discussion is to provide a framework for the examination of trans-jurisdictional water governance issues examined in subsequent chapters in this book and the conclusions that will be drawn in the final chapter.

The meaning of 'trans-jurisdictionality'

It is perhaps easier to explain what 'trans-jurisdictionality' is not, rather than what it is. The term is most commonly used to refer to situations in which water physically crosses some sort of sub-national or national border such as where a river flows from or through the territory of opposite or adjacent states (see, e.g. Shi, 2007). In such cases, the characteristic actually being invoked is that of 'trans-boundary' rather than 'trans-jurisdictional'. While it is true that trans-boundary issues are trans-jurisdictional in nature, our conception of trans-jurisdictionality is broader than trans-boundary, encompassing water problems that cross administrative – and even social – borders at all levels, not only across sovereign state boundaries. In this respect, the term trans-jurisdictional expressly acknowledges water's disrespect for *all* human boundaries: borders between nation states, within nation states, between different bodies of law, between public agencies, between public, private and non-government stakeholders, borders across formal law and informal forms of social control, as well as interactions within and between fresh and saltwater commons. Trans-boundary issues are, therefore, merely a subset of trans-jurisdictional ones.

The trans-jurisdictional nature of water can be characterized along horizontal and vertical axes – the former requiring cooperation between actors at the same level to deal with cumulative impacts from multiple stressors and the latter requiring cooperation between actors at different levels. This characterization assists in delineating some of the key spaces in which legal, regulatory and governance tools need to be mobilized and applied if sustainable water outcomes are to be achieved and the seemingly intractable problem of effective water governance is to be overcome.

Thus, the term 'trans-jurisdictional water governance' is used here to refer to legal and policy arrangements for the conservation, management and use of water resources that cross either functional or geographic administrative borders and either seek to reinforce human-imposed borders, divisions and boundaries or seek to achieve coherence and institutional coordination both between and across different bodies of laws at the local, national, regional and international levels. 'Water governance', for these purposes, is intended simply to refer to 'the intentional shaping of the flow of events' so as to realize desired water outcomes (Parker and Braithwaite, 2003 p. 119; Conca, 2006), as distinct from the concept of 'government' which is defined as a political authority/state auspice (Holley *et al.*, 2012; Norman *et al.*, 2015 p. 3). Our particular concern with 'trans-jurisdictional water governance' accordingly focuses on legal and policy responses to water resources that cross administrative borders, and seek to achieve coherence, institutional coordination and harmony between different bodies of laws at international, regional, national or local levels; between different government and non-government auspices; and between water law and other areas of law (e.g. climate change, biodiversity and land regimes, and/or the law pertaining to the energy and food sectors) (Moynihan, 2014).

Admittedly, this broad concept of trans-jurisdictional water governance may gloss over some theoretical and practical differences between international, national and local tiers. With these limitations in mind, and paying careful consideration to these differences, this book therefore merely utilizes the term 'trans-jurisdictional' as a means of orienting water governance scholars around a structured, albeit limited, understanding and conceptual framework (Holley *et al.*, 2012). The limitations of the framework undoubtedly invite digression and alternative formulations or definitions (as we explore below by developing tentative typologies of trans-jurisdictional water governance). Nevertheless, it is hoped that this framework will enable comparative lessons and insights to be drawn across international, national, regional and local efforts as to the ways in which trans-jurisdictional water resources might be more effectively, efficiently and legitimately governed.

The challenge of trans-jurisdictional water governance

The concept of trans-jurisdictionality introduces complex, multi-dimensional, multi-scalar problems which require complex solutions (Norman *et al.*, 2015). Thus, achieving effective, efficient and legitimate trans-jurisdictional water governance remains a significant challenge at and between the international, regional, national and local level. This section provides an overview of these challenges.

Challenges at the international level

Internationally, state sovereignty remains a fundamental governance challenge for numerous water resources, including approximately 250 trans-boundary river basins, 200 trans-boundary aquifers, various trans-boundary lakes and the oceans, often hindering the development of cooperative forms of governance which engage with, and accommodate the impacts of, decisions in one state on another (Gupta, 2013). This remains the position despite the continual refinement of water rules and practice (see, e.g. the Helsinki Rules, 1966; Berlin Rules, 2004; Convention on the Law of Non-Navigational Uses of International Watercourses; ILC Articles on Transboudary Aquifers, 2008; and other related regional conventions/agreements such as the UNECE Convention on the Protection and Use of Trans-boundary Watercourses and International Lakes) which has sought to ensure the equitable use of international watercourses and their waters to deliver better protection, preservation and management.

Scholars have explored a number of concepts aimed at addressing these challenges including, for example, cooperative sovereignty (Moynihan, 2014), Integrated Water Resources Management (IWRM) (Zeitoun, 2013), and regional cooperation models and agreements (see Eckstein, 2002; Salman, 2007; McIntyre, 2007, 2010; Biswas, 2004; Lim, 2014; Moynihan, 2014) in order to identify the elements necessary to achieve successful trans-jurisdictional water governance. The elements identified have included the need for 'shared visions' among relevant organizations and communities, the importance of 'communities of interests', notions of 'shared sovereignty' and 'boundary spanning' (Bressers and Lulofs, 2010), as well as early and strong engagement of stakeholders, local capacity building and policy development (Servos *et al.*, 2013). Nevertheless, unresolved issues remain. For example, criticisms of IWRM are prolific and conflicts between upper and lower riparian states remain common, fed by geopolitical stakes, a privileging of efficiency and securitization, the dominance of 'basin hegemons', and the evasion of politics and political realities (Zeitoun, 2013). Expanding pressure on international water governance from competing uses and users of water has also seen a host of new actors (such as those in the energy sectors) shaping international water management, including the injection of new ideas on 'horizontal alignment', 'nexus' management and more pluralistic basin governance (Zeitoun, 2013).

Clearly, specific institutional architectures remain necessary for the achievement of collaborative solutions to common problems in trans-boundary and trans-jurisdictional water management, particularly when questions of human rights, indigenous rights and climate change are added to the equation.

In the human rights context, for example, the increasing intersection of international water law and international human rights law has opened up new points of interaction with international water management. The UN Committee on Economic, Social and Cultural Rights' General Comment No. 15 on the Right to Water (UNCESCR, 2002) paved the way for the UN General Assembly to adopt a declaration on the Human Right to Water and Sanitation (United Nations General Assembly, 2010) and the UN Human Rights Council resolution on the human right to access safe drinking water and sanitation (United Nations Human Rights Council, 2010), providing strong political recognition to these rights and creating enforceable responsibilities that can empower the vulnerable, including indigenous peoples (see, e.g. United Nations Declaration on the Rights of Indigenous Peoples (United Nations General Assembly, 2007). Furthermore, progress in the field of water and sanitation supply and management was influenced by the Millennium Development Goals which sought to 'halve, by 2015, the proportion of the population without sustainable access to safe drinking water and basic sanitation' (UN, 2002). While the goal was largely met, in 2015 some 2.5 billion people in developing countries still face physical or economic water shortage (UNDP, 2015). These statistics suggest that much work remains to be done to understand and address the interaction and coordination of human rights regimes and other areas of water law. This includes managing trade-offs between rights and other water interests (such as the environment), as well as coordinating implementation and developing win-win solutions (Gupta, 2013; Pahl-Wolst *et al.*, 2013; Hearns *et al.*, 2014).

Similarly, despite studies of indigenous engagement in water governance (Von der Porten and de Loe, 2013), there are numerous gaps and unresolved issues including how best to acknowledge and address real-world Indigenous goals related to self-determination and nationhood in different contexts (Von der Porten and de Loe, 2013).

The influence of climate change is also expected to compound the trans-jurisdictional governance challenges in the coming decades, requiring more adaptive and cooperative approaches that tap into local knowledge and can deal with water quality and availability concerns (Zeitoun, 2013). In short, the challenge of trans-boundary and trans-jurisdictional water governance at the international level remains a significant one (Gupta, 2013; Tujchneider *et al.*, 2013; Hearns *et al.*, 2014).

Challenges at the regional level

At the regional level, the European Union has been a prominent innovator, developing mechanisms that encourage cooperation between nation states and civil society under its Water Framework Directive and the Marine Strategy Framework Directive (Trubeck and Trubeck, 2006; Keessen *et al.*, 2008). Nevertheless, while many European Union states have delivered on these efforts, a number of studies have highlighted the ongoing challenges of engaging non-government actors and moving beyond planning to implementation. In essence, these studies raise more questions than answers (Pahl Wostl *et al.*, 2010), demonstrating how regional level obstacles still pose significant challenges for the realization of trans-jurisdictional governance.

Challenges at the national level

Similarly at the national level, federal systems, in particular, confront numerous problems facilitating cooperation and power sharing across subnational borders.

Identifying successful institutional designs and harmonizing legal and policy responses between agency 'silos', across local political boundaries and between multiple public and private stakeholders are ongoing challenges. Certainly, concepts such as sustainable development, IRWM, integrated coastal zone management and similar integrated planning approaches have come to represent dominant policy narratives. However, many (if not all) of these approaches continue to come under strong criticism from scholars for their poor record of implementation, their heavy emphasis on technocratic solutions, the exclusion of stakeholders and indigenous peoples, and their failure to achieve integration with energy and other relevant policies (Araral and Wang, 2013; Mukhtarov and Gerlak, 2014). While a number of interesting studies have explored these concepts and concerns, most of this work has been limited to specific non-legal disciplines (e.g. political science, institutions and economics) and often remains at an experiential, rather than at a systematic empirical or coherent theoretical level (Aaral and Wang, 2013).

The challenge of trust, political will and financial commitment

Regardless of the levels of government or the identity of the actors involved, achieving effective trans-jurisdictional water governance requires significant trust, as well as political and financial commitment on the part of law and policy makers and their non-government partners. As a first step, trust is an important factor underpinning the development of effective trans-jurisdictional governance, as is a willingness among different actors to compromise in the interests of workable processes, ease of implementation and overall benefit. Yet trust and willingness to compromise are not givens and may take time to develop. Particularly, where conflict forms the backdrop to relations, such as that exhibited in the long history of violence and underprivilege associated with indigenous–non-indigenous relations in many countries, it may be more difficult to reach agreement on and implement governance approaches in a trans-jurisdictional manner (Bernauer, 2003). Moreover, its effective implementation involves a translation process whereby objectives and goals find their way into actions and decisions. This process usually requires the allocation and availability of adequate capital, including financial and social. Thus, trans-jurisdictional governance can be an expensive approach.

The challenge of privileging statist institutions

In addition, states commonly assume the mantle of organizer or decision maker in trans-jurisidictional water governance because they have been fashioned to do so by history, constitutions and other instruments. However, the focus on the state and the importance of the state's role as the creator of, and the filter for, knowledge can have the effect of marginalizing and trivializing knowledge or ideas from other non-state actors and sources. This has historically been evident in relation to indigenous water knowledge, which has conventionally been excised from the dominant water narrative and which has been overlooked in favour of 'official' knowledge. However, this situation is not limited to indigenous knowledge. Institutional knowledge, or knowledge 'owned' by specific government agencies or departments, may be privileged over that held by, local residents, historians, conservationists, water users and other members of society, who are often the custodians of water stories and water histories which can inform governance. Thus, effective forms of knowledge incorporation, including mechanisms to ensure effective and meaningful public participation in decision-making processes, are needed to ensure the development and implementation of trans-jurisdictional water law and governance (Holley and Sinclair, 2013).

Moreover, the institutions supporting trans-jurisdictional water governance require states to bestow sufficient authority to carry out their functions (such as the authority to carry out water basin management). If they do not, they are at risk of becoming mere coordination entities, which although important, will be limited in capacity (Louka, 2008). A major barrier to empowering such institutions is that states and their various agencies and providers often prioritize their own interests above any joint, shared or trans-jurisdictional agenda or mechanism (including its compliance mechanisms). Indeed, state interests and parochialism often constrain the voluntary nature of compliance with standards that form part of trans-jurisdictional governance arrangements (Connell, 2007, p, 125; Louka, 2008, p. 395). In a competition between an independent, trans-jurisdictional governance entity or institution which takes account of multiple state interests both within and beyond human-imposed boundaries, and the preservation of state autonomy itself, evidence suggests that state autonomy commonly prevails (Louka, 2008, p. 395).

Finally, conventional sectoral, or boundary-driven, compartmentalized approaches to legal and jurisdictional boundaries between, for example, ground- and surface water, local and national authority, urban and rural, water law and energy, health or land law, also pose challenges. While sectoral or boundary-driven approaches provide systematization and some semblance of order, they also lead to fragmentation (see, e.g. Louka, 2008). The challenge in overcoming the complexities and deficiencies of legal, policy and practical fragmentation in the water context is at the root of trans-jurisdictional approaches.

Addressing the challenge of trans-jurisdictional water governance

It will be clear from the foregoing that considerable research attention is needed to provide a comprehensive, empirical, comparative and theoretical understanding of the nature and challenges of trans-jurisdictional water governance. In this respect, four questions can be framed:

- Why does trans-jurisdictional water governance remain such an intractable problem?

- What are the key challenges and confounders to successful trans-jurisdictional water law and policy?
- What opportunities, successes and learning experiences can be gleaned from existing trans-jurisdictional governance practices?
- What does the 'trans-jurisdictionality' of water suggest for the theory and practice of future water governance research?

Existing legal and governance theory does not yet fully address these questions. Rather, in spite of the increasingly vast and dense scholarship on water law and governance, established knowledge on trans-jurisdictional water governance is limited, fragmented and often 'level' specific (international, national, regional, local), providing poor guidance for the future (Öjendal *et al.*, 2010). Little has been assimilated in a systematic way either within or between levels, requiring policy makers and scholars to consult a variety of sources to understand key concepts of trans-jurisdictional governance (Earle *et al.*, 2012). However, as cooperation, integration and coordination become increasingly central to efforts in water governance, it is vital that we learn from trans-jurisdictional experiences at all levels, to better articulate and synthesize standards of good trans-jurisdictional water practice and theory (Öjendal *et al.*, 2010; NCCARF, 2012; Pahl-Wostl, 2013; Lim, 2014).

The chapters in this book seek to contribute to this research agenda by examining existing and emerging mechanisms for trans-jurisdictional water governance in order to assess the efficacy of this approach in contributing to the more sustainable water governance outcomes. The chapters in Part 2 introduce some theoretical underpinnings and discuss selected international approaches and developments in trans-jurisdictional water governance.

In Chapter 2, Godden canvasses the rationale for an 'architecture', or theoretical framework, for trans-jurisdictional water governance, and sketches out some models which, although nascent in form, have emerged in different water law and governance settings. While acknowledging that existing trends towards collaborative and decentralized models could become part of an architecture for trans-jurisdictional water governance, she cautions that it remains unclear what these models can, or will, contribute to effective water governance. Noting that law's attempts to engage with multi-scalar and long-term adaptive planning considerations are nascent, and that the challenges of integrating trans-temporal and trans-spatial factors into the trans-jurisdictional water governance frame remain, she concludes by urging caution about embracing new theories and tools. Rather, she encourages serious and genuine interrogation of the expectations of water governance so that it is possible to identify whether trans-jurisdictional water governance is the best response.

Following Godden's chapter, there is a shift towards discussion of international approaches to trans-jurisdictional water governance. Paisley *et al.* lead the way, with Chapter 3 reviewing trans-boundary international waters governance and the carefully orchestrated model employed by the southern Mekong Basin states. In particular, they highlight the role of the 1995 Mekong

Agreement, cooperation between states and the importance of understanding that the best alternative to a negotiated agreement may be less palatable than what can be gained through agreement. Chapter 4 (van Rijswick) follows, with discussion and analysis of water governance in the European Union where trans-jurisdictional water issues, including 'trans-boundary water governance, human rights and water; water law and consumer law; water and spatial planning; water and agriculture; water and the use of medicines; water and pesticides; water and energy; freshwater and marine protection; and water law and climate change', have been on the agenda for some time. In particular, Chapter 4 demonstrates how external integration is organized, focusing on trans-boundary water management, freshwater supply, water and conservation, and water and internal market legislation. In Chapter 5, Butterly and Techera examine the nexus between protection of marine spaces and indigenous issues arguing among other things for the merits of the World Indigenous Network of Land and Sea Managers (WIN) as a tool of trans-jurisdictional governance.

Part 3 consists of chapters analysing key national attempts at trans-jurisdictional governance, including experiences from the United States of America, New Zealand, South Africa and Australia. In Chapter 6, Martin and Kennedy employ theoretical approaches largely drawn from social and managerial research to discuss intra-national water rivalries, conflicts and collaboration in Australia, South Africa and the Mekong Basin states. In Chapter 7, Thompson focuses on three main types of water fragmentation which exist in the USA, arguing that they obstruct efficient water management, generate distributional disputes, encourage protectionism and undercut conjunctive management of groundwater and surface water. Thompson argues in favour of 'hard restructuring' and top-down defragmentation as particularly effective tools in addressing siloed or fragmented governance, but observes that collective action is a major barrier to such solutions. In Chapter 8 Nelson and Casey also focus on the USA, analysing its groundwater governance arrangements. They acknowledge the 'out of sight – out of mind' approach that so commonly accompanies groundwater, and observe that although crossing jurisdictional boundaries, groundwater has often been left out of traditional legal arrangements for apportioning and managing water across these boundaries. In the spirit of this book, they extend their analysis and discussion beyond interstate compacts and other traditional mechanisms to incorporate unconventional modes of governing trans-jurisdictional groundwater, including approaches 'based on collaborative science, planning and centralised regulatory authorities'.

In the next chapter in this Part, Chapter 9, the focus shifts continents as Kidd analyses examples of trans-jurisdictional water governance in South Africa, predominantly with reference to national, provincial and local spheres and against the backdrop of South Africa's developing nation status. He observes that although the least well resourced, the local sphere has responsibility for several critical aspects of water governance and he argues that the three spheres of government are 'leaving the ball to each other with the result that the ball is dropped'. Ruru, on the other hand, takes up water governance issues in New Zealand, which is a unitary state, in Chapter 10. She explores the intersection of Maori water law, state legislation and related case law, and the common law. She sees reconciliation and public participation as key methods in the quest to govern water effectively across jurisdictions.

Part 4 of the book consists of case studies examining a number of emerging trans-jurisdictional governance challenges from selected national and local jurisdictions. In Chapter 11, Waschka and Gardner examine diffuse source pollution in the Great Barrier Reef, Australia, as 'unfinished business'. They observe that diffuse source pollution is largely unregulated in Australia and that, despite decades of cooperation between the relevant state and federal governments, the problem has not yet been effectively addressed. Their case study is followed by Gray's examination, in Chapter 12, of the possible applicability of trans-jurisdictional approaches to aspects of water governance in the unconventional gas (particularly coal seam gas) context. Observing that trans-jurisdictional governance is a two-way street, such that water governance may impact on the governance of other sectors just as much as it impacts on water, she expresses reservations about whether trans-jurisdictional water governance alone will ultimately prove to be the best solution, despite acknowledging its many (interim) benefits. Meanwhile the case study by McDonald and Foerster in Chapter 13 examines the governance of coastal wetlands, noting that decisions over their use, protection, destruction and management potentially span multiple tiers of government, involve a range of public and private actors, and engage a number of sectors and agencies. With this in mind they argue for a more integrated governance approach and examine the role that trans-jurisdictional governance could play. In Chapter 14 Tan focuses on what the Aboriginal struggle for water in the Australian state of Queensland means in the context of the management of near-natural (wild) rivers, and interrogates what that struggle offers to the development of trans-jurisdictional water governance. In Chapter 15, the final chapter in this Part, the focus shifts back to South Africa where Clark introduces readers to the idea of a constitutionally enshrined right to water and the difficulties of implementing that right in the context of a decentralized system of delivery which in theory, at least, should serve to increase community participation.

In Part 5 we provide a conclusion in which we attempt to synthesize key generalizable lessons and insights that might be learned from the other chapters in the book. However, before doing so, and in order to provide the reader with a more concrete framework in which to appreciate the chapters, it is useful here to outline the range of possible understandings of trans-jurisdictional water governance that are explored by the authors in this book.

Types of trans-jurisdictional water governance

The contributions to this collection employ a range of theoretical and conceptual approaches and identify a range of different trans-jurisdictional water governance 'types', including cooperative, cultural, centralized, non-government-sanctioned, and legal and administratively layered models.

That these different models of trans-jurisdictional water governance rarely operate separately and independently is clear from Thompson's chapter (Ch. 7), where he demonstrates how overcoming various forms of water fragmentation, such as geographical, hydrological and functional fragmentation, may involve engagement with a range of trans-jurisdictional water governance types.

Cooperative models

Paisley *et al.* (Ch. 3) identify what might be termed cooperative trans-jurisdictional water governance in the discussion of the institutionalization of governance arrangements in the Mekong River Basin. Under this model the Mekong River Basin states developed a governance framework which, although driven to some extent by national interests, employed cooperation as a foundational element, becoming its defining feature. Yet, such a cooperative typology is dependent on trust and relationship-building, meaning that it is difficult to employ this type of trans-jurisdictional governance without first having engaged in setting, and then creating, the conditions which nurture such a relationship and allow it to flourish. Those conditions are dependent, among other things, on respectful listening, compromise, sensitive decision making, a willingness to share and an awareness of the *realpolitik*. Such a typology is also commonly dependent on the neutral support of outside entities and actors to assist if/when negotiations break down. As this chapter demonstrates, this type of governance has been shown to work more effectively if the parties have a strong understanding of their legal rights.

Another example of cooperative trans-jurisdictional water governance may be seen in the WIN network (Butterly and Techera, Ch. 5). Like the Mekong case, WIN conceptually embeds elements such as trust, dialogue and sharing as preconditions to the development of an effective governance model that permits enhanced water management across a range of borders and jurisdictions. WIN also serves to demonstrate how different trans-jurisdictional water governance models may interact, in that WIN is a governance experiment that not only embraces the cooperative model but also employs aspects of the administratively layered model (international, national, local and regional boundaries) and the cultural model.

Cultural

Another model of trans-jurisdictional water governance emerges as a result of the significance of different cultural understandings. The cultural classification acknowledges the impact of the interaction of competing cultures – cultures which have been responsible for the development of often vastly different sets of laws, commonly within the same country, such as Maori law, legislation and the common law in New Zealand, which is explored by Ruru in Chapter 10. This type of trans-jurisdictional water governance emphasizes the importance of concepts such as place (especially local place), physicality, spirituality and dialogue to indigenous understandings (see Ruru, Ch. 10; Butterly and Techera, Ch. 5). It is a model that has grown out of the tensions, in colonial countries, between customary indigenous and imposed colonial laws and it aims to develop culturally more inclusive forms of governance, often relying heavily on knowledge exchanges, the significance of respect, the role and capacity of indigenous languages to capture special relationships with water, and other tools which potentially empower indigenous communities such as indigenous tribunals, treaties, special legislation and other agreements. As demonstrated in the chapters by Ruru and Butterly and Techera, this type of governance often also interacts with the legal model discussed below.

Centralized

The centralized model of trans-jurisdictional water governance offers yet another take on governance. As the name suggests, it emphasizes the role of centralized regulatory authorities and agencies in designing and implementing trans-jurisdictional water governance. However, as Nelson and Casey demonstrate (Ch. 8), it may simultaneously operate in conjunction with very different models of trans-jurisdictional water governance that are unconventional and innovative but which rely on non-regulatory governance arrangements.

Non-government-sanctioned

In contrast to the centralized model, the non-government-sanctioned typology of trans-jurisdictional water governance represents a form of governance that is directed by forces outside government (see Norman *et al.*, 2015, p. 3). Under this model, government does not solely direct outcomes. Rather, governance is directed by non-government actors who commonly have a different agenda and set of goals to that of government, but who nevertheless seek to direct behaviours between and across jurisdictions and borders so as to achieve their preferred outcomes. The grassroots-level governance engaged in by resistance groups in the coal seam gas-mining space, which influences and shapes water behaviours and subsequently outcomes, is an example of such governance (Gray, Ch. 12).

Legal

Previous models notwithstanding, the legal classification remains the most traditional and possibly the most well-recognized method of trans-jurisdictional water governance. It is dependent on legal tools such as legislation and the common law engaging with the effects of decisions made in one delineated space or sector on another space or sector. This refers, for example, to incorporation into national law of obligations adopted at the international level, or where water statutes specifically refer to governance in other sectors such as energy, human health, land or other natural resources. This type of trans-jurisdictional water governance operates both vertically and horizontally, employing international, national, regional and local divisions as the lens through which legal and administrative matters are conceived and considered (Van Rijswick, Ch. 4).

Conclusion

In conclusion, it is to be hoped that this edited collection, dependent as it is on unsettled definitions and meanings, multiple theoretical positions and offering far-reaching, diverse, pluralistic and multi-dimensional expositions and case studies of trans-jurisdictional water governance, opens up new possibilities for engagement, imagination, understanding and debate. The old approaches to water governance have not worked. If new ways of seeing old problems and creative alternatives which refashion governance approaches are opened up, novel and effective solutions may also become available.

References

Araral, E. and Wang, Y. (2013) 'Water governance 2.0: A review and second generation research agenda', *Water Resources Management*, 27, 11, 3945–57.

Berlin Rules on Water Resources (Berlin Rules) (adopted 21 August 2004). Seventy-First Report of the International Law Association, 2004.

Bernauer, T. (2003) 'Trans-boundary freshwater management: When and why does it succeed or fail', in S. Castelein (ed.), *From Conflict to Co-Operation in International Water Resources Management: Challenges and Opportunities*, UNESCO-IHE Institute for Water Education Conference, Delft, The Netherlands, pp. 545–83, http://unesdoc.unesco.org/images/0013/001354/135494e.pdf, accessed 30 August 2015.

Biswas, A. (2004) 'Integrated water resources management: A reassessment', *Water International*, 29, 1, pp. 248–56.

Braithwaite, J. (2008) *Regulatory Capitalism*, Edward Elgar, Cheltenham, UK.

Bressers, H. and Lulofs, K. (2010) *Governance and Complexity in Water Management: Creating Cooperation Through Boundary Spanning Strategies*, Edward Elgar, Cheltenham, UK.

Conca, K. (2006) *Governing Water*, MIT Press, Cambridge, MA.

Connell, D. (2007) *Water Politics in the Murray Darling Basin*, Federation Press, Sydney, Australia.

Convention on the Law of Non-Navigational Uses of International Watercourses, 21 May 1997, into force 19 May 2014, 36 *International Legal Materials* 700 (1997).

Convention on the Protection and Use of Trans-boundary Watercourses and International Lakes, 17 March 1992, into force 6 October 1996, 1936 UNTS 269.

Earle, A., Jägerskog, A. and Öjendal, J. (eds) (2010) *Transboundary Water Management: Principles and Practice*, Routledge Earthscan, London, p. 59.

Eckstein, G. E. (2002) 'Development of international water law and the UN Watercourse Convention', in A. Turton and R. Henwood (eds), *Hydropolitics in the Developing World: A Southern African Perspective*, African Water Issues Research Unit, Lynwood, South Africa.

Fisher, L. (2013) 'Environmental law as "hot" law', *Journal of Environmental Law*, 25, 3, 347–58.

Folke, C. (2007) 'Social ecological systems and adaptive governance of the commons', *Ecological Research*, 22, 14–15.

Gupta, J. (2013) 'Global water governance', in R. Falkner (ed.) *The Handbook of Global Climate and Environment Policy*, Wiley-Blackwell, Milton, Australia.

Gray, J. (2006) 'Legal approaches to the ownership, management and regulation of water from riparian rights to commodification', *Transforming Cultures Ejournal*, 1, 2, 64–96.

Hardin, G. (1968) 'The tragedy of the commons', *Science*, 162, 3859, 1243–8.

Hearns, G. S., Henshaw, T. W. and Paisley, R. K. (2014) 'Getting what you need: Designing institutional architecture for effective governance of international waters', *Environmental Development*, 11, 7, 98–111.

Helsinki Rules on the Uses of Waters of International Rivers, 20 August 1966, Fifty-Second report of the International Law Association (1976), 484.

Holley, C. and Sinclair, D. (2013) 'Deliberative Participation, Environmental Law and Collaborative Governance: insights from Surface and Groundwater Studies', *Environmental and Planning Law Journal*, 30, 1, 32–55.

Holley, C., Gunningham, N. and Shearing, C. (2012) *The New Environmetnal Govenance*, Earthscan, Oxford, UK.

Holling, C. S. (ed.) (1978) *Adaptive Environmental Assessment and Management*, Wiley, Chichester, UK.

Hurrell, A. (1995) 'International political theory and the global environment', in K. Booth and S. Smith (eds), *International Relations Theory Today*, Polity Press, Cambridge, UK.

International Law Commission (ILC) (2008) 'Articles on Trans-boundary Aquifers', United Nations, Geneva, Switzerland.

Keessen A. M., van Kempen, J. and van Rijswick, H. (2008) 'Trans-boundary river basin management in Europe: legal instruments to comply with European water management obligations in case of trans-boundary water pollution and floods', *Utrecht Law Review*, 4, 3, 35–56.

Lim, M. (2014) 'Is water different from biodiversity? Governance criteria for the effective management of trans-boundary resources', *Review of European Community and International Environmental Law*, 23, 1, 96–110.

Louka, E. (2008) *Water Law and Policy*, Oxford University Press, Oxford, UK.

McIntyre, O. (2007) *Environmental Protection of International Watercourses under International Law*, Ashgate Publishing, London.

McIntyre, O. (2010) 'International water law: concepts, evolution and development', in A. Earle, A. Jägerskog and J. Öjendal (eds) *Trans-boundary Water Management: Principles and Practice*, Earthscan, Oxford, UK.

Moynihan, R. (2014) 'The rising role of regional approaches in international water law: Lessons from the UNECE Water Regime and Himalayan Asia for Strengthening Trans-boundary Water Cooperation', *Review of European Community and International Environmental Law*, 23, 1, 43–58.

Mukhtarov, F. and Gerlak, A. (2014) 'Epistemic forms of integrated water resource management: towards knowledge versatility', *Policy Sciences*, 47, 2, 101–20.

National Climate Change Adaptation Research Facility, Water Governance Research Initiative (2012) 'Water Governance Research for Transformation', Briefing Paper no 4, National Climate Change Adaptation Research Facility, Southport, Australia.

Norman, E., Cook, C. and Cohen, C. (2015) *Negotiating Water Governance*, Ashgate, UK.

Öjendal, J., Earle, A. and Jägerskog, A. (2010) 'Towards a Conceptual Framework for Trans-boundary Water Management', in A. Earle, A. Jägerskog and J. Öjendal (eds), *Trans-boundary Water Management, Principles and Practice*, Earthscan, Oxford, UK.

Parker, C. and Braithwaite, J. (2003) 'Regulation', in P. Cane and M. Tushnet (eds) *The Oxford Handbook of Legal Studies*, Oxford University Press, Oxford, UK.

Pahl-Wostl, C., Holtz, G., Kastens, B. and Knieper, C. (2010) 'Analyzing complex water governance regimes', *Environmental Science and Policy*, 13, 7, 571–81.

Pahl-Wostl, C., Palmer, M. and Richards, K. (2013) 'Enhancing water security for the benefits of humans and nature – the role of governance', *Current Opinion in Environmental Sustainability*, 5, 6, 676–84.

Roseneau, J. N. (1995) 'Governance in the twenty-first century', *Global Governance*, 1, 1, 13–34.

Salman, M. A. S. (2007) 'The Helsinki Rules, the UN Watercourses Convention and the Berlin Rules: Perspectives on international water law', *Water Resources Development*, 23, 4, 625–40.

Servos, M. R., Munkittrick, K. R., Constantin, G., Mngodo, R., Aladin, N., Choowaew, S., Hap, N., Kidd, K. A., Odada, E., Parra, O., Phillips, G., Ryanzhin, S. and Urrutia, R. (2013) 'Science and management of trans-boundary lakes: Lessons learned from the global environment facility program', *Environmental Development*, 7, 1, 17–31.

Trubek D. and Trubek L. (2006) 'New governance and legal regulation', *Columbia Journal of European Law*, 13, 3, 539–64.

Tujchneider, O., Christelis, G. and van der Gun, J. (2013) 'Towards scientific and methodological innovation in trans-boundary aquifer resource management', *Environmental Development*, 7, 6–16.

United Nations Committee on Economic, Social and Cultural Rights (UNCESCR, 2002) General Comment No. 15 on the Right to Water 29th CESCR sess, Agenda item 3, UN Doc E/C/12/2002/11, United Nations Economic and Social Council, Geneva, Switzerland.

United Nations General Assembly (2007) 'Declaration on the rights of indigenous peoples', GA Res 61/295, UN GAOR, 61st sess., UN Doc A/RES/61/295 (13 September 2007).

United Nations General Assembly (2010) 'The Human Right to Water and Sanitation', GA Res 64/292, UN GAOR, 64th sess., UN Doc A/RES/64/292 (28 July 2010).

United Nations Human Rights Council (2010) 'Human rights and access to safe drinking water and sanitation', HRC Res 15/9, 15th sess., UN Doc A/HRC/RES/15/9 (6 October 2010).

United Nations World Water Assessment Program (WWAP) (2015) 'The United Nations world water development report 2015: Water for a sustainable world', United Nations Educational, Scientific and Cultural Organisation, Paris.

Vogler, J. (2005) 'The European Contribution to Global Environmental Governance', *International Affairs*, 81, 4, 835–50.

von der Porten, S. and de Loe, R. (2013) 'Collaborative approaches to governance for water and Indigenous peoples: A case study from British Columbia, Canada', *Geoforum*, 50, 149–60.

Young, O. (1994) *International Governance: Protecting the Environment in a Stateless Society*, Cornell University Press, Ithaca, NY.

Yu, X. (2011) 'Trans-boundary water pollution management: Lessons learned from river basin management in China, Europe and the Netherlands', *Utrecht Law Review*, 7, 1, 188–203.

Zeitoun, M. (2013) 'Current and future challenges facing trans-boundary river basin management', *WIREs Climate Change*, 4, 331–49.

Part II

Trans-jurisdictional water law governance

Theoretical underpinnings and international approaches

2 Trans-jurisdictional water governance

Law's role at multiple levels

Lee Godden

Introduction

Legal systems over many centuries have variously described and categorized 'water', concurrently shaping through such definitions how water was to be governed. For many legal systems, water is configured by reference to physical boundaries or criteria that focus water law within identified spatial and jurisdictional confines. In reality, water is a substance that is part of a wider ecological cycle, but which also transgresses any number of imposed legal, political and social boundaries. In its multiple forms, surface water, groundwater, overland flow, rivers and water courses, and even as ice, water intersects at numerous scales with human systems that seek to use it, protect it, pollute it, transform it and transfer it. Governance systems have often proved unequal to the task of adequately managing this diversity of boundary crossing. While the possible permutations of trans-jurisdictional governance are many, this chapter offers an initial overview of how systems of trans-jurisdictional water governance might be conceptualized.

The chapter canvasses the rationale for an 'architecture', a theoretical framework for trans-jurisdictional water governance. Various water law systems have approached the question of trans-jurisdictional water governance from different perspectives. For the common law, although water is regarded as a shifting, fugacious substance, it has not confounded development of the riparian doctrine with its spatially referenced rules of allocation and use anchored around the watercourse. In other legal systems, rules governing water access and use are embedded within transactional models organizing economic activity in regulatory spaces defined by reference to institutions and corporations (Bakker, 2012). However, the archetypal form of water law is one which vests water as a resource in territorially identified national or regional governments with detailed statutory regimes prescribing how water is governed (McKay, 2008, p. 48). Typically, water law is associated with policies of water productivity, with territorial control of water seen as integral to nationhood. State-centric schemes for governing water at various levels have become the mainstay in urban and rural areas across the globe. While water has long been conceived at multiple scales and models, from the local catchment to large river basins, the problem of ensuring effective water law and governance that transcends spatial and jurisdictional boundaries has gained recent prominence.

Disputes and conflicts over water between nations, although not a new phenomenon, also focus attention on the need to develop overarching water governance arrangements (Stephens, 2009, pp. 163–95). This requires approaches that transcend the physical and administrative difficulties of dealing with water across multiple political boundaries to maintain flexibility in local implementation, as well as ensuring enforceable standards across jurisdictions (Green *et al.*, 2013, p. 90). These problems are compounded by challenges to long-held 'divisions' in water law, such as that between groundwater and surface water, and the continuation of separate regulatory spheres for freshwater and coastal marine areas. These divisions, once seen as requiring separate water law categories, can be contrasted to many indigenous cultures where any divide between land and water is subsumed within a broader cultural engagement with 'country' (Ruru, 2009).

Similarly, ecological and systemic approaches linking freshwater and coastal marine processes emphasize that water should be governed comprehensively and adaptively. Collaborative and participatory processes for water governance, when aligned to human rights (Gerber and Chen, 2011, p. 23), emphasize that existing state-centric models are limited in their efforts to address ethical and equality considerations. Concerns around environmental sustainability have posed a challenge to state-centric water governance for many decades (Fisher, 2009, ch. 2), but these matters have assumed critical urgency under climate change scenarios.

In international law, the trans-boundary harm concept has been an important impetus for developing water law models that address inter-jurisdictional liability (see Wouters and Chen, 2014). In tandem, hard and soft law bilateral and multilateral agreements which expound collective water objectives rather than political divisions and narrow national interests have proliferated in the public international law sphere (Johns *et al.*, 2010). Within countries, the tensions of regulating water use and access across competing priorities under conditions of increasing scarcity, in the face of uncertainty and escalating cumulative impacts, all indicate that water governance must transcend the narrow accountability of spatially defined resource-based rules to engage with more holistic and dynamic models. In these contexts a trans-jurisdictional concept of water law and governance becomes an imperative (Holley, 2012).

Why is trans-jurisdictional water governance required?

In considering trans-jurisdictional water law and governance, contributors to this collection (Gray *et al.*, 2013) were invited to ponder that:

> Water, in all its forms, is essential to life on earth. However, habitat destruction, biodiversity loss, pollution (including diffuse/non-point source pollution) and increasing human demands for water-use are some of the

> most pressing problems facing the globe's freshwater and coastal marine environments. Despite a plethora of national and international law relating to the protection, management and use of marine and fresh water resources, current solutions to global and domestic water problems remain both complex and problematic.

Recognising the growing complexity of water problems and the limitations of existing legal models, in 2006 Fisher argued that:

> [W]hat is emerging through the system is a notion not simply of management of but rather of governance of water resources. Governance thus seeks to bring together within a coherent framework the interests and concerns of everyone involved and at the same time to impose upon them a sense of responsibility – but perhaps not yet legal liability – for the way in which water resources are managed
>
> (Fisher, 2006, p. 2)

Water governance has gained currency as a concept within many disciplines, including law (Moore *et al.*, 2014). Concurrently, within organizations charged with developing and administering water law, there is increased acknowledgment that effectively addressing multifaceted water 'problems' requires a sophisticated governance approach that can be implemented across scales and jurisdictions. There is awareness of the need for water governance, but uncertainty about what it might entail in terms of approaches, discrete institutions, models and measures. Similarly, acceptance of the need for trans-jurisdictional operation of water governance is growing, although again little convergence on what it encompasses. A theoretical framework therefore can provide 'scaffolding' for elucidating what is comprised by water governance in a trans-jurisdictional context, and how effective trans-jurisdictional models for water governance might be developed.

Within Australia, for example, a 2012 Council of Australian Governments' review of the National Water Commission endorsed the Commission's important work in the institutional landscape as an independent body able to provide national leadership and to enable cross-jurisdictional cooperation in water reform (Rosalky, 2012). Yet, the review noted that governance remains a challenge to sustainable water use and the full realization of the benefits of water reform requires sophisticated institutional and governance arrangements (Rosalky, 2012, p. 57). The National Water Initiative is a set of principles designed to drive water law and policy reform in that country. Water governance, although it might be assumed from a number of key elements, is not given precise articulation (Council of Australian Governments, 2004, art. 24). Similar dilemmas around precisely identifying the content and function of water governance occur at a national, regional and international level. Often, the call for adaptive or resilient governance is a synonym for trans-jurisdictional approaches, drawing inspiration from a variety of socio-ecological constructs that bring with them implicit normative framings and outcomes for water governance (Green, 2011, p. 83).

The lack of agreed definitions for water governance highlights the contested meaning of governance, although this is belied by its widespread adoption. Accordingly, a working definition of what is involved in 'governance', and its relationship with law and regulation in the water sphere, is a useful foundation for any architecture of trans-jurisdictional water governance. Using this platform as a starting point, it is possible to 'explore how to manage and promote legal and policy arrangements for effective trans-jurisdictional water governance on different scales and axes' (Gray *et al.*, 2013). Before moving to the normative question of what constitutes 'effective' trans-jurisdictional water governance, it is necessary to unpack what constitutes water governance.

What is water governance?

Governance originated as a specialist term embedded within a particular theoretical exposition (Jose, 2004) that identified the historical rise of sites of power and control beyond the formal political and legal institutions. It is now applied in many situations to refer broadly to a form of coordination or use of power for defined ends within social systems. In everyday discourse, governance ironically often substitutes for 'government' (Jose, 2004), albeit where governments may 'steer rather than row' and employ a range of regulatory and behavioural measures in addition to formal law. In developing an understanding of trans-jurisdictional water governance, given the diffuse nature of the term and its normative dimensions, it is not simply a matter of deciding what legal and regulatory measures or 'tools' may be appropriate, although these remain important.

Identifying the water 'problem' involves recognizing the multidimensional characteristics of water and its spatial, temporal, political, economic and social interconnections. A starting point is to investigate the framing of existing problems around water governance. Typically the water problem is 'located' in specific physical characteristics, for example, inadequate urban water supply or high levels of river pollution. It is also appropriate to ask what broader vertical and horizontal dimensions and systems are involved beyond the immediate manifestation of the problem. Furthermore, institutional framings that set the boundaries and divisions for managing water, such as catchments, have a critical influence in setting the conditions under which any law, policy, project or decision-making process will operate. Revisiting the framing of water problems provides insights into the appropriateness of the current strategies employed (NCCARF Water Governance Research Initiative, 2012). Systemic and institutional analyses of water that look beyond the immediate 'problem' offer scope for widening legal and policy alternatives around trans-jurisdictional governance. Many existing water law models that emphasize state-centric and 'top-down' regulation have limited capacity to address the system-wide problems that trans-jurisdictional governance potentially is able to address

(Davidson and de Loë, 2014). In this sense, trans-jurisdictional water law models must operate across scales rather than 'at' scales.

Many cross-cutting social, cultural and political conflicts and structural transitions crystallize around water. Typically though, the normative conditions under which water governance models are to operate are not fully articulated; the plurality of actors, institutions and boundaries involved may not be fully delimited; and the social and political goals to be achieved may be unclear, or at least multifaceted (see wicked problems in a climate change context in Ison, 2010). 'Governance experiments' at an international and domestic level that engage with complex water problems signal the need to actively interrogate the permeability of existing jurisdictional boundaries (Holley and Gunningham, 2011). Moves toward collaborative governance may redress some acknowledged limitations of jurisdictional (territorially defined) models. In turn, shifts prompted by collaborative governance theory may have their own limitations for developing an overarching conceptual model for community participation in trans-boundary water management (Holley and Sinclair, 2013, p. 52).

In this context of diverse, emerging governance models, this chapter takes up the challenge to 'lay bare the architecture of trans-jurisdictional water governance' (Gray *et al.*, 2013). However, it is not clear whether a comprehensive pre-existing architecture is already in place, although there has been a noticeable transition from the language of law to governance as the prevailing approach in many areas of environmental and resource management (see Gunningham, 2014). In tandem, there is renewed interest in water planning, in participatory approaches to decision-making (see e.g. Holley, 2010) and in the development of multi-level approaches, such as integrated catchment management (Nelson, 2005; Molle, 2009). However, few existing arrangements approach the type of adaptive and collaborative management across multiple dimensions required to achieve the complexities of trans-jurisdictional (and trans-scalar) water governance. Therefore, this chapter offers an overview as to why a theoretical framework is important, together with an exploration of key concepts, relevant principles, and critical analysis. The chapter remains a preliminary analysis to be refined by the insights offered by contributors to this collection.

Trans-jurisdictional water governance: Locating and framing the problem

If law is to offer particular insights into trans-jurisdictional water governance, it must resolve critical theoretical questions such as the relationship of law and governing institutions for water to the nation state, as well as between the nation state and supra-national institutions. Within countries, the role of the state, industry and the private sector, civil society and local communities in trans-jurisdictional water governance must be probed. This necessarily involves technical legal issues, such as questions of sovereignty as well as less familiar legal analyses that engage with multiple normative outcomes. Lawyers bring particular expertise to trans-jurisdictional water governance but can benefit from drawing upon relevant multidisciplinary scholarship that has long engaged with multi-scalar and trans-boundary concepts to build a robust theoretical framework for trans-jurisdictional water governance (see Frank *et al.*, 2014).

A theoretical framework provides several insights. First, it is important to situate the water governance 'problem' and how the 'entity' to be governed is to be characterized. Typically the policy problem around water is identified as stakeholder conflict over available water supply or water quality. The classic upstream/downstream conflicts in river basins across the world exemplify these simplified types of problems (Boisson de Chazournes, 2013, pp. 22–7; for the Murray-Darling Basin see Connell, 2007, pp. 26–32). Furthermore, the 'problem' often is regarded as associated with the 'physical' qualities of water, rather than with the institutional, organizational and community arrangements in place for managing water. The relative paucity of attention directed to law and institutional practices, and now to governance, has focused attention on narrowly conceived water supply/demand management, as well as on engineering and technical solutions for addressing water scarcity or water degradation. Highlighting the important interface between the 'physical' and the 'social' in managing conflict around water is a useful starting point.

Second, water can be 'located' within many different areas of analysis depending on the trans-jurisdictional issues that arise. The virtual export of water in food, for example, where water 'crosses national boundaries', might be conceived simultaneously as a matter of water resources law, trade law or biodiversity protection law or of human rights obligations under international law. These overlaps are not confined to water governance. International law scholars identify the regulation of transnational and trans-boundary phenomena requiring multi-level and multi-scalar analysis as regime interaction or its obverse as regulatory fragmentation. These paradoxes call for careful 'attention to the processes, principles and actors that can ensure fruitful normative and institutional interplay' (Young, 2011, p. 157). The development of a theoretical architecture for trans-jurisdictional water governance should emphasize interconnection between and within political, jurisdictional and organizational spheres, as well as within legal disciplines. An effective trans-jurisdictional architecture also must engage questions of intergovernmental interaction, negotiation and agreement-making. National constitutional frameworks and international structures, such as treaties, are significant over-arching institutional and regulatory responses to effectuate more specific trans-jurisdictional water governance structures (Johns *et al.*, 2010).

Integrated and systemic approaches

Similarly, at national and intra-national levels, integrated models are required to avoid water law 'silos' (Gray *et al.*, 2013). Systems theory often underpins integrated, multi-level approaches (Godden *et al.*, 2011). While there has been strong integration of systemic approaches into the objectives of water legislation in many countries, the corresponding structures for water management often do not reflect a similar systemic cohesion. Thus a theoretical framework for trans-jurisdictional water governance might consider systemic approaches as an organizing governance principle.

The classic systems model is that of bio-regions where law and decision-making structures align to ecosystemic boundaries and scales rather than existing administrative regions. The construct is embedded within some forms of catchment management. In New Zealand, the Resource Management Act 1989 correlates administrative boundaries for decision making on resources and planning law matters with river basin boundaries. The legislation aligns watersheds with resources and planning decision making, but has not escaped criticism for ineffective water governance (Bramley and McNeill, 2011). Other catchment management models seek to perform a similar integrative function. Such models offer a basis for developing specific trans-jurisdictional water governance architecture where the advantages of managing water at a larger, ecosystem scale potentially can take into account more localized concerns (Holley and Sinclair, 2013, p. 48).

In terms of multi-scalar models, Nelson (2011) identifies how regulation of non-point source water pollution necessarily operates on two structural levels. Specific regulations and institutions that are created to deal with non-point source water pollution problems operate at 'lower levels' in specific geographic areas, with overarching superstructures such as constitutional divisions of governmental powers occurring at a second level (Nelson, 2011). While formal intergovernmental arrangements are significant for developing trans-jurisdictional water governance, so too are more pervasive networks of formal and informal practices. Johns *et al.* discuss the iteration between 'hard law and soft law' that characterizes water law and governance in the Mekong River Basin. They identify the normative network function of soft law as an important factor in the 'highly contested terrain' of the Basin, where insufficient attention has been paid to plural and local knowledge that can inform trans-jurisdictional water management (Johns *et al.*, 2010).

Potentially, developing a theoretical framework for trans-jurisdictional water governance can allow water 'problems' to be 'framed' as dynamic, multi-layered and multi-scalar, where human 'systems' such as law which are designed to manage water across many interfaces constitute both opportunities and constraints (Johns *et al.*, 2010). Models drawn from work on water governance in a climate change era that emphasize resilience are pertinent (McDonald and Styles, 2014). Water law typically predicts future management options on the basis of past data and practice. Such retrospective paradigms 'are no longer adequate for the legislative challenges that relate to complex, cross-scale and interacting social-ecological systems because of their linkages to assumptions of stationarity and uniformitarianism' (Clarvis *et al.*, 2014). Designing legal measures to adequately respond to uncertainty across jurisdictions and scales and to institute legal mechanisms for adjusting water management to changing conditions in the water resource over time is a considerable challenge

(Tan, 2010, pp. 137–8). A comprehensive, trans-jurisdictional model must be oriented toward proactively seeking to reduce stresses to water systems in advance rather than being reactive.

State-centric or collaborative water governance?

Over the past decade, there has been a purported movement away from state-centric models of water management. Policy makers have experimented with flexible regulatory techniques (Braithwaite and Parker, 2004, p. 273), adopting a spectrum of regulatory approaches, economic instruments, behaviour change models and toolkits that are potentially applicable in trans-jurisdictional water contexts. Despite such new tools, the extent to which decision-making has been effectively relocated from the conventional 'top-down' models is more debatable. A theoretical framework may probe how questions of transparency, implementation and enforcement can be accounted for in a trans-jurisdictional context (Holley and Sinclair, 2013, p. 53). The relationships between effective water governance, monitoring and enforcement are questions of particular complexity in a trans-jurisdictional setting (Holley and Sinclair, 2013, p. 53).

Furthermore, a theoretically informed approach to trans-jurisdictional water governance can illustrate how current social technologies for water governance may constrain innovation and change (Wallis and Ison, 2011, p. 4082). These unquestioned organizational models may create path dependencies around existing patterns of water governance that offer limited opportunities for transcending jurisdictions, scales and temporal frameworks (Wallis and Ison, 2011, p. 4082). Overcoming such legal and institutional barriers to developing trans-jurisdictional water governance is a major challenge. Theoretical frameworks raise for consideration the barriers for some groups in terms of justice, human rights and equity in water governance.

Trans-jurisdictional water governance: What legal models?

If legal scholarship is to contribute to trans-jurisdictional water governance, the obvious question is – what law? Simplistically, law can traverse a spectrum from law as principle or right to law as 'top-down' authority. Such differentiation typically maps into the oppositions between collaborative/values-based governance models and state-centric models (derivate of national, sovereignty-based models). The latter relies principally on statutory water law. That model has been dominant in many countries since the late 19th century when statutory water laws emerged, predicated upon vesting water resources in the nation state. These movements were accompanied by a proliferation of public utilities charged with water management.

With respect to Australia:

> . . . the foundations of the legal arrangements for managing water resources in Australia evolved from a common law system, based upon individual

> rights of access to water, to a statutory system, based upon the right of the Crown – now the State – to the use and control of water resources.
>
> (Fisher, 2010, p, 85)

Colonial water laws also displaced Aboriginal and Torres Strait Islander customary law (Godden and Gunther, 2010). The predominance of the state-centric model was confirmed in *ICM Agriculture v. Commonwealth* (2009) 240 CLR 140. Nonetheless, 'in the short history of its modern polity state-based disputes about water distribution have been common' (Rochford, 2013). Over time, there have been attempts to transcend the state-centric model through agreement-based and transactional models. The attempt to deal with trans-boundary water quantity and quality issues by means of the Murray-Darling Agreement (initiated in 1914) is an example, (this Agreement reappears as a trans-jurisdictional mechanism as sch. 1 of the Water Act 2007 (Cth)).

Rochford (2013) identifies international drivers for the emergence of trans-jurisdictional water regulation within nations as

> the existence of multiple constraints on Parliamentary legislative authority by international trade-based treaties creates additional sites from where sovereignty is effectively exercised [which] will likely become normatively entrenched.
>
> (Jenkins, 2002, p. 2)

The EU framework Directives on Water exemplify the situation where 'additional sites of sovereignty' occur beyond nation state-centric management of water resources (Jessel and Jacobs, 2005). This approach combines territorially based water governance modes with specific, extra-territorial forms of sovereignty. Attempts to bring together state-centric or territorially discrete models for water governance with transactional, extra-territorial and collaborative models are being pursued with varying levels of success.

Law in a trans-disciplinary and trans-jurisdictional context

A trans-disciplinary approach aligns well with a trans-jurisdictional water governance framework. Trans-jurisdictional water governance can benefit from incorporating epistemologically aware research that is expressly informed by theoretical and methodological frameworks from non-law or cross-disciplinary traditions (NCCARF Water Governance Research Initiative, 2012).Trans-disciplinarity retains a disciplinary core of knowledge but works 'between, across and beyond' disciplines (Riedy, in NNCARF Water Governance Research Initiative, 2012), transcending the boundaries of traditional research disciplines. It challenges fragmentation, deals with research problems that arise out of complex heterogeneous domains, accepts local uncertainties, continuous collaboration, is action orientated, crosses disciplinary boundaries, but also engages theoretical development and professional practice (Brown *et al.*, 2010, pp. 17–18).

Cross-disciplinary research can assist in identifying general principles for trans-jurisdictional water governance applicable across various scales: temporal, spatial, institutional and cultural (Wallis *et al.*, 2013). The transformation could comprise further engagement with social sciences and humanities to extend the theoretical insights. Although law often presents itself as methodologically neutral, adopting a trans-jurisdictional approach to water governance will require specific engagement with normative objectives and methods to engage meaningfully in a trans-disciplinary approach. Explicit use of theories grounded in other disciplines can assist in managing 'boundary relationships'. As the Australian National Water Commission pertinently observed:

> [w]hile the focus of reform effort to date has been on improving the core structures of water management and on resolving issues of water quantity, there is now a need to more effectively coordinate 'boundary' relationships between water management and related policy interventions.
>
> (National Water Commission, 2012)

Sensitivity to context and local relationships is important in developing trans-jurisdictional water governance. Thus, trans-disciplinary research is needed to embrace collaborative approaches at a number of levels, but with the capacity to evaluate the respective utility of alternative models across multiple scales. Research on agreement-making and collaborative governance, complexity theory, institutional change and 'soft-path' management has been effectively applied to water governance (Brooks and Brandes, 2011).

Integrating water governance and legal models

Determining exactly which theoretical framework to adopt for water governance, and when and how this is to be integrated into law, poses significant challenges. Subsequent chapters in this book provide a range of responses. The foregoing discussion has sought to identify certain commonalities without seeking to pre-empt the choice of relevant theoretical frameworks. This chapter raises, in a preliminary way, key questions with which a theoretical architecture might engage. These issues, which are fleshed out in later chapters in this volume, include:

- the roles and forms of government intervention and institutional design conditions that might best overcome transaction costs to produce successful collaborative processes;
- when and under what conditions trans-jurisdictional cooperation is likely to deliver effective environmental outcomes;
- whether it will be possible to develop principles and practical tools of trans-jurisdictional governance capable of replication and successful application across a variety of water problems; and

- blending trans-jurisdictional governance with state-centric governance understandings and practices to better govern human impacts on fresh and marine water.

New governance, government intervention and water governance

Discussion of government intervention in water governance assumes some transition from dominant state-centric models of water law towards privatized and/or collaborative models. The extent of the transition is contested (see generally, Holley *et al.*, 2012). Nonetheless, the conceptual shift from 'government' to 'governance' has been accompanied by a growing decentralization of regulatory functions and by the emergence of hybrid regulatory forms (Termeer *et al.*, 2011). While the state is no longer the sole entity governing society, such changes have been accompanied by a proliferation in the number of state-related organizations engaged in various modes of regulation (Gunningham, 2009).

The devolution of decision-making power has changed the nature of the legal and regulatory tools that are utilized. A distinction is drawn between more formalized structures of law, where nation states are regarded as the central driver of change and transformation (i.e. a state-centric model) anchored in legislation and public agencies, and emerging 'bottom-up' models. Reliance on market mechanisms, voluntary codes and self-regulation are examples of steering mechanisms utilized by the state to regulate 'at a distance' (Termeer *et al.*, 2011). 'New governance' looks at formal and informal public and private interaction among a wide variety of actors. Rhetorically, new governance suggests that policy may be narrowed to making an 'intervention' using appropriate tools or instruments – one of which may be legislation (Hartley, 2014). However, in the water sector, the view that law is simply the 'add-on' problem solver intervening at points in the water supply and distribution process may require critical investigation in trans-jurisdictional contexts, as 'current solutions to global and domestic water problems remain both complex and problematic' (Gray *et al.*, 2013).

An important aspect of 'new governance' that will require translation into trans-jurisdictional water governance is how to incorporate the pursuit of social objectives, such as sustainability and community value systems, where the 'public interest' no longer sits exclusively in any one nation state or jurisdiction. Some contend that new governance 'just has a greater emphasis on governments giving explicit reasons to justify and gain support for the objectives it seeks to pursue' (Hartley, 2014). A new governance model can lead to 'a "feedback loop" that, combined with robust public participation, can produce better outcomes' (Solomon, 2008, p. 825). The European Union Directive to protect water quality, which involved sharing of best practice ideas and benchmarking across member states, is cited as an example (Scott and Holder, 2006).

The role of formal legal structures as a driver for social change is debated in new governance as informal and extra-legal problem solving is emphasized. Conventional legal approaches and 'new governance' can successfully interact by coupling binding laws with new governance implementation regimes (Solomon, 2008, p. 827). This is apparent in the challenge that governance models pose to the rights-based frameworks for water allocation and use, where norms are strengthened by solving problems through negotiated governance and not just by identifying rights-based positions (Solomon, 2008, p. 827).

In summary, a working definition of 'new' water governance might comprise increasingly decentralized organizational structures for allocating and managing water, with multi-level and multi-spatial points of control or coordination that combine hierarchical and non-hierarchical legal and regulatory frameworks. It will require horizontal, as well as vertical, information flows and subsidiarity in decision making with an emphasis on community collaboration and partnerships. Localized knowledge that is co-generated with government, technical expertise and local knowledge will be required, together with the co-evolution of management systems (Wallis and Ison, 2011). Ideally, there should be an increase in local capacity building, enhanced public participation and accountability and robust monitoring and feedback.

Collaborative environmental outcomes

Trends towards more collaboratively grounded environmental approaches are evident in the water sector. These legal and policy approaches encompass: collaboration, integration, knowledge generation, flexibility, decentralization and participatory and deliberative styles of community decision making (Wallis and Ison, 2011, p. 4082). Such participatory processes increasing blur the line between state and civil society (Wallis and Ison, 2011, referring to Karkkainen, 2002, pp. 189, 200–204 and 224).

Whether ideals of collaborative and adaptive governance are being achieved is problematic. State-centric, formalized water law systems remain the core organizing models, albeit with varying degrees of modification to integrate some consultative and indirect governance approaches. Holley and Sinclair (2013) found mixed results in the Australian context in their study of Australian examples of 'collaborative environmental governance' to assess whether current legal tools and governance architectures are adequate to address pressing water problems. Many 'bottom-up' approaches frequently fail to live up to promise. Full participatory models are normative ideals that are difficult to realize in most water law systems (Holley and Sinclair, 2013, p. 53). Nonetheless, collaborative processes will contribute significantly to 'move normative debate forward on the issue of participatory decision-making' (Holley and Sinclair, 2013, p. 55). Debates that challenge given jurisdictional boundaries, decision-making points and operational scales in water law are vital to giving institutional space for principles such as ecologically sustainable development, and are fundamental to realigning the existing organizational and legal forms toward more environmentally sensitive outcomes. Adaptation to climate change is another important driver for trans-jurisdictional water governance (Terneer *et al.*, 2011). Three key concepts are posited to guide more adaptive governance: organizing connectivity, (re)allocating responsibilities and dealing with controversies (Terneer *et al.*, 2011).

Water is governed at many different levels and scales, often bedevilled by a jigsaw of political and bureaucratic boundaries. Institutions charged with managing water can range from local government authorities to specialized water agencies through to complex intergovernmental and international arrangements, often with little coordination of effort (Fisher, 2009, pp. 21–41). The difficulties created by administrative and institutional 'fiefdoms' are well known. Reforms to institutional arrangements in the Murray-Darling Basin were designed in part to resolve long-standing inter-jurisdictional conflict among the powerful 'fiefdoms' of provincial governments within the Basin (Gardner, 2012).

Examining the nature of the basic unit through which water is governed is necessary if a trans-jurisdictional approach is to be developed. A fundamental construct in water law is that of the river basin. In most jurisdictions, under a state-centric approach, the river basin is regarded as an objectively discernible natural territory or as an ecological entity. In contrast, however, a river basin can be thought of as:

> . . . a bio-geo-physical unit and its associated social actors and institutions. [Such] social-ecological systems typically are regarded as complex and adaptive and delimited by spatial or functional boundaries surrounding particular ecosystems and their problem context.
>
> (Glaser *et al.*, 2008, quoted in McCarthy *et al.*, 2012, p. 306)

Del Moral and Do Ó (2014, p. 343) argue for acceptance of 'the existence of multiple geographies of water, with their overlapping social, economic, political, cultural and physical spaces, and the importance of collaborative and flexible ways of working across the boundaries they entail'.

The complexity and multi-scalar nature of multiple geographies of water rarely fit neat spatial units, although the contours of a river basin reflect significant natural regimes. Accordingly, river basins may constitute better governance units than municipalities, regions, states or nations. Integrated water resource management is often coupled with the idea of natural river basins and watersheds to allow holistic approaches within the catchment scale. It has been the dominant paradigm for water management since the middle of the twentieth century (Molle, 2009, p. 486). Yet, the adoption of river basins as a governance unit can obscure the interface of freshwater with marine areas and even surface–groundwater interactions.

In using the river basin, watershed or catchment as an organizing principle, however, there has been implicit recognition that water requires an integrated approach within the encompassing basin structure. Despite many years of promoting integrated water management and awareness that management must reflect underpinning ecosystems, disparate regimes remain for governing water separated across jurisdictions, geo-spatial and biodiversity regions and across temporal parameters.

Matching organizational arrangements to 'natural systems', such as catchments, can provide some rationalization to the diverse and largely artificial institutional and jurisdictional structures for water management. A study conducted into 29 river basins highlighted the importance of polycentric governance structures (Pahl-Wostl *et al.*, 2012, p. 32). Polycentric structures can embrace trans-jurisdictional as well as multi-scalar and multi-dimensional water governance configurations. Nevertheless, it is doubtful whether a bio-regional approach will be widely adopted despite movements toward catchment management authorities in some jurisdictions. Davidson and de Loë (2014, p. 368; quoting Young, 1992) distinguish between use of boundaries for management, operational and on-ground activity and 'governance', referring to 'the structures and processes by which people in societies make decisions and share power'. Their study indicates that many provincial governments are already using non-boundary governance mechanisms to navigate the complex water system, and integrate issues and solutions. Furthermore, policy transfer across nations and jurisdictions to shorten the policy innovation timeline is a critical response that can assist the development of trans-jurisdictional approaches. Michaels and de Loë (2010) point to successful international policy transfer examples. Increasing recognition of common factors, such as climate change, creates awareness of the need for trans-jurisdictional water governance. An emphasis on collaborative approaches to trans-jurisdictional water governance calls for continual, mutual policy learning and adaptation across jurisdictions (Michaels and de Loë, 2010, pp. 502 and 504).

Yet, Swainson and de Loë (2011) suggest carefully considering the degree of compatibility for the transferability of mechanisms and tools, including the political, institutional, socio-cultural, administrative and biophysical contexts that may make simple institutional transfer across scales and jurisdictions inappropriate. Carefully designed principles and tools for trans-jurisdictional water governance are required.

Principles and practical tools for trans-jurisdictional governance

The international sphere has seen considerable experimentation with tools for trans-jurisdictional water governance. Principles expressed in water law treaties and in the jurisprudence of the International Court of Justice evidence growing awareness of the necessity for trans-jurisdictional governance. At the transnational level, there is a growing focus on soft law and procedural mechanisms in trans-boundary water agreements (Gao, 2014, p. 6). In part, these transitions reflect the recognition that maintaining the emphasis on legal structures is problematic in transnational settings (Gao, 2014, p. 294). Bottom-up, informal processes, particularly consultation and negotiation, are to be preferred (Gao, 2014, p. 295). The increasing use of non-formal legal structures in a trans-jurisdictional setting, nationally or internationally, highlights the difficulty of achieving agreed, yet enforceable, modes of water governance across scales and political dimensions.

Internationally, there are a growing number of procedural rights supporting trans-jurisdictional collaboration embedded in recent water-related treaties (see, e.g. the Convention on the Protection and Use of Trans-boundary Watercourses and International Lakes 1992). Agreement-centred or negotiated models based on multilateral and bilateral treaties have specific limitations in effectively implementing substantive trans-jurisdictional water governance principles, as evident in the decisions of the International Court of Justice involving transnational conflicts over water (Sands and Peel, 2012, pp. 303–41). The decisions commenced in 1997 with *Gabcikovo-Nagymaros Project* ((*Hungary v. Slovakia*) (*Judgment*) [1997] ICJ Rep 7), where the important principle of nations acting to sustain the aquatic environment in 'a state of ecological necessity' was articulated. The principle promised significant potential to govern trans-boundary and transnational water disputes. However, the Court, while articulating these important principles, has been relatively reluctant in directly enforcing them in disputes. The decision in *Pulp Mills on the River Uruguay* ((*Argentina v Uruguay*) (*Judgment*) [2010] ICJ Rep 135) demonstrates the type of pyrrhic victory on procedural grounds that may await nations seeking redress for trans-boundary harm in shared river contexts. Despite considerable potential, the incorporation of agreement-making, consultative and negotiated modes of governance in these extra-territorial conflicts has delivered minimal substantive outcomes.

Indeed,

> [I]t is essential for good trans-jurisdictional water governance at local, national and international levels that parties collaborate, and that they do so in a genuine way. However, existing legal and governance theory has so far failed to provide an appropriate model for achieving this.
>
> (Gray *et al.*, 2013, p. 2)

Accordingly, a trans-jurisdictional water governance framework must facilitate flexibility while still retaining effective institutional forms – a difficult balance.

Blending trans-jurisdictional governance with state-centric law

Water law already utilizes several models adopting transactional (consensual or negotiated) forms that may be adapted to 'top-down' state-centric models for water governance to form the basis for a recalibrated governance model. Drivers outside the conventional water resource sector have been significant.

The growing momentum of human rights discourses around water evidences movements to immediately de-centre the nation state while still requiring its involvement in water governance on particular terms (Staddon *et al.*, 2012, p. 69). Globally and within nations, environmental NGOs, indigenous peoples and disadvantaged local communities figure prominently in rights-based and transactional (agreement-based) claims for inclusion in 'state-centric' structures for water governance. Often, the claim is not just for consultation or inclusion of these groups but also for a redefined participation that seeks to refashion existing power dynamics and jurisdictional boundaries (Ruru, 2009).

Within Australia, transactional modes, including agreement and negotiation frameworks, have been an important vehicle for claims by Aboriginal peoples for a stronger voice in water governance (Morgan *et al.*, 2006). Nonetheless, opportunities for collaborative participation by indigenous and other cultural groups still remain constrained by state-centric models of water governance, primarily the statutory water allocation systems.

Similarly, also in Australia, the progressive inclusion of environmental interests in water law was initially strongly contested (Foerster, 2012). Now, however, the environment as a pervasive 'actor' has achieved a degree of legal formalization within the state-centric water governance process (O'Donnell, 2012). The capacity of legal models for environmental water management to transcend specific jurisdictions to operate across many catchments at a basin scale has yet to be fully tested. Blended governance models, such as the Australian Commonwealth water holder enshrined by legislation as a statutory authority but also a 'market player', represents a blending of both transactional and state-centric models of water governance that already operates across jurisdictions. Groundwater, too, in Australia has started to edge toward its own trans-jurisdictional position based on its physical boundary-crossing functions and the integration of groundwater/surface water approaches. Trans-jurisdictional governance mechanisms in the freshwater/marine environment interface are less common and tend to be largely state-centric in orientation and territorially defined. It is in the interface between freshwater and marine areas that some of the most significant future challenges lie in developing the architecture for trans-jurisdictional water governance. Work on coastal development, driven by climate change adaptation research, offers a promising field (see Bell, 2014).

Finally, of particular pertinence for trans-jurisdictional water governance is the growing number of contractual models of water governance emerging in various regulatory spaces. In these arenas, water governance is expressed in both formal contracts and through normative regimes, corporatization, bench-marking and hybrid models of self-regulation and negotiation. In the water sector within Australia, an example of a blended state-centric and collaborative model is the requirement for community engagement as part of the development of the statutory imposed Murray-Darling Basin Plan (this mechanism has proved highly controversial, see Murray-Darling Basin Authority, 2011). Similarly, in an analysis of the UK Water Act 2003, McGillivray identifies the movement toward to stewardship ideals as supporting more dynamic community-based environmental protections (McGillivray, 2013).

Even so, instances of actual devolution of significant decision making around water management to the community are patchy. Clearly many 'water managers' recognize the need for broader community engagement that transcends the 'consult' model, but show much less agreement on its means of achievement. Moreover, notably absent in many water governance models are the minimal legal safeguards of transparency, such as third party appeals on administrative decision-making.

A collection of studies examining human rights to water have also investigated how 'soft law' and normative 'governance' models align to social and cultural struggles around water (Sultana and Loftus, 2012). In a related vein, the NGO sector has emerged as a significant player in water governance, developing principles such as free, prior and informed consent that are potentially extra-territorial in application but require nation state enforcement. It is in the emergent governance models blending collaborative, rights-based and state-centric models that there are fruitful sources for investigation.

Conclusion: Transactional and trans-jurisdictional models of water governance

In a preliminary way, the discussion has sketched some transactional (normative, collaborative, rights and agreement-based) models and state-centric (territorial and jurisdictional) models that have been developed, albeit often in nascent form, in water law and governance in many different settings. These tools, principles and practices provide a platform for suggesting how existing models and emerging trends toward collaborative and decentralized models might comprise the 'architecture' for trans-jurisdictional water governance.

This rationale for a theoretical framework canvasses the idea of two interacting models of water governance. The first is a model that has its origins in traditional legal sources, including constructs of sovereignty, territory and jurisdiction. Typically it is designated as 'state-centric' and aligned to public authority over a spatially defined water resource. It relies on mechanisms, such as state vesting of water resources and the development of extensive legal and public institutional structures for managing water. The other model is predominantly transactional or agreement-based and, to that extent, is extra-territorial in focus. It can encompass collaborative, contractual, private and civil sector engagement, and even human rights-based claims for crossing boundaries in water governance.

With the shift from government to governance, most legal systems, be they national, international or transnational, are adopting a mixture of models. In terms of effective governance, both have their own particular limitations and gaps in coverage which have driven a search for more effective regulatory, governance and extra-legal mechanisms to deal with trans-jurisdictional water management.

Complex challenges for water law remain around integrating trans-spatial and trans-temporal factors (cumulative impacts, inter-generational equity, enhanced climatic risk, etc.) into trans-jurisdictional water governance. Law is just beginning to examine how to bring multi-scalar and long-term adaptive planning considerations that operate across disciplinary and jurisdictional boundaries into water governance in a comprehensive manner. Importantly, before seizing on new theories and tools, it will be important to examine the expectations of effective water governance and to identify when and where trans-jurisdictional governance is the best response. In this sense, trans-jurisdictional water governance is normative and in the future will be stretched to achieve a growing number of outcomes, including efficiency (reduced transaction costs), equity and sustainability. The challenge will be in articulating a fit-for-purpose 'architecture' to cover the diverse demands upon, and contexts in which, trans-jurisdictional water governance comes into being.

References

Bakker, K. (2012) 'Commons vs. commodities: Debating the human right to water', in F. Sultana and A. Loftus (eds) *The Right to Water: Politics, Governance and Social Struggles*, Earthscan, Abingdon, UK.

Bell, J. (2014) *Climate Change and Coastal Development Law in Australia*, Federation Press, Annandale, Australia.

Boisson de Chazournes, L. (2013) 'Regulation of fresh water uses: Evolution, scope and developments', in *Fresh Water in International Law*, Oxford University Press, Oxford, UK.

Braithwaite, J. and Parker, C. (2004) 'Conclusion', in J. Braithwaite, N. Lacey, C. Parker and C. Scott (eds) *Regulating Law*, Oxford University Press, Oxford, UK.

Bramley, M. and McNeill, J. (2011) 'Up the creek and down the river: In-stream ecological values and property rights under the RMA', in K. Bosselmann and V. Tava (eds) *Water Rights and Sustainability*, New Zealand Centre for Environmental Law, Auckland, New Zealand.

Brooks, D. and Brandes, O. (2011) 'Why a water soft path, why now and what then?', *International Journal of Water Resources Development*, 27, 2, 315–44.

Brown, V., Harris, J. and Russell, J. (2010) *Tackling Wicked Problems through the Transdisciplinary Imagination*, Earthscan, London.

Clarvis, M. H., Allan, A. and Hannah, D. M. (2014) 'Water, resilience and the law: From general concepts and governance design principles to actionable mechanisms', *Environmental Science and Policy*, 43, 98–110.

Connell, D. (2007) *Water Politics in the Murray-Darling Basin*, Federation Press, Annandale, Australia.

Convention on the Protection and Use of Trans-boundary Watercourses and International Lakes, opened for signature 17 March 1992, 1936 UNTS 269 (entered into force 6 October 1996).

Council of Australian Governments (COAG) (2004) 'Intergovernmental agreement on a national water initiative', http://nwc.gov.au/__data/assets/pdf_file/0008/24749/Intergovernmental-Agreement-on-a-national-water-initiative.pdf, accessed 18 May 2015.

Davidson, S. and de Loë, R. (2014) 'Watershed governance: Transcending boundaries', *Water Alternatives*, 7, 2, 367–77.

Del Moral, L. and Do Ó, A. (2014) 'Water governance and scalar politics across multiple-boundary river basins: States, catchments and regional powers in the Iberian Peninsula', *Water International*, 39, 3, 333–47.

Fisher, D. (2006) 'Water resources governance and the law', *Australasian Journal of Natural Resources Law and Policy*, 11, 1, 1–41.

Fisher, D. (2009) *The Law and Governance of Water Resources: The Challenge of Sustainability*, Edward Elgar, Cheltenham, UK.

Fisher, D. (2010) 'Water law, the High Court and techniques of judicial reasoning', *Environmental and Planning Law Journal*, 27, 2, 85–97.

Foerster, A. (2012) 'Emerging environmental water governance in Australia', *Journal of Water Law*, 22, 58–72.

Frank, S., Fürst, C., Witt, A., Koschke, L. and Makeschin, F. (2014) 'Making use of the ecosystem services concept in regional planning – trade-offs from reducing water erosion', *Landscape Ecology*, 29, 8, 1277–91.

Gabcikovo-Nagymaros Project ((*Hungary v. Slovakia*) (*Judgment*) [1997] ICJ Rep 7.

Gao, Q. (2014) *A Procedural Framework for Trans Boundary Water management in the Mekong River Basin: Shared Mekong for a Common Future*, Brill, Leiden, The Netherlands.

Gardner, A. (2012) 'Water reform and the federal system', in P. Kildea, A. Lynch and G. Williams (eds) *Tomorrow's Federation: Reforming Australian Government*, Federation Press, Annandale, Australia.

Gerber P. and Chen, B. (2011) 'Recognition of the human right to water: Has the tide turned?', *Alternative Law Journal*, 31, 1, 21–6.

Glaser, G., Krause, G., Ratter, B. and Welp, M. (2008) 'Human–nature-interaction in the anthropocene: Potential of social–ecological systems analysis', Preparation paper for the annual conference of the German Society for Human Ecology, Sommerhausen, Bavaria, http://dg-humanoekologie.de/pdf/DGH-Mitteilungen/GAIA200801_77_80.pdf, accessed 18 May 2015.

Godden, L. and Gunther, M. (2010) 'Realising capacity: Indigenous involvement in water law and policy reform in south-eastern Australia', *Journal of Water Law*, 20, 5, 243–53.

Godden, L., Ison, R. and Wallis, P. (2011) 'Water governance in a climate change world: Appraising systemic and adaptive effectiveness', *Water Resources Management*, 25, 15, 3971–6.

Gray, J., Holley, C. and Rayfuse, R. (2013), 'Trans-jurisdictional water governance', Overview of workshop, University of New South Wales, Sydney, Australia, Copy on file with author.

Green, C. (2011) 'Mapping the field: The landscapes of Governance', SWITCH Central Management Unit, UNESCO-IHE, Delft, The Netherlands, www.switchurbanwater.eu/outputs/pdfs/W6–1_GEN_RPT_D6.1.1b_Mapping_Landscapes_of_Governance.pdf, accessed 18 May 2015.

Green, O. O., Garmestani, A. S., van Rijswick, H. F. M. W. and Keessen, A. M. (2013) 'EU water governance: Striking the right balance between regulatory flexibility and enforcement?', *Ecology and Society*, 18, 2, 89–98.

Gunningham, N. (2009) 'Environmental law, regulation and governance: Shifting architectures', *Journal of Environmental Law*, 21, 2, 179–212.

Gunningham, N. (2014) 'RegNet Professor Seminar: The Architecture of Environmental Regulation', Australian National University, Canberra, Australia.

Hartley, M. (2014) 'Regulating for groundwater-use efficiency: A toolbox approach based on the experiences of three disparate jurisdictions', *Environmental and Planning Law Journal*, 31, 2, 92–113.

Holley, C. (2010) 'Removing the thorn from new governance's side: Examining the emergence of collaboration in practice & the roles for law, nested institutions & trust', *Environmental Law Reporter*, 40, 7, 10656–868.

Holley, C. (2012) 'Trans-boundary environmental governance, collaboration and the law: Empirical insights from water and natural resource management in inland Queensland, Australia', in R. Warner and S. Marsden (eds) *Trans-boundary Environmental Governance: Inland, Coastal and Marine Perspectives*, Ashgate Publishing, Farnham, UK.

Holley, C. and Gunningham, N. (2011) 'Natural resources, new governance and legal regulation: When does collaboration work?', *New Zealand Universities Law Review*, 24, 3, 309–36.

Holley, C. and Sinclair, D. (2013) 'Deliberative participation, environmental law and collaborative governance: Insights from surface and groundwater studies', *Environmental and Planning Law Journal*, 32, 1, 32–55.

Holley, C., Gunningham, N. and Shearing, C. (2012) *The New Environmental Governance*, Earthscan, New York.

ICM Agriculture v. Commonwealth (2009) 240 CLR 140.

Ison, R. (2010) *Systems Practice: How to Act in a Climate Change World*, Springer, London.

Jenkins, D. (2002) 'Both ends against the middle: European integration, devolution, and the sites of sovereignty in the United Kingdom', *Temple International and Comparative Law Journal*, 16, 1, 1–25.

Jessel, B. and Jacobs, J. (2005) 'Land use scenario development and stakeholder involvement as tools for watershed management within the Havel River Basin', *Limnologica*, 35, 3, 220–33.

Johns, F., Saul, B., Hirsch, P., Stephens, T. and Boer, B. (2010) 'Law and the Mekong River Basin: A socio-legal research agenda on the role of hard and soft law in regulating trans-boundary water resources', *Melbourne Journal of International Law*, 11, 1, 154–74.

Jose, J. (2004) 'Exploring the Genealogy of "Governance" within Australian Political Discourse', paper presented at Australasian Political Studies Association Conference, University of Adelaide, Adelaide, Australia, http://adelaide.edu.au/apsa/docs_papers/Pub%20Pol/Jose.pdf, accessed 18 May 2015.

Karkkainen, B. (2002) 'Collaborative ecosystem governance: Scale, complexity, and dynamism', *Virginia Environmental Law Journal*, 21, 2, 189–243.

McCarthy, D. D. P., Whitelaw, G. S., Anderson, S., Cowan, D., McGarry, F., Robins, A., Gardner, H. L., Barbeau, C. D., Charania, N. A., General, Z., Liedke, J., Sutherland, C., Alencar, P. and Tsuji, L. J. S. (2012) 'Collaborative geomatics and the Mushkegowuk Cree First Nations: Fostering adaptive capacity for community-based sub-arctic natural resource management', *Geoforum*, 43, 2, 305–14.

McDonald, J. and Styles, M. C. (2014) 'Legal strategies for adaptive management under climate change', *Journal of Environmental Law*, 26, 1, 25–53.

McGillivray, D. (2013) 'Water rights and environmental damage: An enquiry into stewardship in the context of abstraction licensing reform in England and Wales', *Environmental Law Review*, 15, 3, 205–24.

McKay, J. (2008) 'The legal frameworks of Australian water: Progression from common law to sustainable shares' in L. Crase (ed.), *Water Policy in Australia: The Impact of Change and Uncertainty*, Earthscan, London.

Michaels, S. and de Loë, R. (2010) 'Importing notions of governance: Two examples from the history of Canadian water policy', *American Review of Canadian Studies*, 40, 4, 495–507.

Molle, F. (2009) 'River-basin planning and management: The social life of a concept', *Geoforum*, 40, 3, 484–94.

Moore, M., von der Porten, S., Plummer, R., Brandes, O. and Baird, J. (2014) 'Water policy reform and innovation: A systematic review', *Environmental Science & Policy*, 38, 263–71.

Morgan, M., Strelein, L. and Weir, J. (2006) 'Authority, knowledge and values: Indigenous nations engagement in the management of natural resources in the Murray-Darling Basin', in Langton, M., Mazel, O., Palmer, L., Shain, K. and Tehan, M. (eds) *Settling with Indigenous People: Modern treaty and agreement-making*, Federation Press, Annandale, Australia.

Murray-Darling Basin Authority (2011) 'Socioeconomic Analysis and the Draft Basin Plan', Murray-Darling Basin Authority, Canberra, Australia, http://mdba.gov.au/what-we-do/basin-plan/development/socio-economic-reports, accessed 18 May 2015.

National Climate Change Adaptation Research Facility (NCCARF) Water Governance Research Initiative (2012), 'Strengthening Water Governance in Australia', Briefing Paper nos 1–4, Monash Sustainability, Monash University, Australia, 2012, http://monash.edu/sustainability-institute/assets/documents/programs-initiatives/wgribrief, accessed 18 May 2015.

National Water Commission (2012) *Australian Environmental Water Management: 2012 Review*, National Water Commission, Canberra, Australia.

Nelson, R. (2005) 'Legislation for ICM: Advancing water resources sustainability?', *Environmental and Planning Law Journal*, 22, 90–123.

Nelson, R. (2011) 'Regulating nonpoint source pollution in the US: Lessons and research paths for Australia', *University of Western Australia Law Review*, 35, 2, 340–84.

O'Donnell, E. (2012) 'Institutional Reform in Environmental Water Management: The New Victorian Environmental Water Holder', *Journal of Water Law*, 22, 73–84.

Pahl-Wostl, C., Lebel, L., Knieper, C. and Nikitina, E. (2012) 'From applying panaceas to mastering complexity: Toward adaptive water governance in river basins', *Environmental Science & Policy*, 23, 24–34.

Pulp Mills on the River Uruguay ((*Argentina v. Uruguay*) (*Judgment*) [2010] ICJ Rep 135.

Resource Management Act 1989 (Cth).

Rochford, F. (2013) 'Water sovereignty and food security', in Q. Farmar-Bowers, V. Higgins and J. Millar (eds) *Food Security in Australia: Challenges and Prospects for the Future*, Springer, New York.

Rosalky, D. (2012) 'Review of the National Water Commission, Council of Australian Governments', Canberra, Australia, http://environment.gov.au/topics/water/australian-government-water-leadership/council-australian-governments-water-reform-1, accessed 18 May 2015.

Ruru, J. (2009) 'Undefined and unresolved: Exploring indigenous rights in Aotearoa New Zealand's freshwater legal regime', *Journal of Water Law*, 20, 5/6, 236–42.

Sands, P. and Peel, J. (2012) 'Freshwater resources', in *Principles of International Environmental Law*, 3rd edn, Cambridge University Press, Cambridge, UK.

Scott, J. and Holder, J. (2006) 'Law and new environmental governance in the European Union', in G. de Búrca and J. Scott (eds) *Law and New Governance in the EU and the US*, Hart, Oxford, UK.

Solomon, J. (2008) 'Law and governance in the 21st century regulatory state', *Texas Law Review*, 86, 4, 819–56.

Staddon, C., Appleby, T. and Grant, E. (2012) 'A right to water? Geographico-legal perspectives', in F. Sultana and A. Loftus (eds) *The Right to Water: Politics, Governance and Social Struggles*, Earthscan, Abingdon, UK.

Stephens, T. (2009) *International Courts and Environmental Protection*, Cambridge University Press, Cambridge, UK.

Sultana, F. and Loftus, A. (eds) (2012) 'Introduction', in *The Right to Water: Politics, Governance and Social Struggles*, Earthscan, Abingdon, UK.

Swainson, R. and de Loë, R. (2011) 'The importance of context in relation to policy transfer: A case study of environmental water allocation in Australia', *Environmental Policy and Governance*, 21, 1, 58–69.

Tan, P. (2010) 'Adaptation measures for water security in a changing climate: Policy, planning and law', in T. Bonyhady, A. Macintosh and J. McDonald (eds) *Adaptation to Climate Change: Law and Policy*, Federation Press, Annandale, Australia.

Termeer, C., Dewulf, A., van Rijswick, H., van Buuren, A., Huitema, D., Meijerink, S., Rayner, T. and Wiering, M. (2011) 'The regional governance of climate adaptation: A framework for developing legitimate, effective, and resilient governance arrangements', *Climate Law*, 2, 2, 159–79.

Wallis, P. J. and Ison, R. L. (2011) 'Appreciating institutional complexity in water governance dynamics: A case from the Murray-Darling Basin, Australia', *Water Resources Management,* 25, 15, 4081–97.

Wallis, P. J., Ison, R.L. and Samson, K. (2013) 'Identifying the conditions for social learning in water governance in regional Australia', *Land Use Policy*, 31, 412–21.

Water Act 2003 (UK).

Water Act 2007 (Cth).

Wouters, P. and Chen, H. (2014) 'Introduction to "The China Water Papers" – trans-boundary water cooperation in Asia with a focus on China', *Water International*, 39, 5, 701–4.

Young, M. (2011) 'Climate change law and regime interaction', *Carbon & Climate Law Review*, 5, 2, 147–57.

3 Trans-boundary waters governance through the prism of the Mekong River Basin

Richard Kyle Paisley, Patrick Weiler and Taylor Henshaw

Introduction

The objective of this chapter is to critically review trans-boundary waters governance in the Mekong River Basin in a context that emphasizes developments over the past four decades.[1] A mere snapshot of governance in the Mekong is inadequate to understand the flow of decisions over time, and a treatment of Mekong governance solely in terms of black letter rules and doctrines, divorced from the social processes that gives them meaning, is misleading. Thus, the effort here is to identify and analyse key events in the interactive processes that have led to the evolution of Mekong River Basin governance.

Covering about half the Earth's surface, 276 watercourses connect the territories of two or more countries (Global Environment Facility, 2013). There are also untold numbers of shared aquifers. International drainage basins[2] have certain characteristics that make their management especially challenging, the most notable of which is the tendency for regional politics to exacerbate the already difficult task of understanding and managing complex natural systems.

Management of international drainage basins is further complicated by the fragmented and underdeveloped nature of international water law. Like international law in general, international water law may take either of two general forms: treaty law or customary international law. The foundational treaty in international water law is the 1997 United Nations Convention on the Law of the Non-navigational Uses of International Watercourses (Watercourses Convention). Although the Watercourses Convention entered into force on 17 August 2014, it is of limited application. However, the Watercourses Convention does reflect the fundamental rules of customary international law applicable in the field (McCaffrey, 2001). These fundamental rules, historically articulated in the 1966 Helsinki Rules,[3] have also been reinforced by the judgement of the International Court of Justice in the *Case Concerning the Gabčíkovo-Nagymaros Project.*

The basic rules that form the foundation of international water law are:[4]

- States agree to use an international watercourse in a way that is 'equitable and reasonable' vis-à-vis other states that are sharing the watercourse.

- States agree to take 'all appropriate measures' to prevent 'significant harm' to co-riparian states.
- States agree to provide 'prior and timely notification' to other international watercourse states concerning any 'new use or change in existing uses' of an international watercourse, together with relevant technical information, and that it will 'consult' with the other international watercourse states.
- States agree to protect ecosystems of international watercourses. (This rule is emerging and does not yet rise to the same level of recognition as the three other basic rules.)

The experience of the Mekong River Basin states provides a fitting illustration of how these principles have coalesced in practice.

Physical characteristics

The Mekong River Basin covers 795,000 km^2 through six countries – China, Myanmar, Lao PDR, Thailand, Cambodia and Viet Nam. The Mekong River is the twelfth longest river in the world (4,350 km) and ranks tenth in terms of total volume (475 billion m^3/year). The Mekong is dominated mostly by a monsoon climate with distinct wet and dry seasons. Most of the total flow volume is delivered to the Mekong from tributaries in the Lower Mekong Basin, with upstream China contributing to over 24% of the total flow from snow melting in the dry season (Mekong River Commission, 2015).

As a result of long-standing regional conflict and various geopolitical barriers, and in spite of 37 years of investigations and planning by the riparian countries, no structures had been built on the Lower Mekong mainstream at the time the 1995 Agreement on the Cooperation for the Sustainable Development of the Mekong River Basin (the Mekong Agreement) was negotiated. China, which contributes 16% of the overall Mekong River flow volume and in which nearly half of the river's length is located, completed its first dam on the river in the year the Mekong Agreement was signed. Since that point, China has completed another three of the 20-plus dams it has planned for the Lancang and Upper Mekong. These hydropower dams have the ability to regulate flows in the wet season and increase river flows during the dry season downstream.

Historical context

The Mekong Agreement is a treaty among Cambodia, Lao PDR, Thailand and Viet Nam, who are described as the 'Lower Mekong River Basin States' (LMRBS), to provide a framework for cooperation in the sustainable development, utilization, conservation and management of the Mekong River Basin waters and related resources. The four parties have all signed and ratified the Mekong Agreement, which entered into force on 5 April 1995.

One of the few legally binding treaties in Southeast Asia (Bearden, 2010), the Mekong Agreement was the outcome of 37 years of regional and supra-regional efforts to cooperatively manage the Mekong River (Öjendal and Jensen, 2012). Throughout this period, extensive planning and feasibility studies were undertaken at considerable cost by the international community. The Lower Mekong became an area of significant interest at the height of the Cold War. This drove investment from the 'Western World' in order to support regional cooperation and development in a strategy to halt the expansion of communist influence in the region.

The early years of the Committee for Coordination of Investigations of the Lower Mekong (Mekong Committee, established by statute in 1957 by the lower Mekong riparians), which was authorized to 'promote, coordinate, supervise, and control the planning and investigation of water resources development projects in the Lower Mekong Basin', led to the development of a network of hydrologic and meteorological monitoring stations for data collection on the mainstream (Molle *et al.*, 2009). When the Pol Pot regime came to power in 1975, Cambodia withdrew cooperative efforts to manage the Mekong. With the lack of a representative government in Cambodia, the Mekong Committee became a three-member (Lao PDR, Thailand and Viet Nam) 'Interim Committee' in 1978 (Hirsch *et al.*, 2006).

Upon Cambodia's reengagement in 1991, the LMRBS set upon a course that, after 21 months of negotiations, produced the Mekong Agreement, which established the Mekong River Commission (MRC) as its regional management body. The MRC was established as the organization responsible for coordinating water resources development in relation to the natural resources and environmental protection in the Lower Mekong River Basin (LMRB) (Browder, 1998). The MRC thus became the replacement mechanism for both the previously established Mekong Committee and the Interim Mekong Committee.

Socio-political context

As a result of extensive planning and feasibility studies conducted during the early stages of the Mekong Committee, the potential benefits of a Mekong Basin-wide agreement were reasonably clear for the LMRBS. Each party had distinct issues that could only be resolved through cooperative development of the LMRB. In brief, Thailand and Viet Nam were interested in reaching agreement on water allocation to promote the larger goal of regional cooperation and solidarity after the Cold War ended in the early 1990s. After decades of regional ideological discord and war, Thailand and Viet Nam did not want water disputes to threaten a more general accord in the region. Moreover, it was recognized that reservoirs in China (which in 1995 were already either under construction or in the advanced planning stage) could significantly increase Mekong flows in the dry season. Considering that water scarcity in the dry season is one of the major water related challenges in the Mekong, the projected increase of flows from the Chinese dams was the catalyst for negotiations of how to share the increased dry season water. In addition, international development agencies were willing to continue funding and providing technical services to the LMRBS to cooperate in managing and developing their common water resources to the MRC. Such assistance was particularly important to Cambodia and Laos PDR, the two least developed states of the group.

Environmental context

Each of the four countries in the LMRB faced issues regarding inconsistent seasonal water flows on the Mekong in the pre-Mekong Agreement stage. Thus, Viet Nam's highest priority was to protect the Mekong Delta, the country's 'Rice Bowl'. The considerable expansion of rice farming over the latter half of the 20th century in Viet Nam required a significant amount of water to be diverted from the Mekong for irrigation purposes. While easily accomplished during the wet season, the Vietnamese needed to maintain at least the existing dry season flows into the Delta to prevent saltwater intrusion and to provide for irrigation. Viet Nam was therefore in favour of dams on the Mekong mainstream if they regulated flows to provide more water availability during the dry season as well as to reduce damaging flooding in the wet season.[5]

For its part, Thailand wanted to ensure access to its share of the Mekong's water. Thailand had plans to expand irrigation to its poor northeastern Issan province through the Khong-Chi-Mun project. Thailand also wanted to divert water from the Mekong into the Chao Phraya Basin in central Thailand. The country's major policy issue was that it did not want these projects, nor any other projects, subjected to prior agreement through the Mekong Committee, which Thailand believed would be unduly burdensome.

Laos PDR's major water concern was to maintain the Mekong in a natural state as a navigable river. Because it was a landlocked country without sufficient land transportation infrastructure – and all major urban centres are located on the Mekong – ensuring a navigable river was vital to Laos PDR. The hydroelectric potential and related revenue that could be generated on the Mekong was also a primary interest to the Laotian government.

Finally, Cambodia's priority was to ensure protection of the Tonle Sap, also known as the Great Lake.[6] Cambodia was concerned that upstream water projects, such as dams in China, Laos PDR and Thailand, would damage the Tonle Sap, which is both central to Cambodian culture as well as of major environmental importance, because it represents one of the largest wetland biodiversity reservoirs in Southeast Asia and is among the most productive fishing grounds in the world.

Development of the Mekong Agreement

Once Cambodia could again participate in efforts to cooperate on multilateral Mekong issues, how this new cooperative platform would operate became a threshold issue. According to Browder:

> When Cambodia's readmission threatened to reactivate the Mekong Committee's constitutional framework – which required approval by the Mekong Committee for all proposed projects – the Thai government called for negotiations to restructure the Mekong regime.
>
> (Browder, 1998, p. 117)

Thailand at this time was lobbying for a new, less stringent system of rules regarding prior notification. This contentious issue resulted in Thailand cancelling a Mekong meeting in Chang Rai in early 1992, when the other participant states proposed rules that would require prior agreement on development (as had been the case under the protocols of the IMC).

In addition, issues over which countries would be involved in the meetings became contentious. For example, Thai delegates were interested in the inclusion of their Chinese and Myanmar counterparts in negotiations. This was opposed by Viet Nam, which wanted a constitution for Mekong River management to be agreed upon by the four downstream countries before moving to management discussions with Myanmar and China, the two countries in the upper reaches of the Mekong River Basin.

The role of the incumbent in the Secretariat office became an issue, as the Thai delegation believed the incumbent was biased in favour of the LMRBS (Browder, 1998). The Secretariat favoured a form of increased cooperation among the four states, which would involve limiting each country's sovereignty over their respective portions of the Mekong. The Secretariat was excluded from the negotiations and reactivation of the Mekong Committee came to a standstill.

The United Nations Development Programme (UNDP) was heavily invested in a Southeast Asia regional management project and was committed to its success. To save the stalled negotiations, the UNDP provided neutral assistance in facilitating a solution to the impasse among the parties by proposing an informal consultation, which took place in Hong Kong in October 1992 (Radosevich, 2009). The success of that meeting led to an historic follow-up meeting in Kuala Lumpur in mid-December that year. At this meeting, the four parties drafted the key points that formed the basis for a future framework for Mekong cooperation, which was officially expressed in a Communiqué and Guidelines after a meeting of the Mekong Working Group (MWG)-I in Hanoi in February 1993. The UNDP agreed to provide technical and financial assistance in formulating a new agreement, and a noted international water law expert was selected by the four countries to serve as the Senior Advisor for negotiating and drafting the document to be approved by the four participating LMRBS.

This 'mandate' of the MWG served to guide the preparation of the draft Mekong Agreement among the parties, along with the various subsequent papers and discussions. The Guidelines contained many important provisions of common interest and mutual acceptance to the parties. Recognizing that 'certain elements of cooperation already exist' that may need redefining, six elements for the future framework of coordination were set out: (1) a set of principles for the sustainable utilization of water resources of the Mekong River system; (2) an institutional structure and mechanism for coordination; (3) a definition of the functions and responsibilities of the structure and mechanism; (4) the legal basis for the governance and financial operation of the structure and mechanism; (5) future memberships of the structure; and (6) management of the structure (Radosevich, 2009).

In October 1992, the LMRBS decided against including China and Myanmar in the negotiations. However, the four lower riparians did agree to negotiate a new framework of cooperation that would be conducive to eventual Chinese and Myanmar (Burmese) participation (Browder, 1998). The negotiation and drafting of the new agreement took nearly two years. There were moments when it was unclear whether an agreement would be realized. However, each participant country recognized that the best alternative to a negotiated agreement ('BATNA') was far less desirable than reaching a formal agreement. This scenario likely encouraged the four countries to make the necessary concessions in order to reach mutual consensus for a cooperative framework on the Mekong.

Absentee riparian countries and organizations

China and Myanmar, whose territories are included in the Upper Mekong Basin, are not parties to the Mekong Agreement. In 1996, as provided for in the Mekong Agreement and explained below, China and Myanmar became official 'Dialogue Partners'. Accordingly, they may send representatives to a special one-day annual fall meeting of the MRC Joint Committee (JC). Should China and Myanmar wish to become parties to the 1995 Agreement – which many believe is necessary for the 1995 Agreement to fully realize its purpose and set goals (Wolf and Newton, 2007; Backer 2007) – they have the potential to do so. As provided for in the Mekong Agreement (art. 39): '[a]ny other riparian State, accepting the rights and obligations under this Agreement, may become a party with the consent of the parties'.

In addition to Member States and Dialogue Partners, some international organizations have the right to attend and participate in JC and Council meetings. The Asian Development Bank, the Association of Southeast Asian Nations, the International Union for Conservation of Nature, the UNDP, the United Nations Economic and Social Commission for Asia and the Pacific, the World Bank and the World Wildlife Fund all possess observer status. Donors have the opportunity of dialoguing with the MRC through the Donor Consultative Group (DCG), established by the MRC to facilitate donor interactions and coordination.

The negotiation process

The negotiations that led to the Mekong Agreement were guided by the principles of customary international water law and the rights and needs of each of the four riparian countries. The four countries established a MWG consisting of five representatives from each country (including at least one from each country's respective Ministry of Foreign Affairs). The MWG would hold five 'formal' meetings ('flags, neckties, seating arrangements and minutes kept') and two 'informal' ('no flags, no ties and no minutes') technical drafting group meetings that would prove crucial to the success of the process (Radosevich, 2009).

The MWG needed to negotiate a new international framework that would protect the fundamental water resource interests of each Member State. This meant that no party should benefit at the expense of another, meaning that all parties 'win' or at least do not 'lose' by cooperating.

Main issues that arose in the negotiations

Compensation

The first major issue that arose in the negotiations was with respect to art. 8 in the Draft Agreement on compensation for damages. The Thai delegation was primarily concerned with the use of the term 'compensation', because such a term would trigger the need for parliamentary approval and ratification. Instead, the Thai delegation suggested that the document should use the term 'responsibility for damages'. The other three country delegations were opposed to Thailand's proposal, but were persuaded that the phrase would in essence mean the same thing. Thus, the terminology suggested by Thailand was adopted.

Dispute resolution

The second major issue in the negotiations focused on the Mekong Agreement's dispute resolution mechanisms. Lao PDR officials proposed that the disputes that remained unresolved through either the MRC or by the respective governments should be referred to the International Court of Justice (ICJ) or to binding arbitration. The Thai government was opposed to this proposal, viewing it as a restriction on their bargaining power vis-à-vis Thai developmental goals. The matter was resolved in art. 35, where without naming the ICJ, if the parties could not resolve the matter through their governments, they would proceed according to the principles of international law, i.e. through the United Nations system and the ICJ.

Minimum flows

Article 6 of the Mekong Agreement would cover several issues related to water levels throughout the year. Provision A of art. 6 mandates a minimum level of natural dry season flow. The Cambodian delegation was able to ensure agreement on an acceptable natural reverse wet season flow (in Provision B) that would be sufficient to sustain Tonle Sap through the dry season until the flow would reverse to the Mekong River. The parties were unable to come to an agreement on a formula to set the minimum flow during the dry season, so the issue was tabled to be determined by the MRC at a later date.

The minimum dry and wet season flows in the agreement would address the Vietnamese concerns about saltwater intrusion, the Laotian concerns about navigability in the summer and the Cambodian concerns about maintaining water levels in Tonle Sap. The Thai government was willing to make concessions in order to foster greater regional cooperation and goodwill. Thai representatives also realized that the forthcoming Chinese dams would increase the amount of flow in the dry season, thus making it a more palatable concession for Thailand.

Power to review

The most difficult issue to resolve in the negotiations related to the power to review proposed water uses (art. 5). Laos PDR, Cambodia and Viet Nam were in favour of a system wherein mainstream projects and diversions of water from the Mekong would be subject to prior agreement by the MRC, as had been the case during the IMC era. However, Thailand was adamantly opposed to other states having veto power over its proposed projects in the Mekong, and accordingly made this issue the primary concern of their delegation. The Thai delegation instead favoured the approach incorporated in the International Law Commission (ILC) Draft Articles on the Law of the Non-navigational Uses of International Watercourses (ILC Draft Articles), which required states to notify and consult with states that potentially could be harmed by a proposed project, and if necessary, to negotiate an agreement with the party/parties. The Thai delegation was further interested in the ILC Draft Articles because it would be more palatable for the Chinese to join the MRC, which the Thai delegation keenly favoured.

In order to make progress on this issue, the Vietnamese delegation proposed a conceptual framework matrix for art. 5. Cambodia was successful in getting the Tonle Sap classified as a tributary, which did not concern the Upper Riparian states considering its downstream position. Viet Nam knew that it was of crucial concern, but its delegation was preoccupied with what it perceived to be more immediate projects in Thailand.

The intent of prior consultation and agreement was not difficult to reach consensus on. However, the intent of notification was disputed between Thailand, who wanted the provision to include 'without any discussion necessary', and Viet Nam, which found this troubling. In the absence of consensus, this section of the definition was eventually left out of the Mekong Agreement.

Throughout negotiations, Thailand wanted to forgo the requirement of prior agreement in dry season flows, as opposed to the other three states, which were in favour of requiring such an agreement. After a number of negotiating sessions on this issue, the sides were nearing agreement until a Thai newspaper article mentioned a Thai Cabinet plan to draw water out of the Mekong for use in Thailand without the approval of the other states. This caused a stir in Vietnam, who emboldened their demands, and caused a slight impasse in negotiations. Nonetheless, Laotian, Cambodian and Vietnamese delegations were subsequently able to come to a consensus on a proposal that satisfied the Thai delegation. The agreement satisfied the Thai delegation because it would not threaten proposed Thai projects, and it limited the scenarios where prior agreement was required before a project could proceed. For the LMRBS, it ensured that projects which threatened their interests at the most vulnerable time (i.e. during the dry season) needed prior agreement.[7]

Key factors leading to successful negotiations

The historical relationships among the LMRBS were a major factor in the successful outcome of the negotiations that produced the Mekong Agreement. The participant countries had nearly 37 years of interaction, which evolved into what was recognized as the 'Spirit of Mekong Cooperation' (Radosevich, 2009). This experience helped to create a strong desire and sense of commitment among the countries to continue their joint efforts to reach an agreement. The LMRBS also had the benefit of shared data and information collected over the course of the life of the MC and IMC.

The ILC Draft Articles, the Helsinki Rules, various other international water treaties, and customary international law provided a solid legal foundation that was influential throughout the course of the negotiations. This platform became a fall-back position when the parties could not come to an agreement on particular issues. Reliance on these international legal norms was aided by the fact that some members of the MWG had received training on international law through their participation in a Legal Studies Group in the Mekong Secretariat. Other key members of the MWG were active members of the ILC's Working Group for the formulation of *the Watercourses Convention*, and the Senior Advisor for the negotiations was a recognized international water law expert.

Several negotiation approaches were utilized to help develop consensus in the negotiations. For example, in order to disengage the participants from their beliefs and perceptions, a 'one-nation basin' scenario was adopted as a hypothetical theoretical framework. Under this scenario, the LMRB was assumed to be one nation and discussions were to be conducted within a shared nation framework. Each participant alternated roles with other representatives in the discussions to create greater empathy at the negotiating table, in an effort to work toward an optimum and equitable solution to the utilization and protection of the water and related resources. National boundaries were subsequently overlain in a later scenario to adjust interests. This process helped the participants to better understand each other's interests and concerns (Radosevich, 2009).

Another useful negotiating approach utilized by the parties, to their great advantage, was to discuss and evaluate various options about the kind of agreement and institution that they believed would be most suitable, and then proceed to the second stage of discussions about the options relating to the range of objectives, principles, specific issues and processes that would be included in the agreement, and ultimately implemented by the established organization. The shared objective in these negotiations was to seek the highest common mutually acceptable option.

Key elements of the Mekong Agreement[8]

Dispute resolution

The Mekong Agreement provides four steps for resolving disputes. The MRC has responsibility for addressing any disagreement first, pursuant to arts. 18.C and 24.F. If the MRC cannot resolve the dispute at this stage, the issue is to be referred to the respective governments, to be resolved diplomatically (art. 35). If this also proves unsatisfactory, countries may refer the issue to a third party for mediation, such as an international organization, an international professional group or an individual. Finally, countries may refer to international law principles (art. 35).

Financing

The MRC's budget consists of contributions from Member States on an equal basis (art. 14), though this refers primarily to the administrative costs for running the MRC. The MRC relies heavily on international donors to fund its projects. Development Partners have pledged more than US$20 million for the 2016–2020 Strategic Planning period for the MRC (Mekong River Commission, 2015). In recognition of this essential donor involvement, art. 41 of the Mekong Agreement specifically acknowledges the contributions of the United Nations and the international community, and expresses the desire to continue these relationships.

Data and information exchange

Data and information exchange among the LMRBS had been ongoing for 40 years in the lead-up to the MRC. However, the Mekong Agreement served to cement this process. The responsibilities related to data and information sharing, exchange and harmonization were split among members of the JC and the Secretariat. The JC is to 'regularly obtain, update and exchange information and data necessary to implement this Agreement' and to 'conduct appropriate studies and assessments for the protection of the environment and maintenance of the ecological balance of the Mekong River Basin' (art. 24). The Secretariat is directed to '[m]aintain databases of information as directed' (art. 30). In this sense the Mekong Agreement only calls for data collection in the most general terms. There remains room for improvement, especially with regard to forecasting models (Schmeier, 2011).

Flexibility

Designing a flexible agreement was seen as imperative to those involved in the negotiations. Thus, the purpose of the MRC was to establish a cooperative institutional framework that would be robust enough to make operational decisions under a variety of future conditions. Conscious of the need to maintain ecological integrity as well as to ensure territorial sovereignty, the provisions of the Mekong Agreement that concern the use of Mekong River waters offer considerable flexibility for varying objectives, while maintaining the protection for the fundamental interests of the LMRBS through its procedural requirements (Schmeier, 2011). For example, art. 5 of the Mekong Agreement, which refers to 'reasonable and equitable utilization' of the Mekong, prescribes a dynamic operational decision-making process rather than a static legal agreement. Countries are to provide notification[9] or prior consultation[10] to the JC of an intended use for the Mekong and its tributaries depending on the season/use, or to come to a prior agreement if such a proposed use is within the most sensitive scenarios. This kind of prior consultation enables all parties to assess and mitigate, or to minimize the harmful effects of each other's actions. Though this mechanism places some constraints on individual country behaviour, it is a useful compromise that reflects both sovereign interests and hydrological perspectives.

Of particular importance is the recognition by the parties, throughout the Mekong Agreement, that the use of waters governed by the Mekong Agreement excludes natural domestic and minor uses of water that do not have a significant impact upon the mainstream flows. This eliminates the doubts and anxieties about unreasonable restrictions of the Mekong Agreement upon riparian rights and normal uses by domestic inhabitants.

The approach of art. 5 of the Mekong Agreement is based upon the universally accepted reasonable and equitable utilization doctrine and the need to maintain certain flows on the mainstream as provided in art. 6. The purpose of this legal framework is to provide a realistic means for cooperation by all riparians who are provided with the practical opportunity to maximize options for trade-offs, exchanges or releases from other sources (i.e. lower tributaries from the same riparian country) to protect the flow levels, while optimizing the sharing and use of waters during the two seasonal variations. This also takes inter-basin impacts on the basin of origin into consideration. In theory, it provides certainty, security and flexibility to all riparians under what may be termed a 'super-fairness' doctrine. Article 37 provides that the Mekong Agreement may be 'amended, modified, superseded or terminated' by mutual agreement of all parties, thereby contemplating and allowing for flexibility in the Mekong Agreement to address future needs.

Implementation and monitoring

Structure of the MRC

There are three organs of the MRC: the Council, the JC and the Secretariat. The Council meets once annually and is responsible for overall MRC governance. It consists of one member from the ministerial or cabinet level from each Member State. The JC is responsible for the implementation of Council decisions and policies, and it functions as a management board. The JC is comprised of one person appointed by each country who is recruited from a rank that is no less than a Head of Department level. The Secretariat provides technical, logistic and administrative assistance to the other MRC organs, and is directed by a Chief Executive Officer (CEO) appointed by the Council. The MRC CEO is responsible for day-to-day MRC operations.

Other functional entities supplement the work of the three main MRC bodies but are not structures of the MRC. The National Mekong Committees (NMCs) coordinate activities at national levels, and link the MRC Secretariat to national ministries and agencies. As mentioned, China and Myanmar are Dialogue Partners to the MRC. China currently provides water level data during flood season from stations located within its boundaries on the Upper Mekong. The DCG provides funding for MRC activities and programmes within the Mekong. It meets with the MRC following Council meetings to discuss strategies and donor harmonization.

Monitoring mechanisms

As the main operational organ of the MRC, the JC is responsible for the implementation of the Mekong Agreement and the Basin Development Plan (BDP), including periodic ongoing assessments of their successes. Article 26 of the Mekong Agreement requires the JC to prepare rules for, inter alia, improving monitoring mechanisms for intra-basin use and diversions from the mainstream. In 2003, MRC countries agreed to a new Procedure for Water Use Monitoring, which is intended to provide an adaptive and comprehensive framework for the implementation of water use and diversion monitoring.[11] It defines the roles of the JC, the Secretariat and the NMCs within this new Monitoring System. A set of guidelines on the Implementation of the Procedures for Water Use Monitoring, developed by a technical committee, provides further input into the operation of the Monitoring System. These procedures are supplemented by Guidelines that form the internal rules of the MRC to implement art. 26. However, some critics view the precise legal statuses of these procedures and guidelines to be unclear, which has weakened their influence on state behaviour (Hirsch *et al.*, 2006).

Assessment of the Mekong Agreement

Implementation

Three months after the draft Mekong Agreement had been finalized, all four countries had ratified. However, in contrast to the relatively rapid rate at which the Mekong Agreement was ratified, the MRC has attracted criticism from some parties because implementation of MRC activities began slowly. The BDP, the main vehicle through which activities in the Mekong are planned and carried out in the Mekong Agreement, did not commence operation until 2001.[12] Many critics also see the MRC as reactive rather than proactive in its approach, particularly with respect to economic development. As a result, there is a perception that the work of the MRC has not necessarily resulted in sufficient tangible benefits for the Member States.

Monitoring

The MRC performs a fair amount of environmental monitoring on the Mekong. The MRC routinely monitors water quality and also publishes annual flood reports that provide hydrological statistics for the flood seasons as well as analysis and recommendations for disaster management. The MRC monitors the successes of the BDP and Strategic Plan through regular reports. In addition to the more general Annual Reports, the MRC publishes yearly Work Programmes which are used for both monitoring and planning purposes (Mekong River Commission, 2015). Monitoring of project progress is done through regular meetings that are held by the JC, as well as through consultations with stakeholders, technical experts and donors.

The data- and information-sharing system is widely considered to be the most effective output of the MRC, and the expansion of the system into the Chinese portions of the Mekong will further build upon this success.

Stakeholder involvement

While the MRC processes are open to external participation, MRC engagement with NGOs and other stakeholders could be considerably improved. Several factors have impeded effective stakeholder involvement in MRC processes. Out of respect for national sovereignty, the NMCs operate autonomously from the MRC. Since each member country allows varying degrees of domestic political freedom, the MRC is powerless to ensure sufficient stakeholder involvement in its activities to satisfy the aspirations of some of its NGO critics.

The nature of external stakeholder involvement within MRC processes is also unclear. MRC documentation is predominately in English, which has meant that much of this reporting output is inaccessible to local stakeholders. This has resulted in significant pressure from a number of NGOs and donor organizations to improve the reporting system. In response, the MRC has enacted formal mechanisms of participation within its governance practices and decision-making institutions (Sneddon and Fox, 2006). The MRC has improved its accessibility to the public through regional Stakeholder Consultation fora on Hydropower and the BDP.

Though China and Myanmar remain Dialogue Partners rather than MRC members, they have increased their level of technical cooperation with the MRC, and have continued to attend all major MRC regional meetings. China continues to provide flood data to the MRC during the rainy seasons, and is a full participant in the MRC Strategic Environmental Assessment of the Mainstream Dams. As China continues to alter the flow of the Mekong through its development of new hydroelectric dams, it is crucial to work in cooperation with the MRC for the overall success of the region.

Sustainability of financing

The MRC's reliance on external donors weakens its capacities in several ways, including slowed implementation of the Mekong Agreement. For example, after the MRC was established in 1995, it identified the creation of the BDP as a top priority. However, the MRC was unable to obtain sufficient funding for initiating the BDP from outside donors until 2001. Furthermore, most external partners donate funds for specific projects, rather than ongoing operational funding. This has created the impression that the MRC is merely a vehicle for managing external donor grants rather than a multilateral political water governance organization.

At times donors have held views that conflict with those of the MRC regarding the future role of the MRC in the Mekong. Conflict in particular has arisen over the primary objectives of the MRC. Donors have been primarily focused on the sustainable management of the MRB, which has conflicted with Member State priorities to develop more hydropower and a de-emphasis on the negative ecological effects that such power generation may create. The concern is that the strong donor presence in the MRC's work may encourage a weaker commitment from riparian countries.

Summary

Although the MRC appears to have made progress with respect to many objectives, it has nevertheless been criticized by both Member States and the international community for not being sufficiently effective. Member States find the work of the MRC to be lacking in tangible results, particularly with respect to economic development and the BDP. Since the formation of the MRC, the BDP has become the centrepiece of the MRC's work and is seen by both the Mekong governments and international donors as a 'primary rationale for the institutional existence of the MRC' (Sneddon and Fox, 2006). However, many critics deem progress on the BDP to be very slow, and that the BDP is not aligned well enough with the actual needs and aspirations of the Member States. The MRC itself noted that delays in the BDP's development might have contributed to perceptions that the MRC procedures are restraining rather than facilitating mechanisms. In particular, the BDP has experienced difficulty in anticipating and guiding the discussion on mainstream dams, an area that has become a focal point of the MRC efforts at present.

One source of the difficulty in the operation of the MRC is its institutional structure, which often pits the interests of donors in the Secretariat against those of the Member States in the Council. Generally speaking, the Secretariat has served as a liaising body between the MRC and donor countries, and in the minds of critics, too often the work of the Secretariat has reflected the policies of donor organizations, whereas the work of the JC and Council has represented the interests of the member countries. When the donors approach the Secretariat with their project assistance plans (with which the JC and Council might not agree), the Secretariat is put into the position wherein it must convince the JC and the Council that donor priorities for project funding will not trump Member State economic development goals, particularly those involving the development of more hydroelectric dams. International donors, while not always possessing a unified perspective, have not necessarily been opposed to the further construction of dams. However, they have been adamant that these dams be built in conformance with recommendations of the World Commission on Dams. The different viewpoints on priorities in development may lead to high-profile disagreements. The controversy surrounding the Nam Theun 2 Dam project has prominently illustrated this internal conflict (Hensengerth, 2009).

One of the major successes of the MRC has been the development of the Water Utilization Programme (WUP), which gives effect to the water allocation framework of the original agreement. Since 2000, this Global Environment Facility (GEF)-funded programme has been a key contributor to the success of the organization. Project activities of this programme have included basin modelling and knowledge base programmes, environmental and trans-boundary analyses, rules for water utilization and management, and institutional strengthening (Mekong River Commission, 2015). Some of the accomplishments from this programming include: (1) the ability to assess impacts of proposed developments, to monitor, tailor and minimize negative impacts and ensure more consistent dry season flow (in dam development); (2) good working relationships developed through the experiential learning-based drafting processes of several of the main water utilization and technical guidelines (i.e. data and information sharing); and (3) spurring of institutional growth of various multilateral working groups that are needed to implement the procedures developed under the plan.

As the WUP illustrates, the MRC's major strength has been its ability to generate shared data and information regarding the Mekong. This includes information about water quality and flow monitoring as well as a flood forecasting system, development scenarios, technical directions for scenario analysis, a comprehensive hydropower database and other technical databases (Mekong River Commission, 2015). The improvement in water monitoring has increased environmental awareness within agencies and local authorities in the region. For example, the Fisheries Programme of the MRC has conducted extensive research that is providing valuable information that needs to be considered in the planning of future developments. This shared information flowing from the work of the MRC is the essential basis to build science-based fishery management plans needed for effective management of this resource on which so many people in the basin are dependent. The approach has been to coordinate the programmes of each Member State to help them implement sustainable fisheries management and development at the local, national and regional levels (Mekong River Commission, 2015).

Conclusions

In the 37 years prior to the Mekong Agreement, the LMRBS were able to cooperate to varying degrees on the management of the Mekong. This history served to facilitate negotiations of the MRC. In addition to this 'Mekong Spirit' (of cooperation), the negotiations leading to the Mekong Agreement were assisted by: (1) a highly experienced UNDP consultant that facilitated negotiations and helped the parties to find common ground; (2) delegations that had a strong grasp of international law which served as an important starting point for negotiations; and (3) the BATNA for each country was less palatable than what could be gained through agreement. As a result of these factors, the countries were willing to make the necessary concessions to accommodate each delegation's primary interests and eventually reach agreement. Considering the lack of other multilateral bodies operating in the region, and the breadth and importance of the issues in discussion, the four countries were able to reach consensus in a remarkably short amount of time.

While the Mekong Agreement was critical to providing a legal and institutional framework to facilitate development in the Basin, it is the actual programmes developed through this framework that will be the real determinant of the success of the Mekong Agreement. Some of the major successes of the MRC have been in the areas of fisheries, navigation, flood management and, especially, the WUP. Major conflicts have occasionally presented themselves between the respective national governments, who have been pursuing large-scale hydroelectric development, and the donors, who appear to be more concerned with the ecological issues arising from these proposed projects.

There has also been a relative dearth of public participation in the functioning of the MRC, and this absence of consultation has become an area of particular focus for improvement by the MRC. This issue is especially concerning when considering the segments of society that have the most to lose from large-scale hydroelectric development are often the ones with no say in these matters.

One of the biggest remaining concerns for the MRC is that China and Myanmar, the two riparian countries in the upper reaches of the Mekong, are not members of the MRC. In spite of an increasing amount of information being shared between the MRC and these two official Dialogue Partners, more cooperation is needed to manage the Mekong more effectively. With facilitative support from regional organizations, the LMRBS entered into a cooperative dialogue with the two upstream riparians, China and Myanmar – The Greater Mekong Subregion Economic Cooperation Program Strategic Framework (Leb, 2013).

As China is exhibiting a growing interest in engaging as a responsible citizen in the international community, it is reasonable to predict that China will increase its participation in the MRC. However, China will need to be persuaded that the MRC will be sensitive to the best interests of all the riparian countries (including China) in the way that the MRC facilitates the management and development of the Mekong (Lee, 2015).

A major obstacle to embracing this more inclusive view of the role of China and Myanmar in the work of the MRC is the objections being voiced by the donor countries that have supported the MRC in the early stages of its life. Over time the MRC may need to become more independent of the influences from outside the region in order to do a better job performing its optimal future role inside the region.

There is no doubt that the LMRBS could all benefit from increased cooperation with China, especially in relation to the potential changes in climate and the increased dry season flows that the Chinese dams may provide for their downstream neighbours. The institutional memory of how the four original members of the MRC forged the Mekong Agreement might provide some key lessons about how to proceed with an increased role for China and/or Myanmar in the work, and ultimately the membership and governance of the MRC.

Notes

1 This chapter draws on hitherto unpublished research conducted by the Global Transboundary International Waters Initiative at the University of British Columbia in Vancouver, Canada, pursuant to a grant from the Global Environment Facility (GEF). The assistance of Mauria Curlier and Theresa Etmanski in helping to produce the case study upon which this chapter is based is gratefully acknowledged. This case study was peer reviewed by well-known Mekong experts George Radosevich, Olivier Cogels, John Dore, John Metzger and Pech Sokhem, whose advice and assistance is also gratefully acknowledged.

2 The terms 'international drainage basin', 'international river' and 'international watercourse' are used synonymously throughout this chapter. Not all academic scholars necessarily agree that the terms may be so used interchangeably.

3 The *Helsinki Rules* (Campioni Consolidation) and the International Law Association (1966, 1987) *Commentary to the Helsinki Rules on the Uses of the Waters of International Rivers*, ILA Report of the 52nd Conference, Helsinki, Finland, pp. 484, 484–505, arts. J-XI, 4. Coming from the non-governmental International Law Association (ILA), the *Helsinki Rules*, a predecessor to the *1997 UN Watercourses Convention*,

are not intergovernmentally authoritative, technically speaking. However, they very much reflect many years of research by a representative body of renowned international law experts, and therefore come easily within the terms of art. 38(1)(d) of the Statute of the International Court of Justice.

4 These rules are generally utilized by international organizations. For example, the World Bank has a number of documents reflecting these principles: Bank Operational Policies (OP 7.50): Projects on International Waterways; Bank Procedures (BP 7.50): Projects on International Waterways; Bank Good Practices (GP 7.50): Projects on International Waterways.

5 Viet Nam's stance in regards to dams has since changed. At the time of writing, Viet Nam was in favour of imposing a 10-year moratorium on mainstream dams.

6 The Tonle Sap Lake and River is a tributary to the Mekong mainstream, and during the wet season, high water flows actually reverse the direction of the river to fill the Lake where it is stored until the dry season, when the river flow is again reversed. In this process, the Lake expands in area by a multiple of five, from 2,000 km^2 to 10,000 km^2, and temporarily stores 15–20% of the Mekong's annual flow of 475 billion m^3.

7 Inter-basin use involves the transfer of water out of the Mekong River Basin into another river basin, while intra-basin implies that the water is used within the Mekong River Basin.

8 See UNDP-GEF International Waters Project (2011) 'International waters: Review of legal and institutional frameworks', International Waters Learning Exchange & Resource Network, Washington, DC, *http://iwlearn.net/publications/misc/governing-marine-protected-areas-getting-the-balance-right-main-report-lower-resolution-2mb,* accessed 24 June 2015.

9 Ch II Definitions of Terms, defines Notification as: 'Timely providing information by a riparian to the Joint Committee on its proposed use of water according to the format, content and procedures set forth in the Rules for Water Utilization and Inter-Basin Diversions under Article 26'.

10 Ch II Definitions of Terms, defines Prior Consultation as: 'timely notification plus additional data and information to the Joint Committee . . . that would allow the other member riparians to discuss and evaluate the impacts of the proposed use. . . . Prior consultation is neither a right to veto . . . nor the unilateral right to use water by any riparian without taking into account the other riparians' rights'.

11 Ch II Procedures for Water Use Monitoring.

12 The Final Report of the BDP Phase 1 project noted that decision-making processes turned out to be 'lengthier than originally anticipated', which was a factor contributing to the plan's extension.

References

Agreement on the Cooperation for the Sustainable Development of the Mekong River Basin, signed 5 April 1995, 2069 UNTS 3 (entered into force 5 April 1995).

Backer, E. (2007) 'The Mekong River Commission: Does it work, and how does the Mekong Basin's geography influence its effectiveness?', *Südostasien Aktuell*, 26, 4, 31–55.

Bearden, B. (2010) 'The legal regime of the Mekong River: A look back and some proposals for the way ahead', *Water Policy*, 12, 798–821.

Browder, G. (1998) 'Negotiating an international regime for water allocation in the Mekong River Basin', unpublished PhD dissertation, Stanford University, Palo Alto, CA.

Campione Consolidation of the International Law Association Rules on International Water Resources (Helsinki Rules) (adopted 20 August 1966).

Convention on the Law of the Non-Navigational Uses of International Watercourses, opened for signature 21 May 1997, UN registration no I-52106 (entered into force 17 August 2014).

Gabčíkovo-Nagymaros Project (Hungary/Slovakia) (Judgment) [1997] ICJ Rep 375.

Global Environment Facility (2013) 'GEF-6 Programmatic Directions', Global Environment Facility, Washington, DC, https://thegef.org/gef/GEF6-Programming-Directions, accessed 20 June 2015.

Hensengerth, O. (2009) 'Trans-boundary river cooperation and the regional public good: The case of the Mekong River', Contemporary Southeast Asia: A Journal of International and Strategic Affairs, 31, 2, 326–49.

Hirsch, P., Jensen K. M., Boer, B., Carrard, N., FitzGerald, S. and Lyster, R. (2006) 'National interests and trans-boundary water governance in the Mekong', Australian Mekong Resource Centre and Danish International Development Assistance, Sydney, Australia.

International Law Commission (1994) 'Draft Articles on the Law of the Non-navigational Uses of International Watercourses', *Yearbook of the International Law Commission*, 2, 2, 89–135.

Leb, C. (2013) *Cooperation in the Law of Trans-boundary Water Resources*, Cambridge University Press, New York.

Lee, S. (2015) 'Benefit sharing in the Mekong River Basin' *Water International*, 40, 1, 139–52.

McCaffrey, S. (2001) 'International water law for the 21st century: The contribution of the UN convention', *Water Resources Update*, 118, 11–19.

Mekong River Commission (1995) 'Agreement on the Cooperation for the Sustainable Development of the Mekong River Basin' (the Mekong Agreement), Vientiane, Laos and Phnom Penh, Cambodia.

Mekong River Commission (2015) 'Mekong River Commission', Mekong River Commission, Vientiane, Laos and Phnom Penh, Cambodia, www.mrcmekong.org, accessed 20 June 2015.

Molle, F., Foran, T. and Flach, P. (2009) 'Introduction: Changing waterscapes in the Mekong region – historical background and context', in F. Molle, T. Foran and M. Kakaonen (eds) *Contested Waterscapes in the Mekong Region – Hydropower, Livelihoods and Governance*, Earthscan, London.

Öjendal, J. and Jensen, K. (2012) 'Politics and Development of the Mekong River Basin: Trans-boundary Dilemmas and Participatory Ambitions', in J. Öjendal, S. Hansson and S. Hellberg (eds) *Politics and Development in a Trans-boundary Watershed*, Springer, New York.

Radosevich, G. (2009) 'Mekong River basin, agreement & commission case study', NEGOTIATE Case Studies, International Union for Conservation of Nature, Gland, Switzerland.

Schmeier, S. (2011) 'Resilience to climate change-induced challenges in the Mekong river basin: The role of the MRC', Working paper no 61810, World Bank, Washington, DC.

Sneddon, C. and Fox, C. (2006) 'Rethinking trans-boundary waters: critical hydropolitics of the Mekong Basin', *Political Geography*, 25, 181–202.

Wolf, A. and Newton, J. (2007) 'Case Study Trans-boundary Dispute Resolution: the Mekong Committee', Program in Water Conflict Management and Transformation, Corvallis, OR, http://trans-boundarywaters.orst.edu/research/case_studies/Documents/mekong.pdf, accessed 24 September 2015.

4 Trans-jurisdictional water governance in the European Union

Marleen van Rijswick

Introduction

This chapter focuses on trans-jurisdictional issues in European Union (EU) water law. The topic of trans-jurisdictional water governance in the EU has been the focus of detailed work across a range of issues, including trans-boundary water governance; human rights and water; water law and consumer law; water and spatial planning; water and agriculture; water and the use of medicines; water and pesticides; water and energy; freshwater and marine protection; and water law and climate change. This chapter combines insights from the author's earlier research and shows how external integration (integration in other policy fields) in the EU is organized, with a focus on EU practices in the field of trans-boundary water management, freshwater supply, water and nature conservation, and water and internal market legislation.

Three trans-jurisdictional issues in the EU's water law are discussed. The first is the geographical and hydrological scope of water management, often related to trans-boundary river basins, which require specific cooperation mechanisms which are not available in EU water law. A second is the need for a common understanding of definitions, concepts and legal obligations in EU Member States, which is necessary for trans-boundary river basin management and a level playing field. This common understanding is needed because each EU Member State has an individual obligation to comply with EU law as well as a shared responsibility to reach the goals of European water law. The third topic discussed is the fragmented and sometimes conflicting legal frameworks in the field of water governance and adjacent policy fields. Water and nature conservation, freshwater supply and land use, and water and the authorization of medicines are used as examples. The analysis explores how best to deal with these trans-jurisdictional issues, paying attention to internal integration (integration within the ambit of water law) within the broad scope of water management and ways to improve external integration (integration in other policy fields).

The chapter is divided into seven parts: Part 1: EU water law: towards an integrated, adaptive ecosystem-based approach; Part 2: multi-level governance of trans-boundary river basins; Part 3: objectives for trans-boundary rivers – a need for a common understanding of the main concepts; Part 4: water management and nature conservation; Part 5: freshwater supply as a trans-jurisdictional issue; Part 6: water management and product policies – inadequate powers due to fragmented legal frameworks; Part 7: conclusion and the way forward.

EU water law: Towards an integrated, adaptive ecosystem-based approach

EU water law developed as a result of many individual water problems. At the beginning in the 1970s, before the Water Framework Directive (WFD) (EC 2000) entered into force at the end of 2000, the focus was on the protection and improvement of water quality as part of European environmental law. A top-down approach, with regulation and standard setting at the European Community level was used. With the coming into force of the WFD plans and programmes, combined with more policy freedom for the Member States became the main instrument to facilitate adaptive water management (van Rijswick and Havekes, 2012; van Rijswick and Keessen, 2012; van Rijswick *et al.*, 2010). Although the EU requires effective legal protection against decisions concerning the aforementioned instruments, a strong litigation position for citizens depends upon the national legal systems in the numerous Member States. In, for example, the Netherlands, citizens have a stronger litigation position when using clear regulation and standard setting because there is no administrative law appeal possible in the case of plans and programmes.

However, by 2000 a new approach to water management was introduced by the WFD. The purpose of the Directive is, according to art. 1, the establishment of a framework for the protection of inland surface waters, transitional waters, coastal waters and groundwater which prevents further deterioration and protects and enhances the status of aquatic ecosystems and, with regard to their water needs, terrestrial ecosystems and wetlands directly dependent on the aquatic ecosystems. The WFD promotes sustainable water use based on the long-term protection of available water resources, and aims to realize the enhanced protection and improvement of the aquatic environment. It should contribute to mitigating the effects of floods and droughts, and thereby contribute to the provision of a sufficient supply of good-quality surface water and groundwater. Finally, it aims to achieve the objectives of relevant international agreements (e.g. the Helsinki Treaty). These objectives immediately make clear that trans-jurisdictional issues lie at the heart of EU water governance.

The river basin management approach, the most important new element, introduced shared responsibilities for all Member States sharing a transnational river basin. This development of an ecological or ecosystem approach required regional physical circumstances to be taken into account. It is now up to the Member States or river basin authorities to formulate ecological goals and standards, and to choose the necessary accompanying instruments. The protection of waters as an ecosystem and the aim to provide users with water services can mainly be deduced from EU water directives (van Rijswick, 2011). Water governance is regulated by means of several EU water directives and is strongly related to regulation in other policy fields. Most water directives have a twofold scope: protecting human health and protecting the environment.

The new approach also urged further integration with other policy fields, including environmental policy, nature conservation, land use, agriculture, product policy, climate change and energy policies. The protection against flooding (EC 2007), protection of and the sustainable use of the marine environment (EC 2008) and the sustainable use of water resources has each fundamentally changed the scope of European water law. Both the substantive content of EU water law and the instrument of framework directives changed, offering more flexibility and policy discretion for Member States and a greater focus on proceduralization and public participation, bringing about a new era in water management and law (van Holten and van Rijswick, 2014).

EU water governance accordingly became more multi-levelled, with regulation at the international level, the EU level, the Member State level and the regional and local levels. Cooperation was required between the EU and Member States, Member States and non-member states, regional authorities, authorities in different policy fields (multi-sector governance), and between governmental authorities and non-governmental organizations, stakeholders and the public (multi-actor governance). This occurred through defining goals and standards, the development of river basin management plans and programmes of measures, the use of exemptions, and in the executive phase, the taking of measures (van Rijswick *et al.*, 2010).

In the following sections, specific attention will be paid to three examples that illustrate the trans-jurisdictional nature of modern EU water law, followed by a detailed discussion on how to shape external integration.

Multi-level governance of trans-boundary river basins

The WFD opts for the coordination of administrative arrangements and the realization of environmental goals by way of a *river basin approach*. This means that Member States have to divide their territory into districts based on river basins (art. 3). Every river basin requires a drawing of the characteristics of the river basin district, a review of the environmental impact of human activity and an economic analysis of the water use (art. 5).

Appropriate administrative arrangements, including the identification of the competent authority, must be made for each river basin district. In the case where a river basin covers the territory of more than one Member State, that river basin should be assigned to an international river basin district. The Member States must ensure that the requirements of the WFD in regard to achievement of the environmental objectives are met, and in particular that all programmes of measures are coordinated for the international river basin district as a whole. This obligation demonstrates the first important trans-jurisdictional issue. It is a noble goal of the WFD, but as highlighted by earlier research, the EU does not offer specific instruments to shape this cooperation (van Rijswick *et al.*, 2010).

A Common Implementation Strategy has been developed in which EU Member States work towards coherent implementation in Member States' territories. Because the EU does not have many specific instruments for trans-boundary cooperation, it is advised or allowed to use the traditional treaty system as an instrument for trans-boundary cooperation. Examples include the use of the Rhine Treaty and the International Rhine Commission, the Meuse Treaty and the Danube Treaty. International cooperation often takes place by virtue of multilateral conventions. These conventions regulate cooperation between states, by stipulating obligations for the parties, determining principles to be taken into account in trans-boundary cooperation, and guiding coordination in implementing the conventions. However, there are some difficulties in and limits to the use of multilateral conventions. Hey and van Rijswick (2011, pp. 227–49) state that '[e]stablished case law of the European Court of Justice illustrates that the court regards treaties that have a link with European law as part of the European legal system and it interprets and applies them accordingly'. Based on analysis of the European Court of Justice's case law, they find that European law takes precedence over international treaties within the legal order of the Community. This entails the requirement that agreements and decisions taken within the framework of such treaties should be compatible with European law.

Keessen *et al.* (2008) argue that, where trans-boundary cooperation does not lead to intended results, one may consider the use of instruments provided by European law to enforce compliance or to participate in administrative procedures (participation, complaints, appeals) in neighbouring countries. A Member State can hold another Member State responsible if the latter fails to take adequate measures – for example, by failing to combat water pollution. Moreover, a Member State can call on the European Commission's assistance when consultations fail to result in a joint approach to water pollution or floods. Although mediation by the Commission does not result in binding arrangements, the Commission can bring a matter before the European Court of Justice to demand compliance, and if need be, demand a penalty and penalty payments. Member States will often not institute proceedings due to non-compliance by virtue of art. 227 of the Treaty on the Functioning of the EU (EU Treaty) after mediation by the Commission, because this is a politically sensitive thing to do. Finally, they discuss the option of civil actions as a last resort to recover the cost incurred from trans-boundary pollution by the responsible Member State, although this may also hamper future cooperation. Actions by the Commission are therefore the most likely option in the event of failing cooperation. An example of this is the letter the European Commission sent to the Netherlands, in which the Commission forced the Netherlands to take care of full implementation of the agreements made in the river basin management plan for the Rhine river basin, in particular with regard to the opening of the Haringvliet dam to enable fish to swim upstream. This letter was sent after a request from the other Member States that share the Rhine river basin.

Objectives for trans-boundary rivers: a need for a common understanding of the main concepts

An interesting feature of modern European water law is that it aims to achieve a specific 'status'. Groundwater and surface water in a river basin should hold 'good status' under the WFD, while the Floods Directive stipulates that flood risks should be manageable in order to reduce the risk as well as the consequences of a flood for humans, the environment, cultural heritage and the economy. The objective of the Framework Directive Marine Strategy (EC 2008) is the 'good environmental status' for marine regions. The obligations arising from the European water directives are, in principle, aimed to achieve a certain result for each individual Member State (art. 249, EC Treaty). In this section, I focus on common understandings of important concepts in the WFD, which are essential for European trans-jurisdictional river basin management.

The environmental objectives of the WFD can be found in art. 4. These relate to surface waters, groundwater and protected areas, and call for a 'good' water status to be reached by 2015. This can be seen as one of the most important substantial elements of the protection regime. Good surface water status means that the status achieved by a body of surface water is at least ecologically and chemically 'good'. Good ecological status may vary depending on the kind of water. For artificial and heavily modified surface water it will suffice to reach a good ecological potential.

Most environmental objectives in art. 4 are further elaborated in *environmental quality standards*, which aim to give substance to the notion of 'good' water status. For rivers, lakes, coastal waters and transitional waters, a good surface water condition, including good ecological and good chemical status, has to be achieved. Achieving a good chemical status is mandatory for all waters, including artificial and heavily modified surface water bodies. Article 4(3) enumerates the conditions under which bodies of surface water may be designated as artificial or heavily modified. For these surface waters, the environmental objectives are less stringent. It is sufficient that good chemical status and good ecological potential are realized. Groundwater must also have a good status, which means that both the quantitative as well as the chemical status must be good. Any significant and sustained upward trend in the concentration of any pollutant resulting from the impact of human activity should be reversed, in order to progressively reduce the pollution of groundwater and to ensure the balance between abstraction and recharge of groundwater.

The Directive requires the establishment of a register of all areas lying within each river basin district that have been designated as requiring special protection under specific Community legislation for the protection of surface water and groundwater, or for the conservation of habitats and species directly depending on that water source. The register or registers shall include all bodies of water identified as waters used for the abstraction of drinking water; areas designated for the protection of economically significant aquatic species; bodies of water designated as recreational waters, including areas designated as bathing waters under the Bathing Water Directive (EC 2006); nutrient-sensitive areas, including areas designated as vulnerable zones under the Nitrates Directive (EC 1991a), and areas designated as sensitive areas under the Urban Waste Water Directive (EC 1991b); and areas designated for the protection of habitats or species where the maintenance or improvement of the status of water is an important factor in their protection, including relevant Natura 2000 sites designated under the Habitat Directive (EC 1992) and the Birds Directive (EC 1979). The register summary, required as part of the river basin management plan, shall include maps indicating the location of each protected area and a description of the community, national or local legislation under which they have been designated. Although it should be highly appreciated that the WFD emphasizes a direct relationship between WFD obligations and requirements following from adjacent policies, the following trans-jurisdictional issue demonstrates the complexity of the approach taken by the EU.

Water management and nature conservation

The WFD requires that Member States shall achieve compliance with any standards and objectives for protected areas by no later than 2015 (art. 4). Important protected areas are designated under the regime of the Birds and the Habitats directives, but in these nature conservation directives no time limits to achieve the goals are foreseen. Where more than one of the objectives relates to a given body of water, the most stringent shall apply. There is in fact a large physical overlap between the water bodies that fall under the protective regime of the WFD and the Natura 2000 sites (de Smedt and van Rijswick, 2015). This applies in particular for the Lowlands of the Netherlands and Flanders, where a large part of the surface water and the terrestrial ecosystems dependent on them are designated as Natura 2000 sites. There are also strong analogies between the objectives of protection, as the WFD aims for the protection of a healthy aquatic ecosystem and the Habitats Directive has the same goal, albeit with a focus on the preservation and protection of habitats and species. Since both directives aim to protect the same aquatic ecosystem, questions related to the cumulative application of the protective regime arise. After analysing both directives, I conclude that the complexity of the WFD is caused by the wording of the environmental objectives, its references to an extensive number of annexes, and the application of the accompanying derogation regime. The many cross-references to other Directives (relevant to water), such as the Habitats Directive, increases interpretation and application problems since the most stringent objective determines the level of protection to be achieved. This may lead to diverse implementation strategies in the several Member States, problematic cooperation in trans-boundary river basins and, finally, non-achievement of the goals.

Although the objectives of the WFD must be achieved by 2015, this time limit may be extended under the conditions mentioned in the Directive (art. 4(4)). One of these conditions is that no further deterioration of the status of water bodies occurs. However, there is still a great deal of uncertainty regarding this principle of no deterioration, leading to different outcomes in several Member States. Recently, however, more clarity has been given by the European Court of Justice in *Bund für Umwelt und Naturschutz Deutschland e.V. v. Bundesrepublik Deutschland* (ECJ,C-461/13) (van Rijswick and Backes, 2015).

It is also possible for Member States to pursue less stringent environmental objectives if waters are so profoundly affected by human activity, or their natural condition is such that the achievement of the general objectives would be infeasible or disproportionately expensive (art. 4(5)). This is only possible under certain circumstances. Article 4(6) offers the possibility of an exception in the case of a temporary deterioration only, while art. 4(7) mentions several grounds for exceptions in the case of new modifications to the physical characteristics, or when the failure to fulfil obligations is the result of new, sustainable human development activities.

To conclude, the complex relationship between nature conservation legislation and the legal regime for protected areas under the WFD raises three issues: (1) how to deal with the clear time limits of the WFD and the lack of time limits under the Habitats Directive; (2) how the 'most stringent' requirement should be treated; and (3) does the derogation regime apply (in full) within these areas? To date, the Court of Justice has not yet handed down a decision on these issues.

Freshwater supply as a trans-jurisdictional issue

The supply of freshwater is an important aspect of water management. Climate change can cause an increasing shortage of water, with consequences for water safety, drinking water supply, agriculture, forestry, fisheries, tourism, health care, buildings, industrial and transport infrastructure, the energy sector (including energy infrastructure), the environment and nature. The protection of sufficient water for food production and economic activities is also based on the WFD, while the protection of water necessary for well-functioning ecosystems is based on the WFD, the Birds Directive and the Habitats Directive. The urge to guarantee freshwater supply is becoming increasingly clear due to, among other things, the effects of climate change. However, freshwater supply and the allocation of freshwater are hardly regulated under EU water law; therefore we also have to look at non-binding European policy instruments.

Fair allocation of scarce water among the various functions within a river basin is becoming more and more important (van Rijswick 2015). Various policy initiatives have been developed for that purpose. The sustainable and balanced use of natural resources is one of the objectives of EU environmental policy. Dealing with drought and the prevention of water scarcity is indeed mentioned, but is unfortunately not further elaborated in the WFD because its emphasis is mainly on the protection and improvement of water quality. In addition to a climate change adaptation strategy at the European level (European Commission, 2013), a strategy for water scarcity and drought has been developed (European Commission, 2007). For the time being no new legislative instruments will be developed because the Member States nowadays prefer non-binding strategies above legislation at the EU level. This gives them more policy freedom.

The protection and allocation of freshwater is elaborated at the European level in the strategy for water scarcity and drought, which aims to achieve greater water efficiency and better management of the demand for water. This is considered a task for the Member States, and should be carried out by measures of adaptation, sustainable practices, greater water savings, fiscal instruments, monitoring systems and adapted risk management tools. The most important elements in this strategy for water scarcity and drought show the importance of trans-jurisdictional water governance issues, such as integration between different sectors and actors – for example (1) the improvement of land-use planning as a driver for water use, since many economic activities take place in areas where there is insufficient water available; (2) the integration of water-related concerns in other policy areas; (3) active promotion of the use of market-based instruments (European Commission, 2000); (4) the amendment of river basin management plans to identify areas with water scarcity, to include drought risk management plans with observatories, warning systems and civil protection (art. 13, para. 5 WFD); and (5) legislative proposals on water-saving, such as the development of standards for water-using devices, including water efficiency criteria in performance standards for buildings and the use of certification systems.

Formal, legally regulated aspects of freshwater supply focus on the protection and improvement of the available amount of water. The actual allocation of scarce water among the different functions is mainly set out in the plans and programmes. The relevant mandatory measures in the programme of measures (art. 11, s. 3, WFD) show the trans-jurisdictional nature of the freshwater supply, since they include measures from the Environmental Impact Assessment Directive, the Habitats Directive, the Birds Directive and the Bathing Water Directive. In order to ensure that the right amount of water is in the right place, there is also an obligation to include measures concerning the abstraction of freshwater and the impoundment of fresh surface water, including registers of water abstractions and a requirement of prior authorization for abstraction and impoundment (art. 11, s. 3(e)). Prior authorization is also required for the artificial recharge or augmentation of groundwater bodies (art. 11, s. 3(f)).

If the failure to achieve these objectives is caused by problems that cannot be solved at the level of the Member States – for instance, if other states take measures within a river basin which obstruct sustainable and efficient water use in the whole river basin, or take measures as a result of which there is not enough water within a river basin for other states – then a Member State can bring the problem to the European Commission (art. 12). Notwithstanding the Commission's recommendations in the strategy for water scarcity and drought, it is unclear to what extent Member States may impose far-reaching requirements for saving water on products, buildings and infrastructure, without coming into conflict with the EU internal market requirements (e.g. Chapter 3, 'Prohibition of Quantitative Restrictions Between Member States' in the Treaty on the Functioning of the European Union).

We can conclude that EU environmental law and policy offer instruments to facilitate the protection of freshwater resources and to prevent water shortages. But again we see a fragmented legal framework that may hamper effective protection and that strongly relies on national initiatives by Member States, effective cooperation in trans-boundary river basins and the effective external integration of several policy domains.

However, clarity is needed on the question of how far Member States have the power to set far-reaching product requirements without infringing the internal market rules or to regulate land use, which is outside the competence of the EU.

Water management and product policies: inadequate powers due to fragmented legal frameworks

The combination of, on the one hand, the geographical area approach, which requires Member States to cooperate with other Member States (and third countries, if need be), and on the other hand, an objective that has to be met by each Member State individually, creates a fragile balance. It is not clear whether Member States can be held responsible for not achieving objectives over which they do not have total control. One reason may be inappropriate cooperation, as described above. Another reason will be discussed in this section; that is, the inability of EU Member States to take the measures required in other policy fields, because they lack (adequate) powers to do so. This is mainly the case in the field of agricultural policy (D'Hondt 2003), fisheries policy (Wakefield 2009) and the admission of products, for example drugs and pesticides (Montforts *et al.*, 2004, 2006; van Rijswick and Vogelezang-Stoute, 2008). In the case of inadequate power, Member States have to rely on initiatives taken by Community institutions.

Keessen *et al.* (2010b, p. 1) speak of a clash between the internal market and environmental legislation, and discuss as an example of trans-jurisdictional water governance the clash between medicines and water regulation. The regulatory challenge that lies at the basis of this clash is the integration of environmental concerns into internal market legislation. The challenge is to limit water pollution caused by the use of medicines that benefit from free movement on the internal market. This study analyses both water legislation as well as the legislation that regulates the allowance of medicines on the internal market, and offers valuable insights on the role of regulation and governance techniques in reconciling regulatory challenges under EU law, in particular in the context of environmental law and internal market law.

European medicines legislation completely harmonizes the regulation of placing medicines on the internal market. It provides for authorization procedures to establish the quality, effectiveness and safety of a medicine, and for a pharmaco-vigilance system to evaluate these aspects once medicines are on the market and are being used. Marketing authorization is required before a medicine can enter the market of an EU Member State. Either a national competent authority issues an authorization decision for its territory (the so-called decentralized procedure) or the Commission issues an authorization decision for the entire European territory on the basis of advice from the European Medicines Agency (EMA, formerly called EMEA) (the so-called centralized procedure) without any further implementing acts by the Member States being required (Keessen, 2009). When a Member State issues a marketing authorization, other Member States can use the mutual recognition procedure.

In 2003, an environmental assessment was introduced in the European medicines legislation to understand and mitigate the environmental risks of medicines, which may foster external integration between the two policy domains. However, the environmental assessment is limited because it is only mandatory for medicines that entered the market after 30 October 2005, when the environmental assessment was introduced. Another problem is that it is not obligatory to make the environmental information public and there is uncertainty about commercial sensitivity. Also important is the fact that the environmental information hardly plays a role in the authorization of medicines. The environmental risks are excluded from playing a role in the risk–benefit balance of medicines for human use and, in the case of medicines for veterinary use, they only carry limited weight. The lack of coordination between water and medicines regulation diminishes the usefulness of the environmental risk assessment of medicines and may place the achievement of good chemical or ecological water status at risk (Vos and Janssen, 2005). The health benefits of medicines may continue to override environmental concerns, just like their free movement may continue to override any perceived needs for setting stricter national standards. Thus, the question is how the environmental impact of medicines can be minimized, remaining as much as possible within the present, European regulatory framework.

Unfortunately, the fragmented European legal system does not provide for a ready-made solution to settle conflicts between European legislative acts. Despite the obligation to integrate environmental concerns into other policies, the problematic relationship between water and medicines legislation provides an example of how difficult this integration is. One reason for this inadequate coordination is that there is no ranking between Regulations and Directives, with the exception of framework and subordinate directives such as the WFD and the Groundwater Directive; although even in that case there is no formal ranking, as daughter directives remain separate directives. Just like the objectives they pursue, they are equal in principle.

Thus, a Member State with water pollution caused by medicines or pesticides or other pollutant substances that are, because of full harmonization, allowed on the internal market, should find a solution to limit the pollution caused by the substances or products at a level below the environmental quality standards prescribed by EU or national law, respecting internal market legislation. To tackle the described conflicts between several fragmented regimes, a combined approach is suggested. Keessen *et al.* (2010b, pp. 47–1452) describe four ways to reduce the occurrence of clashes between Directives and Regulations, of which three are based on Beijen's (2010) research. The first suggestion is to group European legislation into framework legislation and daughter legislation to gain a good overview of the relevant legislation. This also contributes to achieving consistency within the policy fields of European legislation. The second suggestion is to use thematic strategies to ensure consistency not only within policy fields but also between policy fields. The third suggestion is to include provisions referring to flanking policies that are necessary to achieve its objectives in environmental framework directives. A practical way to do so is to include an obligation to take the objectives of the framework directive into account, and a derogation clause that can be invoked if flanking policies render it impossible to achieve the objectives of the framework directive. However, the WFD already contains several provisions that link it to other policy fields, including the suggested provisions mentioned above, which may strengthen their effect. Finally, more attention to transparency and public participation may facilitate solving conflicts between fragmented legal frameworks. It is therefore suggested to use governance techniques in addition to formal legislative instruments, such as by better involving stakeholders in the case of medicines, e.g. drinking water companies and veterinarians. Their involvement could be achieved by employing governance instruments: making information publicly available and creating opportunities for public participation (Kjaer, 2004). These governance instruments can also be used to involve various administrative authorities in decision making when their involvement is beyond their competences but is of interest to them.

External integration and appropriate coordination mechanisms among several policy fields

From the above three examples, it becomes clear that achieving water management goals is strongly dependent on measures and regulations adopted by other policy fields. The WFD recognizes this need and therefore one can find many cross-references throughout the directive. Although the WFD does not explicitly require external integration, this so-called external integration is important to achieving 'good' status, since activities regulated in other policy fields can have a positive or negative impact on water quality.

There are various interpretations of the integration principle that lead to diverse results. The integration principle is established in art. 11 TFEU, which prescribes that environmental protection requirements must be integrated into the definition and implementation of the Community's policies and activities. For instance, taking into account environmental objectives of the WFD could either mean that objectives are merely taken into consideration when a decision is made, or that the authorities are bound by these objectives because they have to be complied with. In the second scenario, objectives will be reflected in the decision and non-compliance is only permitted when there are good reasons (D'Hondt, 2003, pp. 80–110). Depending on the interpretation of the integration obligation, either a strong or weak internal and external integration approach will result.

In a comparative study of Member States' approach to internal and external integration, the following results became clear (Keessen *et al.*, 2010): while integrated water legislation may facilitate internal integration (integration within the ambit of water law), the issue of external integration (integration in other policy fields) remains unanswered. Since EU water law does not provide competent authorities with the necessary instruments to achieve good water quality, the achievement of the environmental objectives of the WFD will also require action outside the ambit of water management. The question that arises is – in which way do water or environmental objectives influence the decision making of authorities, acting outside the ambit of water law, such as in the field of land use, nature conservation and medicine, including veterinary medicines (see above)?

It appears that internal integration is widely found within the relevant national laws of the investigated Member States. This means that the water authorities have to take water quality norms and standards into account when they make plans and issue permits. However, 'taking into account' does not necessarily mean 'complying with'. For example, in Romania, the water authorities may authorize a discharge under specific conditions when the quality of the receiving water exceeds the quality norms. The same appears in Spain, where decisions concerning water may diverge from the water plan as long as there are good reasons for this. In the Netherlands, it is also possible for permits to be issued even though the surface waters in which discharges take place exceed the quality norms, particularly when the water authorities consider that this can be compensated.

External integration mainly takes place at the level of spatial planning. In general, it appears that water management plans or water quality objectives are required to be taken into account when spatial plans are being drafted in all investigated Member States, except in Spain where it is the other way around. In Luxembourg, external integration is not organized in the absence of links between various pieces of legislation. Both the Netherlands and Belgium have organized external integration with the introduction of a water test, which serves to integrate water issues into spatial planning. In Belgium, the water test is also used in other policy areas. However, in both countries the authorities merely use the water test to address water safety concerns and not to address water quality concerns. In Italy, river basin management plans (RBMPs) bind other authorities, which are required to ensure that water issues are integrated into other policy areas. In France, spatial plans must be compatible with RBMPs although no enforcement mechanisms are foreseen. In German legal literature it is stated that the management plan is binding on administrative authorities, but this has no direct effect on the rights and obligations of individuals and compliance with the plan, and cannot therefore be enforced by administrative authorities before the courts. Individuals can, however, challenge a management plan incidentally – in other words by alleging that an administrative act is unlawful because the management plan on which the act was based is unlawful (Breuer, 2004). Outside the ambit of spatial planning, the water quality objectives may also be taken into account; for instance, in Romania the Ministry of Administration and Administrative Reform takes the objectives into account when it promotes processes concerning water infrastructure. Yet, it remains to be seen to what extent the water management plans or water quality objectives will be taken into account by other authorities in other European countries.

External integration can also take place by cooperation between the various stakeholders. In the Netherlands, cooperation occurs in an informal manner, which differs between water authorities (Keessen *et al.*, 2010; van Rijswick and Havekes, 2012). In Denmark, the external integration of water policy and nature conservation policy is ensured through both a legal link, as both policies are based on the same Act, and an organizational link, both in the preparation and in the execution phase (Keessen *et al.*, 2010). Regional Environmental Centres are responsible for the development of both RBMPs and Natura 2000 management plans, while each municipality has to develop a municipal action plan containing the measures required take to implement these plans. While many measures will mutually reinforce each other, irrespective of whether they are taken for nature conservation or water management, it is also possible for conflicts to arise. For instance, the re-creation of wetlands and the re-establishment of natural hydrology in a river valley will retain nutrients, thus reducing eutrophication and creating a more natural environment around the river, but it will also threaten the survival of some Natura 2000 habitats and species adapted to the artificial hydrology. This conflict may be avoided by the re-creation of new, similar Natura 2000 habitats to compensate for the flooded habitats. It is, however, unclear whether this is allowed under EU nature conservation law.

The integration of water and agriculture can also take place in water legislation, as in France. Water pollution by nitrates constitutes a serious problem in the French river basin district of Loire-Brittany. In order to reduce the nitrate pollution of waters, local authorities can impose the use of fertilizer-free buffer zones for the reduction of nitrates by farmers in the sub-RBMP.

In England and Wales, a general integration provision obliges all authorities to give consideration to the approved RBMPs and any supplementary plan in exercising their functions, so far as they affect a river basin district. This has resulted in a link between water and agriculture, which is created in both water policy and agricultural policy. The RBMPs take the Regional Rural Development Frameworks into account, which are developed by regional organizations under the coordination of regional government offices and contain, inter alia, the agreed environmental priorities. In addition, the Environmental Agency, working with the Department for Environment Food and Rural Affairs and Natural England (in Wales, known as Natural Wales), has set up the Catchment Sensitive Farming programme. This programme offers a range of instruments, including advice and incentives, targeted support in priority catchments and, where required, enhanced regulation. It encourages voluntary action by farmers to tackle diffuse water pollution. By contrast, a specific link between water policy and spatial planning is absent.

It becomes clear from the experiences described above that the internal and external integration programmes of the WFD differ widely. This may not come as a surprise in view of the absence of any specific provisions to that effect in the WFD, but it does undermine the achievement of effective trans-jurisdictional water governance and may hamper the effective protection of European waters.

Conclusion and the way forward

Several EU trans-jurisdictional water governance issues have been identified that require urgent attention in European water law. The EU takes a progressive and integrated approach in its water legislation. Building on experience in classical water regulation, aimed at tackling water pollution with dangerous substances, the scope of water regulation has been broadened to include integrated water resource management and the combined protection of human health, ecosystems and the environment. This chapter has highlighted the introduction of new regulatory approaches with a strong emphasis on planning, adaptive water management and public participation.

This chapter has emphasized that trans-jurisdictional water governance in Europe relates to three key topics. The first is the *scope* of water management, often related to trans-boundary river basins. This raises specific problems for instruments shaping cooperation, jurisdiction and conflict resolution. We see mixed approaches in tackling these problems, ranging from informal cooperation, the use of international law instruments, including treaties, to the development of new European instruments for trans-boundary cooperation. Conflict resolution between EU Member States is strongly influenced by the leading role of the European Court of Justice. The recent development of arbitration-style conflict resolution in European Commission water directives may help to solve some problems, but because the European Commission lacks appropriate powers, except for starting infringement procedures, this is only a small step forward.

A second important point to tackle is *the consistent use of definitions and concepts* across Member States. This can be a problem even when only one Directive is being implemented, despite the fact that the EU should be one legal order. For the trans-jurisdictional governance of river basins this is crucial. For this reason, the EU has developed a Common Implementation Strategy. Nevertheless, it should be noted that this Strategy is not legally binding and that in the end it is the task of the Court of Justice to provide clarity on the scope and meaning of obligations, definitions and concepts.

The third problem to be tackled concerns *fragmented legal frameworks* that may hamper effective water management. Each policy field has its own goals, competent authorities, regulatory and governance approaches. At the same time the protection of water resources is highly dependent on measures taken in other policy fields: land use, energy and nature conservation or policies – such as agriculture or product policies, which are often fully harmonized within the legal framework of the internal market. In these cases, water authorities or Member States lack appropriate powers to realize the goals set in the field of water management. Three examples of fragmented and conflicting legal frameworks were discussed: water and nature conservation, freshwater supply and land use, and water and the authorization of medicines.

Internal integration within the broad scope of water management, as the EU has done with the WFD, offers great benefits. However, external integration with other policy fields is highly diverse, due to the policy field at stake, the legislative frameworks within the EU, the numerous Member States and the importance that is given to water management in relation to other interests.

One way of dealing with fragmented legal frameworks is to place stronger focus on the general principles of EU law – and environmental principles in particular. The fact that the EU has made the integration principle an important objective has not been of great practical value thus far. Furthermore, in the case of a lack of coordination, mechanisms improving transparency and providing information on environmental effects may be useful. However, it has become clear that even in the relation between nature conservation and water management, where goals overlap to a great extent, it is difficult to create an integrated approach for protected areas because of the use of different concepts, definitions and time lines. Therefore explicit and clear coordination mechanisms and a shared understanding of concepts, definitions and obligations are essential.

References

Beijen, B. A. (2010) *De kwaliteit van milieurichtlijnen. Europese wetgeving als oorzaak van implementatieproblemen*, Boom Juridische uitgevers, The Hague, Netherlands.

Breuer, R. (2004) *Öffentliches und privates Wasserrecht*, 3rd edn, Beck Verlag, Munich.

Bund für Umwelt und Naturschutz Deutschland e.V. v. Bundesrepublik Deutschland (C-461313) (2015) OJ C 294, pp. 3–4.

de Smedt, P. and van Rijswick, H. F. M. W. (2015) 'Nature conservation and water management: One battle?', in C. Born, A. Cliquet, H. Schoukens, D. Misonne, and G. van Hoorick (eds) *The Habitats Directive in its EU Environmental Law Context, European Nature's Best Hope?*, Routledge, London.

D'Hondt, N. (2003) *Integration of Environmental Protection into other EC Policies. Theory and Practice*, Europa Law Publishing, Groningen, Netherlands.

EC 1979, Council Directive 79/409/EC of 1997 on the conservation of wild birds [1979] OJ L 103/1, last amended by Directive 97/49/EC amending Council Directive 79/409/EEC on the conservation of wild birds [1997] *OJ* L 223/9.
EC 1991a, Council Directive 91/676/EEC of 1991 concerning the protection of waters against pollution caused by nitrates from agricultural sources [1991] *OJ* L/375.
EC 1991b, Council Directive 91/271/EEC of 1991 concerning urban waste-water treatment [1991] OJ L/135.
EC 1992, Council Directive 92/43 of 1992 on the conservation of natural habitats and of wild fauna and flora [1992] OJ L 206/7, last amended by Council Directive 97/62/EC adapting to technical and scientific progress Directive 92/43/EEC on the conservation of natural habitats and of wild fauna and flora [1997] OJ L 305/42.
EC 2000, Directive 2000/60/EC of the European Parliament and of the Council of 23 October 2000 establishing a framework for Community action in the field of water policy, [2000] OJ L/327, pp. 1–73.
EC 2006, Directive 2006/7 of 2006 of the European Parliament and of the Council of 15 February 2006 concerning the management of bathing water quality and repealing Directive 76/160/EEC [1992] OJ L/64, pp. 37–51.
EC 2007, Directive 2007/60/EC of the European Parliament and of the Council of 23 October 2007 on the assessment and management of flood risks, [2007] OJ L/288, pp. 27–34.
EC 2008, Directive 2008/56/EC of 2008 of the European Parliament and of the Council of 17 June 2008 establishing a framework for community action in the field of marine environmental policy (Marine Strategy Framework Directive) [2008] OJ L164/19.
European Commission 2000, Communication on Pricing policies for enhancing the sustainability of water resources, COM (2000) 477, final.
European Commission 2007, Communication Addressing the challenge of water scarcity and droughts in the European Union, COM (2007) 414, final.
European Commission 2013, a EU Strategy on adaptation to climate change, COM (2013) 216, final.
Hey, E. and van Rijswick, H. F. M. W. (2011) 'Transnational water management', in O. Jansen and B. Schöndorf-Haubold (eds) *The European Composite Administration*, Intersentia, Cambridge, UK.
Keessen, A. M. (2009) *European Administrative Decisions. How the EU Regulates Products on the Internal Market*, Europa Law Publishing, Groningen, Netherlands.
Keessen, A. M., van Kempen, J. J. H. and van Rijswick, H. F. M. W. (2008) 'Trans-boundary river basin management in Europe, legal instruments to comply with European water management obligations in case of trans-boundary water pollution and floods', *Utrecht Law Review*, 4, 3, 35–56.
Keessen, A. M., van Kempen, J. J. H., van Rijswick, H. F. M. W., Robbe, J. and Backes, C. W. (2010a) 'European river basin districts: Are they swimming in the same implementation pool?', *Journal of Environmental Law*, 22, 2, 197–222.
Keessen, A. M., Freriks, A. and van Rijswick, H. F. M. W. (2010b) 'The clash of the Titans: The relation between the European water and medicines legislation', *Common Market Law Review*, 47, 5, 1429–54.
Kjaer, A. M. (2004) *Governance*, Cambridge University Press, Cambridge, UK.
Montforts, M. H., van Rijswick, H. F. M. W. and de Haes, H. A. (2004) 'Legal constraints in EU product labeling to mitigate the environmental risk of veterinary medicines at use', *Regulatory Toxicology and Pharmacology*, 40, 3, 327–35.

Montforts, M.H. M. M., van Rijswick, H. F. M. W., Freriks, A. A., Keessen, A. M. and Wuijts, S. (2006) *The relation between product registration and water quality legislation: medicines, veterinary medicines and feed additives*, The National Institute for Health and the Environment, Bilthoven, Netherlands.

Scott, C. (2009) 'Governing without law or governing without government? New-ish governance and the legitimacy of the EU', *European Law Journal*, 15, 2, 160–73.

Treaty on the Functioning of the European Union, opened for signature 7 February 1992 (2012), OJ C 326, pp.47–390 (entered into force 1 November 1993).

van Holten, S. and van Rijswick, H. F. M. W. (2014) 'The consequences of a governance approach in European environmental directives for flexibility, effectiveness and legitimacy', in M. Peeters and R. Uylenburg (eds) *EU Environmental Legislation: Legal Perspectives on Regulatory Strategies*, Edward Elgar, Cheltenham, UK, pp. 13–47.

van Rijswick, H. F. M. W. (2011) 'The status of consumers in European water regulation', *European Journal of Consumer Law*, 1, 115–48.

van Rijswick, H. F. M. W. (2015) 'Mechanisms for water allocation and water rights in Europe and the Netherlands – lessons from a general public law perspective', *Journal of Water Law*, 24, 3/4, 141–9.

van Rijswick, H. F. M. W. and Backes, C. W. (2015) Groundbreaking Landmark Case on Environmental Quality Standards?, *Journal for European Environmental and Planning Law*, 12, 361–375.

van Rijswick, H. F. M. W. and Havekes, H. J. M. (2012) *European and Dutch Water Law*, Europa Law Publishing, Groningen, Netherlands.

van Rijswick, H. F. M. W. and Keessen, A. M. (2012) 'The implementation of the first generation river basin management plans in Dutch water law, lessons from the past, lessons for the future', *Romanian Journal of Environmental Law*, Suppl., 262–73.

van Rijswick, H. F. M. W. and Vogelezang-Stoute, E. M. (2008) 'The influence of environmental quality standards and the river basin approach taken in the Water Framework Directive on the Authorisation of Plant Protection Products', *European Energy and Environmental Law Review*, 4, 78–89.

van Rijswick, H. F. M. W., Gilissen, H. K. and van Kempen, J. J. H. (2010) 'The need for international and regional trans-boundary cooperation in European river basin management as a result of new governance approaches in EC water law', *ERA Forum*, 11, 1, 129–57.

Vos, J. H. and Janssen, M. J. M. (2005) *Options for emission control in European legislation in response to the requirements of the Water Framework Directive*, Report 601300003, Rijksinstituut voor Volksgezondheid en Milieu, Bilthoven, Netherlands.

Wakefield, J. (2009) 'Fisheries: A Failure of Values', *Common Market Law Review*, 46, 2, 431–70.

5 Critical linkages

Trans-jurisdictional approaches to advancing Indigenous marine governance

Lauren Butterly and Erika Techera

Introduction

Over the last two centuries, marine spaces have been used, and resources harvested, for economic gain with little or no regard for their sustainability or spiritual and intrinsic value. Slowly it has become apparent that both marine living resources and the marine environment are being degraded. Legal responses initially focused on issues of sovereignty over territorial waters and freedom of the high seas, with little consideration given to the issue of any reciprocal obligations. It was not until international attention was drawn to specific issues such as marine pollution and over-fishing that global interest turned to conservation. Simultaneously, a global emphasis on environmental responsibilities emerged, and later still, calls for more participatory forms of environmental governance. These historical developments provide the background to contemporary marine environmental governance.

The increasing involvement of Indigenous peoples in marine resource management and growing interest in Indigenous environmental governance have been influenced by a number of international developments, two relating to international law and the another two involving the creation of fora for discussion at the global level. The first development was the recognition, in the 1992 Convention on Biological Diversity (CBD), of the significance of traditional knowledge and practices. The second was the creation of the United Nations Permanent Forum on Indigenous Issues (UNPFII) in 2000, as a mechanism through which information and advice on Indigenous issues is provided to UN agencies and a wide range of Indigenous groups are able to express their voice at the international level. The third development was the adoption, in 2007, of the United Nations Declaration on the Rights of Indigenous Peoples (UNDRIP), which includes provisions relating to 'lands, territories, waters and coastal seas and other resources' (including 'conservation and protection' as well as 'productive capacity') and practising culture and customary practices (arts. 25–34). Then, in June 2012, establishment of the World Indigenous Network of Land and Sea Managers (WIN) was announced at the UN Conference on Sustainable Development (Rio+20).

While there is literature on the first three of these developments, the newly formed WIN has as yet received little academic attention. The WIN is a 'network that brings together Indigenous and local community land and sea managers to share their knowledge and practices in managing ecosystems, protecting the environment and supporting sustainable livelihoods' (WIN, undated(a)). In this sense, it is a knowledge-sharing network. At this stage, WIN is achieving this through conferences which bring together land and sea managers, side-events at conferences organized by other organizations (such as the UNPFII, the UN Conference on Small Island Developing States, the CBD and the International Union for the Conservation of Nature (IUCN) World Parks Congress), online interaction through the WIN website and social media, WIN community 'dialogues' and exchanges between land and sea managers involved in WIN ('knowledge exchanges') and a level of financial and technical support to implement 'knowledge' learnt through exchanges (WIN, 2014b).

Initially, the WIN was launched, funded and supported by the Australian Government. However, at the inaugural WIN conference in May 2013, it was 'handed over' to the UNDP Equator Initiative (Burke, 2013).[1] At this stage, there is little public information about the governance structure of the WIN (beyond the creation of an advisory committee that 'guides WIN's strategic focus and direction' (WIN, 2014b)) and, it seems that this is a practical element that is currently under consideration. This is therefore a unique time to be considering the ways and means for Indigenous peoples to continue to strengthen their involvement in marine governance. The developments which we will consider have seen the meeting of a regime of Indigenous peoples' rights (in the broad sense, not narrowly in terms of actionable rights) with fora in which to provide expert advice to international bodies and an inclusive conservation framework involving knowledge sharing, enhancing biodiversity conservation, improving social cohesion and increasing economic opportunities to alleviate poverty. Nevertheless, while international instruments and institutions have furthered global recognition of Indigenous rights, translating this at regional and local levels remains a challenge. There is an 'implementation gap' and Indigenous peoples must take a leading role in the implementation of these rights (Davis, 2013).[2] This has also been specifically identified with respect to Indigenous environmental management, including of marine areas (Firestone *et al.*, 2005; Techera, 2010). The WIN may be the missing link in communicating and promoting bottom-up approaches of Indigenous communities on a global scale. It therefore has the potential to facilitate the operationalization of the international framework, by drawing global attention to ways and means by which different Indigenous peoples and jurisdictions have implemented marine governance approaches, rights and conservation initiatives.

This chapter first considers the broad background to Indigenous peoples' participation in marine conservation and then explores the global developments identified above. This analysis is focused on drawing out the links (and potential links) between the developments across jurisdictions; in particular, how the WIN has the potential to feed local experiences, knowledge and processes 'up the chain' to inform the implementation of certain articles of the CBD and the UNDRIP, to enrich the work of the UNPFII and to share horizontally across the world success stories relating to Indigenous marine governance.

Indigenous peoples and marine conservation

Indigenous peoples generally approach nature and its management in a way that is different – socially, spiritually, economically and ecologically – from Western societies. However, international approaches to environmental conservation and management have historically reflected the dominant ethical perspectives of Western societies. Geographical landscapes have been viewed as 'physical rather than metaphysical spaces', which are 'inanimate, non-living and non-feeling' (Kwaymullina and Kwaymullina, 2010, p. 201). This disconnectedness from nature is said to stem from beliefs regarding the separation and superiority of humans over nature: '[s]pace was something for the humans to tame and subdue, to conquer and make their own' (Kwaymullina and Kwaymullina, 2010, p. 201). This belief system was particularly emphasized in relation to the marine environment, due to the notion of the ocean as 'the commons' and the 'outsider perceptions' of 'unused or inefficiently used resources' (Armitage 2008, p. 10). The result of this Eurocentric relationship with the environment has tended to be over-exploitation, environmental degradation and loss of marine biodiversity.

As the world turned its attention to marine degradation, industrialized nations applied Western conventional approaches to land management to ocean areas (Rothwell and Vanderzwaag, 2006). These were centralized, top-down, 'fortress conservation' measures involving the declaration of protected areas quarantined from human inhabitants or use. These conventional approaches were transposed into the marine environment, with little analysis of their potential efficacy or broader implications (Kearney *et al.*, 2013). One example of a marine 'version' of this style of governance is the 'no-take' marine protected area. If used bluntly, such tools and mechanisms take no account of the complex relationships between people and marine areas that extend beyond subsistence needs to cultural and spiritual connections. While it is clear that many Indigenous communities utilized a similar tool for marine management, this was in the context of broader Indigenous marine governance.[3] In Western contexts, marine protected areas were often implemented with little or no community consultation or ongoing involvement in management (see, e.g. Firestone *et al.*, 2005, p. 277). Furthermore, rather than assisting, it has been argued that 'dedication to the fencing model [protected areas] in form, but not in substance, has often furthered environmental degradation', including in the marine environment (McAllister, 1999, p. 199). Therefore, top-down approaches to marine management cannot be said to automatically aid conservation.

It is important to reflect on the impact more broadly that 'top-down' approaches to governance have had on Indigenous communities. Richardson (2009, p. 337) notes the 'evictions' of Indigenous communities to make way for 'wilderness' (and the irony of this given that European colonizers 'plundered

Indigenous lands'). In the marine context, Indigenous groups have referred to the issue of 'eviction' and also the frustration of watching environmental degradation by the non-Indigenous community (see, e.g. Ngarrindjeri Tendi, 2006, p. 13). This has resulted in a lack of confidence in non-Indigenous management techniques, which has in turn led to strong calls from Indigenous communities to secure Indigenous involvement in management of marine areas. As noted by Scott (2006, p. 19):

> Indigenous peoples struggle to be re-empowered to continue what we have done for millennia – to interact with our sacred lands and waters as part of the sacred circle of life and in doing so to promote biodiversity and sustainable development.

More broadly, principles of sustainable development have also led to support for bottom-up approaches. However, despite the drive for bottom-up approaches, national governments will always have a role to play in creating the context for bottom-up approaches (or stymieing them) (McAllister, 1999, p. 209). National governments have an impact on the formal domestic recognition of Indigenous rights within their jurisdiction. The foundational principle of Indigenous rights is the right to self-determination, which can be said to include rights such as the practice and revitalization of cultural traditions, participation in decision making, conservation and protection of the environment, development of lands and resources and the enjoyment of means of subsistence. As a consequence of the lack of recognition of Indigenous rights by nation states, in the 1970s, Indigenous peoples turned to international law (Anaya, 2004, pp. 56–7; Davis, 2012, p. 31). This has resulted in a number of institutions and instruments specifically relating to Indigenous peoples. These include the Convention Concerning Indigenous and Tribal Peoples in Independent Countries, UNPFII, UNDRIP and, most recently, WIN. There are also other international legal instruments which make mention of the rights of Indigenous peoples, particularly in relation to the environment, such as the CBD. In a marine context, however, there has been no guidance on how to facilitate Indigenous participation in marine management. Nevertheless, an examination of the relevant international institutions and legal instruments provides evidence of the linkages between these developments, both on a practical and theoretical level.

Convention on Biological Diversity (1992)

Prior to the adoption of the UNDRIP, the CBD 'emerged as one of the most important international instruments for recognising and protecting [Indigenous] traditional knowledge and rights to natural resources' (Fourmile, 1998, p. 14). The CBD continues to play an important role in the environmental context; it can be argued that it currently plays a unique linking role in relation to the UNPFII, UNDRIP and WIN. The focus of the CBD is conservation (generally by nation states), not rights of self-determination for Indigenous peoples.

However, it has come to be a major focus for advancing Indigenous peoples' claims to the environment (Richardson, 2001), primarily because the places where Indigenous peoples live often coincide with areas of high biodiversity. It has therefore been argued that it is 'in relation to management of biodiversity that the most important claims of Indigenous peoples are made in international law making' (Richardson, 2001, p. 7).

The CBD provides a framework and general measures for conservation and sustainable use of living resources. It recognizes in its Preamble, 'the close and traditional dependence of many Indigenous and local communities embodying traditional lifestyles on biological resources'. Biological diversity is defined in art. 2 as including marine and aquatic ecosystems and living organisms include those of the marine environment.[4] The CBD has three broad aims identified in art. 1: (1) the conservation of biological diversity; (2) the sustainable use of its components; and (3) the fair and equitable sharing of the benefits of the utilization of genetic resources. The more recently adopted Nagoya Protocol on Access to Genetic Resources and the Fair and Equitable Sharing of Benefits Arising from their Utilization to the Convention on Biological Diversity (Nagoya Protocol) is also of particular relevance to Indigenous peoples, as its purpose is to implement the third of the core objectives: fair and equitable sharing of benefits. The Preamble of the Nagoya Protocol 'notes' the UNDRIP and affirms 'that nothing in this Protocol shall be construed as diminishing or extinguishing the existing rights of Indigenous and local communities'. There are also a number of articles (arts. 7, 12(4) and 5(2)) relating to Indigenous peoples, including ensuring that access to traditional knowledge is based on the prior and informed consent of Indigenous and local communities.

The CBD 'inspires' action rather than creates enforceable obligations. Phrases such as 'in accordance with particular conditions and capacity' are used throughout the Convention. However, its wording embraces the cultural diversity and traditional knowledge of Indigenous peoples. The CBD Strategic Plan for Biodiversity 2011–2020 and its Aichi Biodiversity Targets, which were revised and updated at the 10th Conference of the Parties (CoP 10) in 2010,[5] (CBD, 2010) included Target 18:

> By 2020, the traditional knowledge, innovations and practices of Indigenous and local communities relevant for the conservation and sustainable use of biodiversity, and their customary use of biological resources, are respected, subject to national legislation and relevant international obligations, and fully integrated and reflected in the implementation of the Convention with the full and effective participation of Indigenous and local communities, at all relevant levels.

Furthermore, the Executive Secretary of the Secretariat of the CBD, Braulio De Souza Dias recently noted that 'Indigenous peoples and local communities are important actors in achieving the three goals of the Convention' (Dias, 2013a).

The link between Indigenous rights and the CBD has received attention from academics, including in the legal field (Birnie *et al.*, 2009, pp. 626–8; Desmet, 2011, pp. 126–34; Morgan, 2011, pp. 113 and 136). The crucial articles in relation to Indigenous peoples are arts 8(j) and 10(c), which state:

> 8(j) Subject to its national legislation, respect, preserve and maintain knowledge, innovations and practices of Indigenous and local communities embodying traditional lifestyles relevant for the conservation and sustainable use of biological diversity and promote their wider application with the approval and involvement of the holders of such knowledge, innovations and practices and encourage the equitable sharing of the benefits arising from the utilization of such knowledge, innovations and practices.
>
> 10(c) Protect and encourage customary use of biological resources in accordance with traditional cultural practices that are compatible with conservation or sustainable use requirements.

Article 8(j) does not use the term 'Indigenous peoples', and arguably the reference to 'communities' 'restricts the rights of Indigenous peoples' (Desmet, 2011, p. 131). The requirement for the community to be 'embodying traditional lifestyles' also reflects a certain perspective of Indigenous peoples' roles in environmental governance (Richardson, 2009, pp. 340–2; Desmet, 2011, p. 132). Only knowledge and customary uses that are compatible with conservation and sustainable use are protected (Desmet, 2011, pp. 132–3), although, in the context of the CBD, this is unsurprising. Furthermore, the phrase 'subject to' national legislation, seen in both the Aichi Biodiversity Targets and art. 8(j), leaves states with discretion, and means the CBD does not contain 'any explicit non-derogable requirement' for consent or participation of Indigenous peoples (Richardson, 2001, p. 9).[6]

The 'fundamental importance of ensuring the full and effective participation of Indigenous communities in the implementation of art. 8(j)' has been acknowledged by the Parties to the CBD (CBD, 2000). In 1998 a Working Group on art. 8(j) and Related Provisions (Working Group) was established to promote dialogue between Indigenous and local communities, Parties to the CBD, and other stakeholders (CBD, 1998).[7] The Working Group is 'an instrument to translate the provisions of the Convention into reality' (Scott, 2006, p. 17). Since its inception, the Working Group has produced the Akwé: Kon Voluntary Guidelines for assessments relating to developments likely to impact on lands and waters traditionally used by Indigenous and local communities (CBD, 2004); adopted the Tkarihwaié:ri Code of Ethical Conduct in relation to cultural and intellectual heritage of Indigenous communities (CBD, 2011); and provided input into the finalization of the Nagoya Protocol (CBD, undated; Desmet, 2011, pp. 132–3). Furthermore, it has established a funding programme to enable Indigenous communities to participate in meetings (Dias, 2013c). Although the Working Group has many achievements, some (potentially inherent) limitations have been identified, including 'limited support to help Indigenous and local communities in drafting their own development and biodiversity conservation plans' and 'lack of capacities for local communities' (Scott, 2006, p. 19).

Even with the potential limitations noted above, the opportunity for Indigenous peoples' involvement in implementation of the CBD is clear and there has been active participation, particularly by Indigenous peoples from Australia (Richardson, 2001, p. 9; Firestone *et al.*, 2005, p. 251). On a practical level, Morgan (2011, p. 113) notes that diverse mechanisms exist under the CBD for Indigenous peoples to participate in various meetings, including at the Working Group on Article 8(j) and Related Provisions. These have been utilized and the Parties to the CBD have 'shown an unprecedented willingness to work with 'Indigenous and local communities . . . to ensure that the Parties observe the Convention's requirement that they respect [Indigenous] biodiversity-related knowledge' (Fourmile, 1998, p. 14) and Indigenous peoples 'have been active participants attempting to influence the work of the parties in the development of their procedural and substantive decisions' (Quiggan, 2008, p. 46). This is relevant to the WIN given that the chair of the current WIN Advisory Committee is a representative of the Secretariat of the CBD (WIN, 2014b). We can see that although the CBD text, in particular, may have inherent limitations due to its focus on state actors and environmental conservation rather than Indigenous rights, the practical impact of Indigenous voices has the potential to take the CBD in a different direction.

In October 2012, the 'potential contribution' the WIN could make in continuing to progress the implementation of art. 8(j) was acknowledged at the 11th COP (CBD, 2012, para.10). The Executive Secretary of the CBD then circulated the report of the WIN Conference to the Working Group, noting the potential contribution the WIN could make to 'input into Article 10(c)' and 'linking Indigenous expertise' (Dias, 2013b). One of the main outcomes of the Working Group meeting was:

> Governments are encouraged to provide support to the recently founded World Indigenous Network, now housed with the Equator Initiative of the United Nations Development Programme, to permit the development of a global network for knowledge and cultural exchanges for conservation and sustainable use of biological diversity.
>
> (CBD Secretariat, 2013)

This acknowledgement from the Working Group is significant. The emphasis on the ability of the WIN to link 'Indigenous expertise' evidences the gap that the WIN can fill by facilitating such sharing of ideas and approaches. This interaction also serves to strengthen the potential importance of the CBD to Indigenous rights more broadly going forward.

The Parties to the CBD and the Working Group have also in recent years participated in a dialogue with the UNPFII. The Executive Secretary of the

CBD noted that in-depth dialogue at the ninth session of the UNPFII led to an 'increased interest' by the UNPFII in the work of the CBD (Dias, 2012, p. 1). The UNPFII now puts forward recommendations and voices concerns to the CBD Secretariat, and these are then considered by the Parties and the Working Group (see, e.g. UNPFII, 2008, para.81). In addition, the dialogue includes the UNPFII drawing attention to relevant reports and welcoming new developments. For example in the tenth session of the UNPFII, the adoption of the Tkarihwaié:ri Code of Ethical Conduct was welcomed, but concerns were raised about the fact that some of the aspects of the code were voluntary (UNPFII, 2011, paras 22 and 23; Dias, 2012, p. 9). Again demonstrating the links between the CBD and the UNPFII, the Executive Secretary of the CBD noted that the Tkarihwaié:ri Code of Ethical Conduct was developed in response to recommendations from the second session of the UNPFII (Dias, 2012, p. 9). In answer to the voluntary nature, the Executive Secretary noted that the mandate of the Working Group was to develop guidelines which would remain voluntary; however, they carry ethical and moral weight as they are adopted by the Parties to the Convention and these parties must report to the Secretariat (Dias, 2012, p. 9).

The links identified among the CBD (and the institutions related to its implementation), the WIN and the UNPFII demonstrate that although the CBD is not focused on Indigenous rights, the Working Group, the UNPFII and the WIN are having an impact on implementation. Clearly (and as demonstrated above), this is not to suggest that there is no tension between the mechanisms of the CBD and, in particular, the UNPFII. The UNPFII also has a mandate to discuss environmental issues but from a different perspective and through different voices. Nonetheless, the connections referred to above demonstrate a willingness to integrate agenda where appropriate.

A global voice and forum – UNPFII

Lobbying by Indigenous peoples led to the establishment of the UN Working Group on Indigenous Populations in 1982, which was the 'first human rights mechanism established to consider Indigenous issues' (Davis, 2008, p. 444). From about the mid-1970s onwards, Indigenous communities fostered networks on both a national and international scale (Morgan, 2011, pp. 62–5). Cheaper air travel, new forms of communication and the rise of the Internet have been crucial for facilitating trans-jurisdictional dialogue among geographically dispersed Indigenous peoples (Morgan, 2011, p. 63). These networks often formed due to the limitations of working within a 'state-based' system and, therefore, the trans-jurisdictional nature of such networks enabled strategies to bypass the state (Morgan, 2011, pp. 62–5).

In 2000, the UNPFII was established by Resolution E/2000/22 of the UN Economic and Social Council (ECOSOC). Pursuant to this resolution, the UNPFII provides expert advice and recommendations to the ECOSOC, raises awareness and promotes integration of activities relating to Indigenous issues within the UN and prepares and disseminates information on Indigenous issues.[8] The UNPFII is significant at an international level in that it includes non-state actor representatives. It is the 'first UN body to be constituted with a membership inclusive of Indigenous peoples' representatives, who meet on an equal basis with their governmental counterparts' (Morgan, 2011, p. 30). Furthermore, some of the members who are government-appointed are also Indigenous.[9] Sixteen Members, who are all independent experts appointed for three-year terms, make up the UNPFII. Half of the Members are nominated by governments and the other half are nominated by Indigenous organizations. All Members are nominated on a regional basis and the regions are different for government and Indigenous peoples' representatives (UNPFII, undated(b)). The UNPFII meets once a year for a 10-day session and is attended by the Members as well as representatives of Governments, other UN bodies and intergovernmental, non-governmental and Indigenous organizations. The ability of Indigenous organizations to attend, and apply to speak on issues on the agenda, is another unique factor of the UNPFII. This affords an opportunity for local communities to provide information directly to the UNPFII Expert Members and to share with other organizations. There is also the opportunity to host 'side events' during the UNPFII, discussed further below.

The mandate of the UNPFII pursuant to the ECOSOC Resolution is to provide advice and recommendations on Indigenous issues relating to six areas: economic and social development; culture; the environment; education; health; and human rights. In the marine context, the UNPFII has considered the issue of fishing, which spans issues of environment, food security, cultural practices, economic opportunities and self-determination (UNPFII, 2009, p. 9). In relation to the 'environment' more broadly, the UNPFII specifically identifies as a focus 'the implementation gap' (UNPFII, undated(a)). In particular, the UNPFII notes the lack of 'Indigenous voices' being heard in decision making (UNPFII undated(a)). This appears to speak to some of the 'gap' that the WIN is aiming to fill. The focus of the WIN is on scaling-up local, bottom-up approaches rather than necessarily being concerned with implementation at a 'national level'; although, as discussed above, in some ways the two are intertwined. The WIN certainly does seek to address the marginalization of Indigenous voices so that local opinions are heard on a global level. The WIN presented two main 'side-events' at the 2014 UNPFII annual session (13th session); however, there seem to be no further links between the two organizations (although, it should be noted that the CBD, UNEP and UNDP participate in the UNPFII annual forum as UN agencies). It seems that this is a result of the manner in which the WIN was formed.

As noted above, at the end of the WIN conference in May 2013, the Australian Government 'handed' over management of the WIN to the UNDP Equator Initiative. The UNDP's submission to the 2013 session of the UNPFII did not mention the WIN (UNDP, 2013), nor was it mentioned in the UNDP's 2014 submission (UNDP, 2014). To address the UNPFII's focus on addressing the 'gap' and facilitating and implementing opportunities for

Indigenous peoples' voices in relation to environmental rights (including ocean issues), a clearer articulation and engagement of WIN with the UNPFII would provide useful exposure and avenues for discussion. Before turning to these issues it is useful, however, to briefly consider the UNDRIP. Article 42 states that the UNPFII, alongside other UN bodies, as well as states, 'shall promote respect for and full application of the provisions of this Declaration and follow up the effectiveness of' the UNDRIP. The UNDRIP thus provides another important trans-jurisdictional development.

UNDRIP

The UNDRIP was adopted by the UN General Assembly in 2007. It includes the right to self-determination (art. 3) in the context of sustainable development (arts. 20, 23 and 32). Self-determination is the cornerstone of the UNDRIP, which itself is a 'very clear exercise in translating the right to self-determination from international law into the domestic context' (Davis. 2008, p. 461). In translating the right to self-determination in the context of sustainable development, UNDRIP provides specific articles relating to decision making, including in relation to natural resources. Article 18 states that Indigenous peoples have the right to participate in decision making 'in matters that would affect their rights'. While art. 25 articulates the right to 'maintain and strengthen' spiritual relationships with 'traditionally owned . . . lands, territories, waters and coastal seas and other resources', UNDRIP does not make specific reference to marine areas in other provisions.[10] Articles 26 and 32 provide that Indigenous peoples have the right to lands, territories and resources that they have traditionally owned and the right to determine and develop priorities and strategies for the development or use of these lands, territories and resources. The word 'resources' is not defined and may include marine resources.

Furthermore, art. 29(1) states:

> Indigenous peoples have the right to the conservation and protection of the environment and the productive capacity of their lands or territories and resources . . .

The two themes of 'conservation and protection' and 'productive capacity' demonstrate the complexity of Indigenous governance within the environmental context (Butterly, 2013, pp. 5–6). Self-determination can be linked to rights to protect the environment for future generations and to development in accordance with Indigenous communities' 'own needs and interests' (Preamble, UNDRIP). Self-determination also includes securing subsistence and engaging freely with traditional and other economic activities, which are particularly relevant to the marine space (art. 20, UNDRIP).

There has been much academic commentary on the UNDRIP more broadly (see, e.g. the overview of the commentary in Davis, 2008). Returning to its fundamental nature, Davis (2013, p. 11) submits that the 'binding/non-binding legal debate is a distraction and obfuscates other uses of the UNDRIP'. She suggests that Indigenous communities need to make the instrument work for them, and in that sense, take ownership and 'put meat on the bones of the UNDRIP' (Davis, 2013, p. 13). In taking ownership, Davis (2013, p. 11) states:

> [I]t is necessary in conversations about the UNDRIP to turn our minds to what we can do, as Aboriginal and Torres Strait Islanders communities, to implement the UNDRIP; then work in partnership with the state (the state as a 'junior partner') to meet those aspirations.

Clearly, the UNDRIP does not provide a 'checklist' for Indigenous rights (or indeed, aspirations) in relation to marine management. Nor should it aim to. However, its terms can (and should) be used by communities to guide their implementation. It has been suggested that the UNDRIP provides an opportunity to link global and local levels, and to think globally but act locally in terms of implementation (Stavenhagen, 2011, pp. 165–6). Thus, it may be seen as a valuable development if it facilitates and promotes cross-jurisdictional linkages and networks. As will be explored below, the 'language' of the WIN is not of rights; rather it is more closely aligned with the language of biodiversity conservation and sustainable development. However, there is no doubt the overarching themes of the WIN contribute to recognition of self-determination. The WIN fills a theoretical governance gap which can feed into local implementation through networked approaches to sharing knowledge across jurisdictions. It also fills a practical gap in that the focus should not be on more international law, but rather implementation through people and programmes.

World Indigenous Network

The second part of art. 29 of the UNDRIP states that nation states 'shall establish and implement assistance programs for Indigenous peoples for such conservation and protection'. The WIN may broadly be seen to fit into this. As previously noted, Australia launched the WIN at the Rio+20 conference in June 2012. Australia committed to develop the initial WIN design and to host the inaugural WIN Conference. The early state partners of the WIN were Brazil, New Zealand and Norway, and other countries were urged to join (WIN, undated(a)).

The inaugural WIN Conference was held in Darwin, Australia in May 2013 (WIN, undated(a); Burke and Snowden, 2013). The timing of the Conference clashed with the annual session of the UNPFII. As suggested above, it may be that the Australian Government's early steering of the process meant that interacting with the UNPFII was not a priority at that time. The presenters at the WIN Conference were from Indigenous communities around the world. Generally, the presentations were about sharing experiences from local Indigenous community perspectives from various jurisdictions. An example of a presentation on marine management at the conference was the 'Nobodyz

Trash – Indigenous Marine Debris Network' (WIN, 2013b). This is an initiative of the Gumma Indigenous Protected Area (which sits within the southern section of the Gumbaynggirr Nation in New South Wales, Australia) in partnership with domestic and international bodies and international Indigenous communities. The Indigenous Marine Debris Network aims to address plastic marine pollution based on a worldview of long and continuing ecological processes and, in doing so, to maximize involvement by developing new initiatives and opportunities. WIN was used as an opportunity to draw attention to their bottom-up approach, workshop and gather ideas from other conference delegates and also to encourage participation by new communities.

Other presentations included a discussion forum on art. 10(c) of the CBD and how land and sea managers can be linked locally, nationally and globally (WIN, 2013a). This relates back to the 'potential' of WIN to contribute to the implementation and interpretation of the CBD. In this way, the conference also provided local communities with a broader 'update' of the work going on at the international level. The videos of many of the sessions are available on the Internet, and this was a deliberate move to encourage Indigenous people around the world to become involved and continue to stay involved.[11]

James Anaya (2013), the then United Nations Special Rapporteur on the Rights of Indigenous Peoples, stated during an interview about the WIN:

> [M]y hope is that the World Indigenous Network can eventually contribute to concrete steps towards securing rights of Indigenous peoples to lands and territories in a way that is genuinely beneficial to all concerned, to all society. WIN . . . I think if it can do that, contribute with their knowledge to sustainable development for everybody while at the same time securing their own rights . . . it can make for a very positive development.

Anaya prominently uses the language of rights. In doing so, it seems that he identified the potential of the WIN in relation to the implementation gap. The documentation associated with the WIN uses the 'language' of local implementation. The WIN seeks to bring 'together Indigenous peoples and local communities' land and sea managers to share stories, knowledge, cultural experiences and ideas to better manage ecosystems, protect the environment and support sustainable livelihoods' (Burke and Snowdon, 2013). Currently, the WIN has three 'overall' aims: (1) 'better conservation of biological diversity and sustainable use of natural resources'; (2) 'improved social cohesion'; and (3) 'increased economic opportunities and the alleviation of poverty' (WIN, 2013f, p. 3). These themes were put forward when the Australian Government had 'carriage' of WIN, indicating that the choice of language may have been influenced by the agenda of the then government. WIN uses the phrase 'Indigenous Peoples and Local communities', which is similar to the dual phrase used in CBD ('Indigenous and local communities'), but with the important distinction of the use of 'Indigenous peoples'. However, the WIN, both in name and in practice, appears to be focused on Indigenous peoples.

The Australian Government also committed to developing the design of the WIN. To that end, a WIN scoping paper was produced (WIN, 2013c). The introduction to the scoping paper emphasized that the Australian Government had been assisted by a working group of Indigenous peoples from around Australia and an International Reference Group (WIN, 2013c, p. 4). The scoping paper covers the framework of WIN (including what topics it should cover), methods of engagement (online, conferences and exchanges), structure (funding, host organizations and working groups) and seeking ideas about what organizations WIN might form links with. This scoping paper was open for consultation prior to, during and after the conference. A consultation report was produced from oral consultations at the conference and written responses received (WIN, 2013f). One issue this consultation report noted from the collective submissions was that as well as being a knowledge-sharing network, WIN also needs to be an advocacy network for the rights of Indigenous peoples (WIN, 2013f, p. 5). However, in order to influence international actors and states, WIN needed to first find 'a collective voice' (WIN, 2013f, p. 5). The consultation paper also made specific reference to how to provide funding for WIN, including ideas such as approaching governments and companies (in affected regions) and examples such as auctioning artworks to raise funds for specific communities to attend exchanges.

At the close of the WIN Conference, the Australian Government 'handed over' the management of the initiative to the UNDP Equator Initiative (Burke, 2013). Broadly, the Equator Initiative aims to bring together the UN, 'governments, civil society, businesses and grassroots organizations to recognize and advance local sustainable development solutions for people, nature and resilient communities' in 'an international knowledge-sharing network' (UNDP Equator Initiative, undated). As noted by the Director of the UNDP Environment and Energy Group:

> UNDP recognizes and embraces its responsibility to learn from what is working on the ground and to assist governments in the process of taking local and Indigenous community successes to scale.
>
> (Burke, 2013)

This draws together themes of process and influencing nation states, but not necessarily rights.

WIN is now in a transition phrase from the Australian Government to the Equator Initiative. However, further details about governance structures still appear to be a work in process. Some of the WIN documents note that the Equator Initiative is only an 'interim' host of the WIN (WIN, 2013f, p. 1). There is further reference to an 'Interim WIN Committee led by a chair from the CBD' (WIN, 2014a). It is clear that the governance arrangements of the WIN going forward are crucial to its potential success. Furthermore, the arrangements must bring together local and international elements.

As well as bringing land and sea managers together at the conference, the WIN programme includes exchanges between Indigenous communities to share knowledge (WIN, undated(b)). This is a key part of the WIN as it deals with enforcement mechanisms. One example is the reciprocal exchange between Torres Strait rangers and rangers in the Arnavon Islands of the Solomon Islands in April/May 2013 (WIN, 2013d). Rangers in the Torres Strait have a particular focus on turtles and dugongs and, although they do not have compliance powers in relation to the turtle and dugong management plans, they ensure the community is aware of their responsibilities (WIN, 2013e). The rangers in the Solomon Islands work in the Arnavon Community Marine Conservation Area. As part of their programme they are involved in marine resource and species management and spatial mapping for management, as well as a range of related sustainable financial initiatives such as seaweed harvesting (WIN, 2013d). Currently, there is limited information available about the outcomes and responses to these WIN ranger exchanges. However, they have the potential not only to share ideas about marine management, but also about implementation processes (e.g. processes for setting up a marine conservation area).

Clearly WIN will facilitate sharing of knowledge between communities, and this has already started to take place. This is a form of participatory network governance. WIN has the potential to be a 'clearinghouse mechanism' whereby best practice models of environmental governance can be shared and communicated across jurisdictions. This is important in the absence of any UN agencies undertaking a comparative analysis of Indigenous marine governance mechanisms and developing 'model' laws to provide guidance and build capacity (although such model laws would need to be flexible, given the variety of situations) (Techera, 2010, p. 197). However, WIN also has the capacity to 'add value' to the other instruments and institutions identified in this paper.

The Parties of the CBD Working Group on art. 8(j) and WIN itself have already identified the potential of WIN to provide input into the implementation of the CBD. WIN also has the ability to 'put meat on the bones' of the UNDRIP and enrich the work of the UNPFII (Davis, 2013, p. 11.). In relation to the latter, what is currently missing is a more formal link 'up the chain' from WIN to UNPFII. This may be a factor of timing, given that WIN was only launched in 2012 and the first major conference held in 2013. It also appears to have been impacted by the Australian Government's involvement in this process. WIN seems to have been launched at Rio+20 without consultation with UNPFII or any other international Indigenous organization. While the Australian Government should be praised for supporting and providing initial funding for WIN and the WIN Conference, such consultation is vital. Moving forward, the importance of these links should be recognized and acted upon and the resolution of the WIN's governance arrangements will be crucial to future success.

Conclusion

The increasing involvement of Indigenous peoples in marine resource management, and interest in Indigenous environmental management, has been heavily influenced by the four developments discussed above. At the global level, these advances contribute to a comprehensive regime comprising instruments which support Indigenous peoples' rights with fora in which to be heard, and an inclusive conservation framework and knowledge sharing of on-the-ground experiences between Indigenous communities. Therefore, given the recent formation of the WIN, this is a unique time to be considering the ways and means for Indigenous peoples to continue to strengthen their involvement in marine governance.

The above discussion and analysis demonstrates the importance of identifying and implementing new approaches in order to achieve the goals of existing instruments and institutions. In other areas of environmental governance a level of stagnation is evident; despite growing awareness of environmental problems, positive advances have not been made either in terms of environmental outcomes nor governance developments. While all of the international instruments and institutions considered in this chapter facilitate the global achievement of Indigenous rights, the challenge is operationalization and in this regard there is a greater need to involve Indigenous peoples in the 'implementation' of these rights.

It is in this context that the emergence of the WIN is most exciting. By sharing experiences of Indigenous communities across the globe, it facilitates a cross-jurisdictional networked approach to local marine management. If it can 'feed' information up to the international level, it will cross another governance boundary – from local to global. It may thus become one of the first initiatives to advance both horizontal and vertical linkages to address environmental concerns. There is little doubt that marine environmental governance is a problematic area, often raising issues that have no direct parallels in terrestrial and freshwater governance. Nonetheless, it is with a level of positivity that Indigenous peoples can look to recent developments in the CBD, UNDRIP, UNPFII and WIN and the future that cooperative initiatives may hold.

Notes

1 In this context, Rosie Cooney, Chair of the IUCN Sustainable Use and Livelihoods Specialist Group who attended the WIN Conference noted that: 'There was some surprise that more consultation had not taken place on this, and many present had no information about what the Equator Initiative was' (Cooney, 2013).

2 The phrase 'implementation gap' in relation to Indigenous rights and the environment (but also more broadly) is used both in UN and academic literature (for example, UNDESA, 2009; Allen, 2011, pp. 253–4; Morgan, 2011, p. 156 and following; Stavenhagen, 2011, pp. 165–6;).

3 For an analysis of some traditional approaches see, for example, Johannes, 1978 and Techera, 2012.

4 Also note Article 22, which requires parties to implement the CBD 'with respect to the marine environment consistently with the rights and obligations of States under the law of the sea'.
5 The Conference of the Parties is the governing body of the CBD.
6 Furthermore, all of the clauses in Article 8 are subject to the introductory phrase: 'Each Contracting Party shall, as far as possible and as appropriate . . .'.
7 There is also an ad hoc '10c network' (WIN, 2013a).
8 It should be noted that the UNPFII is one of three UN bodies that is mandated to deal specifically with Indigenous peoples' issues. The other two are the Expert Mechanism on the Rights of Indigenous Peoples and the Special Rapporteur on the Rights of Indigenous Peoples.
9 For example, Professor Megan Davis who is Government nominated by Australia is Indigenous.
10 Although we note that art. 32 references 'water'.
11 It has been noted that: '. . . a focal point of the WIN has been its website, with over 90,000 page views from more than 170 countries. During the Conference week itself, over 19,000 page views occurred from 79 countries'. (UNDP Equator Initiative, 2013b).

References

Allen, S. (2011) 'The UN Declaration on the Rights of Indigenous Peoples and the limits of the International Legal Project' in S. Allen and A. Xanthaki (eds) *Reflections on the UN Declaration on the Rights of Indigenous Peoples*, Hart Publishing, Oxford, UK.

Anaya, S. J. (2004) *Indigenous Peoples in International Law*, 2nd edn, Oxford University Press, Oxford, UK.

Anaya, S. J. (2013) *Interview at the World Indigenous Network Conference (2013)* (Video), 28 May, http://unsr.jamesanaya.org/videos/interview-at-the-world-Indigenous-network-conference-2013, accessed 19 September 2014.

Armitage, D. (2008) 'Governance and the commons in a multi-level world', *International Journal of the Commons*, 2, 1, 7–32.

Birnie, P., Boyle, A. and Redgwell, C. (2009) *International Law and the Environment*, 3rd edn, Oxford University Press, Oxford, UK.

Burke, T. (2013) *WIN takes to the international stage*, media release, 27 May, Department of Environment, Canberra, Australia, http://environment.gov.au/minister/archive/burke/2013/mr20130527.html, accessed 19 September 2014.

Burke, T. and Snowdon, W (2013) *WIN Conference to tackle big environment issues*, media release, 25 May, Department of Environment, Canberra, Australia, http://environment.gov.au/minister/archive/burke/2013/mr20130526.html, accessed 19 September 2014.

Butterly, L. (2013) 'Changing tack: *Akiba* and the way forward for Indigenous marine governance', *Australian Indigenous Law Review*, 17, 1, 2–22.

Convention Concerning Indigenous and Tribal Peoples in Independent Countries, adopted 27 June 1989, C169 (entered into force 5 September 1991).

Convention on Biological Diversity (CBD), adopted 5 June 1992, 1760 UNTS 79 (entered into force 29 December 1993).

CBD (1998) COP 4 Decision IV/9, 'Implementation of Article 8(j) and related provisions', Fourth Ordinary Meeting of the Conference of the Parties to the Convention on Biological Diversity, Bratislava, Slovakia, https://cbd.int/decision/cop/default.shtml?id=7132

CBD (2000) COP 5 Decision V/16, 'Article 8(j) and related provisions', Fifth Meeting of the Conference of the parties to the Convention on Biological Diversity, Nairobi, Kenya, https://cbd.int/decision/cop/default.shtml?id=7158

CBD (2004) 'Akwé: Kon Guidelines', Secretariat of the Convention on Biological Diversity, Montreal, Canada, https://cbd.int/doc/publications/akwe-brochure-en.pdf, accessed 19 September 2014.

CBD (2010) COP 10 Decision X/2, 'Strategic Plan for Biodiversity 2011–2020', Tenth Meeting of the Conference of the Parties to the Convention on Biological Diversity, Nagoya, Japan, https://cbd.int/decision/cop/?id=12268

CBD (2011) 'Tkarihwaié:ri Code of Ethical Conduct', Secretariat of the Convention on Biological Diversity, Montreal, Canada.

CBD (undated) 'Outcomes of Article 8(j)', Secretariat of the Convention on Biological Diversity, Montreal, Canada, http://cbd.int/traditional/outcomes.shtml, accessed 19 September 2014.

CBD Secretariat (2013) 'Press release: Biodiversity meeting affirms key role of traditional knowledge in implementing UN biodiversity convention', Secretariat of the Convention on Biological Diversity, 12 October, http://cbd.int/doc/press/2013/pr-2013-10-12-8j-en.pdf, accessed on 19 September 2014.

Cooney, R. (2013) *World Indigenous Network*, International Union for the Conservation of Nature, Gland, Switzerland, http://web.archive.org/web/20150421170255/http://iucn.org/about/union/commissions/sustainable_use_and_livelihoods_specialist_group/sulinews/issue_5/sn5_win/, accessed 19 September 2014.

Davis, M. (2008) 'Indigenous struggles in standard-setting: the United Nations Declaration on the Rights of Indigenous Peoples', *Melbourne Journal of International Law*, 9, 2, 439–71.

Davis, M. (2012) 'To bind or not to bind: the United Nations Declaration on the Rights of Indigenous peoples five years on', *Australian International Law Journal*, 19, 17–48.

Davis, M. (2013) 'Community control and the work of the National Aboriginal Community Controlled Health Organisation: putting meat on the bones of the UN DRIP', *Indigenous Law Bulletin*, 7, 7, 11–14.

Desmet, E. (2011) *Indigenous Rights Entwined with Nature Conservation*, Intersentia, Cambridge, UK.

Dias, B. F. dS. (2012) *Recommendations Arising from The Ninth And Tenth Sessions of the United Nations Permanent Forum on Indigenous Issues to the Convention On Biological Diversity: A Concise Summary Of Actions Taken*, Convention on Biological Diversity, Montreal, Canada, http://un.org/esa/socdev/unpfii/documents/2012/session-11-scbd.pdf, accessed 19 September 2014.

Dias, B. F. dS. (2013a) Message of the Executive Secretary, Dr. Braulio F. De Souza Dias on the Occasion of the International Day of the World's Indigenous Peoples: 9 August (2013), Convention on Biological Diversity, Montreal, Canada, http://cbd.int/doc/speech/2013/sp-2013-08-09-ilc-en.pdf, accessed 19 September 2014.

Dias, B. F. dS. (2013b) *Report Of The International Conference For Indigenous Peoples And Local Community Land And Sea Managers and The Resulting Network: 23 August (2013)*, Convention on Biological Diversity, Montreal, Canada, http://cbd.int/doc/meetings/tk/wg8j-08/information/wg8j-08-inf-12-en.pdf, accessed 19 September 2014.

Dias, B. F. dS. (2013c) *Statement by Mr. Braulio Ferrera De Souza Dias, Executive Secretary Convention On Biological Diversity To The Eighth Meeting Of The Ad Hoc Open-Ended Working Group On Article 8(J) And Related Provisions: 7 October (2013)*, Convention on Biological Diversity, Montreal, Canada, http://cbd.int/doc/speech/2013/sp-2013–10–07–8j-en.pdf, accessed 19 September 2014.

ESC Res 22, UN ECOSOC, 45th meeting, UN Doc E/2000/22 (28 July 2002).

Firestone, J., Lilley J. and Torres de Noronha, I. (2005) 'Cultural diversity, human rights, and the emergence of Indigenous peoples in international and comparative environmental law', *American Universities International Law Review*, 20, 219–92.

Fourmile, H. (1998) 'Using prior informed consent procedures under the Convention on Biological Diversity to protect Indigenous traditional ecological knowledge and natural resource rights', *Indigenous Law Bulletin*, 4, 16, 14–17.

Johannes, R. E. (1978) 'Traditional marine conservation methods in Oceania and their demise', *Annual Review of Ecological Systems*, 9, 349–64.

Kearney, R., Buxton, C. D. and Farebrother, G. (2013) 'Australia's no-take marine protected areas: appropriate conservation or inappropriate management of fishing?', *Marine Policy*, 36, 5, 1064–71.

Kwaymullina, A. and Kwaymullina, B. (2010) 'Learning to read the signs: law in an indigenous reality', *Journal of Australian Studies*, 34, 2, 195–208.

McAllister, S. (1999) 'Community-based conservation: restructuring institutions to involve local communities in a meaningful way', *Colorado Journal of International Environmental Law and Policy*, 10, 195–225.

Morgan, R. (2011) Transforming Law and Institution: Indigenous Peoples, the United Nations and Human Rights, Ashgate, Burlington, VT.

Nagoya Protocol on Access to Genetic Resources and the Fair and Equitable Sharing of Benefits Arising from their Utilization to the Convention on Biological Diversity, adopted 29 October 2010, UN registration no A-30619 (entered into force 12 October 2014).

Ngarrindjeri Tendi, Ngarrindjeri Heritage Committee and Ngarrindjeri Native Title Management Committee (2006) *Ngarrindjeri Nation Yarluwar-Ruwe Plan: Caring for Ngarrindjeri Sea Country and Culture*, Department of Environment, Canberra, Australia, http://environment.gov.au/Indigenous/publications/pubs/ngarrindjeri-scp-2006–1.pdf, accessed 19 September 2014.

Quiggin, R. (2008) 'Indigenous knowledge and the Convention on Biological Diversity', *Australian Journal of Indigenous Education*, 37S, 46–55.

Richardson, B. J. (2001) 'Indigenous peoples, international law and sustainability', *Review of European Community & International Environmental Law*, 10, 1, 1–12.

Richardson, B. J. (2009) 'The ties that bind: Indigenous peoples and environmental governance', in B. J. Richardson, S. Imai and K. McNeil (eds) *Indigenous Peoples and the Law: Comparative and Critical Perspectives*, Hart Publishing, Portland, OR.

Rothwell, D. and Vanderzwaag, D. (2006) 'The sea change towards principled oceans governance', in D. Rothwell and D. Vanderzwaag (eds) *Towards Principled Oceans Governance: Australian and Canadian Approaches and Challenges*, Routledge, London.

Scott, J. (2006) 'Protecting traditional knowledge and the Convention on Biological Diversity', *Indigenous Law Bulletin*, 6, 20, 17–19.

Stavenhagen, R. (2011) 'Making the Declaration on the Rights of Indigenous Peoples work: the challenge ahead', in S. Allen and A. Xanthaki (eds) *Reflections on the UN Declaration on the Rights of Indigenous Peoples*, Hart Publishing, Oxford, UK.

Techera, E. J. (2010) 'Legal pluralism, indigenous people and small island developing states: achieving good environmental governance in the South Pacific', *Journal of Legal Pluralism*, 42, 61, 171–204.

Techera, E. J. (2012) *Marine Environmental Governance: From International Law to Local Practice*, Routledge, Abingdon, UK.

United Nations Department of Economic and Social Affairs (UNDESA) (2009) 'State of the World's Indigenous Peoples Report', UNDESA, New York, http://un.org/esa/socdev/unpfii/documents/SOWIP/en/SOWIP_web.pdf, accessed 19 September 2014.

(UNDP) (2013) 'Contribution to the 12th session of the UN Permanent Forum on Indigenous Issues', United Nations Development Program, New York, http://un.org/esa/socdev/unpfii/documents/2013/agencies/2013_UNDP.pdf, accessed 19 September 2014.

UNDP (2014) 'Contribution to the 13th Session of the UN Permanent Forum on Indigenous Issues', United Nations Development Program, New York, http://un.org/esa/socdev/unpfii/documents/2014/undp.pdf, accessed 19 September 2014.

UNDP Equator Initiative (undated) 'What we do?', UNDP, New York, http://equatorinitiative.org/index.php?option=com_content&view=article&id=47&Itemid=849, accessed 19 September 2014.

UNDP Equator Initiative (2013a) 'WIN Dialogues at UNCCD COP-11', United Nations Development Program, New York, http://equatorinitiative.org/index.php?option=com_content&view=article&id=785:win-dialogues-namibia&catid=192:2013&Itemid=732#tuesday-17-october-2013, accessed 19 September 2014.

UNDP Equator Initiative (2013b) 'WIN Lunch Discussion', United Nations Development Program, New York, http://equatorinitiative.org/index.php?option=com_content&view=article&id=784:brown-bag&catid=192:2013&Itemid=894, accessed 19 September 2014.

United Nations General Assembly (2007) 'Declaration on the rights of indigenous peoples' (UNDRIP), GA Res 61/295, UN GAOR, 61st sess., UN Doc A/RES/61/295 (13 September 2007).

(UNPFII) (undated(a)) 'Environment', United Nations Permanent Forum on Indigenous Issues, New York, http://undesadspd.org/IndigenousPeoples/ThematicIssues/Environment.aspx, accessed 19 September 2014.

UNPFII (undated(b)) 'Permanent Forum: Origin and Development', United Nations Permanent Forum on Indigenous Issues, New York, http://undesadspd.org/IndigenousPeoples/AboutUsMembers.aspx, accessed 19 September 2014.

UNPFII (2008) 'Report on the seventh session (21 April-2 May 2008)', United Nations Permanent Forum on Indigenous Issues, New York, http://undesadspd.org/IndigenousPeoples/UNPFIISessions/Seventh.aspx, accessed 19 September 2014.

UNPFII (2009) 'Report on the eighth session (18–29 May 2009)', United Nations Permanent Forum on Indigenous Issues, New York, http://un.org/esa/socdev/unpfii/documents/E_C_19_2009_14_en.pdf, accessed 19 September 2014.

UNPFII (2011) 'Report on the tenth session (16–27 May 2011)', United Nations Permanent Forum on Indigenous Issues, New York, http://undesadspd.org/IndigenousPeoples/UNPFIISessions/Tenth.aspx, accessed 19 September 2014.

WIN (undated(a)) 'About', World Network of Indigenous Peoples and Local community Land and Sea Managers, New York, http://winlsm.net/about, accessed 19 September 2014.

WIN (undated(b)) 'Knowledge Exchanges', World Network of Indigenous Peoples and Local community Land and Sea Managers, New York, http://winlsm.net/exchanges, accessed 19 September 2014.
WIN (2013a) 'CBD side-event: Expanding and strengthening an Indigenous and local community network on customary sustainable use (CBD Article 10c)', World Network of Indigenous Peoples and Local community Land and Sea Managers, New York, http://worldIndigenousnetwork.net/win-conference-darwin-2013/cbd-side-event-expanding-and-strengthening-an-Indigenous-and-local-community-network-on-customary-sustainable-use-cbd-article-10c, accessed 9 September 2014.
WIN (2013b) 'Nobodyz Trash – Indigenous Marine Debris Network', World Network of Indigenous Peoples and Local community Land and Sea Managers, New York, http://winlsm.net/nobodyz-trash, accessed 19 September 2014.
WIN (2013c) *Scoping Paper*, WIN, New York (copy of source with author).
WIN (2013d) 'Solomon Islands – Australia', World Network of Indigenous Peoples and Local community Land and Sea Managers, New York, http://worldIndigenousnetwork.net/win-exchanges/solomon-islands-australia, accessed 19 September 2014.
WIN (2013e) 'Solomon Islands – Australia Part 2', World Network of Indigenous Peoples and Local community Land and Sea Managers, New York, http://winlsm.net/solomon-islands-australia-part-2, accessed 19 September 2014.
WIN (2013f) 'WIN Scoping Paper Consultation Outcomes', World Network of Indigenous Peoples and Local community Land and Sea Managers, New York, http://winlsm.net/wp-content/uploads/2013/12/WIN-Consultation-Summary.pdf, accessed 19 September 2014.
WIN (2014a) 'WIN at Permanent Forum on Indigenous Issues', World Network of Indigenous Peoples and Local community Land and Sea Managers, New York, http://winlsm.net/events/win-at-permanent-forum-on-Indigenous-issues, accessed 19 September 2014.
WIN (2014b) 'WIN: Connecting Indigenous and Local Community Land and Sea Managers', World Network of Indigenous Peoples and Local community Land and Sea Managers, New York, http://winlsm.net/wp-content/uploads/2013/12/WIN-brochure-Final-Single-Pages.pdf, accessed 19 September 2014.

Part III

Trans-jurisdictional water law and governance

National approaches

6 Intra-national rivalries

A submerged aspect of trans-boundary water governance

Paul Martin and Amanda Kennedy

Introduction

A growing realization of the need to shift from a focus upon instruments to a governance[1] systems approach is evident in the Rio+20 declaration 'The World We Want', the IUCN 'Natural Resources Governance' framework, UNEP's Environmental Governance sub-programme and in the UNEP/WGEA MoU (4/90 implementation – GEO5).[2] Despite more than 2000 international legal instruments and 110,000 national laws and regulations,[3] resource governance outcomes fall short of being sustainable. Environmental law has to respond to political, social and economic system conditions that are less than optimal, including insufficient resources or political commitment, or the inability of society to change to the degree that is necessary. Effective law has to be designed and implemented based upon these realities.

This chapter concerns water rivalries, in particular how the law interacts with other non-law variables to generate, shape or manage conflicts, and how these impact upon the effectiveness, efficiency and fairness of water policy. In international trans-boundary water law, conflict has been given a lot of attention because of the serious consequences of nation state conflicts and because the parties have equal sovereign status (Wolf, 1997; Waslekar and Futehally, 2013).[4] However, in intra-national, settings rivalry issues do not seem to be given sufficient attention.

Whenever there is a structure that defines 'them' and 'us', an organizational boundary is created. The points of interaction with other groups are the focus for trans-boundary conflict (Coase, 1937; Oliver, 1993; Santos and Eisenhardt, 2005). Given diverse rivalrous issues, relevant organizational types include corporate firms, political entities, non-government organizations and social groups. Different disciplines use different terms for collective bodies. Two are 'organizations' (mainly used in social and managerial research) and 'firms' (mainly used in organizational economics and corporate strategy). We will use 'firms' as the descriptor because our focus is strategic rivalry, conflict and collaboration (Chen, 1996).[5]

Organizational theories of the firm explain why firms emerge, the nature of the firm in terms of transaction costs and agent behaviour (Coase, 1937;

Holmström and Roberts, 1998). The organizational boundary definition of 'us' versus 'them' has behavioural implications demonstrated by tribal conflict and xenophobia. In the strategy literature increasing attention has been focused on rivalry and firm boundaries, in particular the challenge of effective collaboration. The nation state and the federal state have many of the fundamental characteristics of the firm, with similar issues of rivalry and collaboration.

With inter-firm rivalries come impediments to transacting including distrust, positional bargaining, information asymmetries and complexities affecting bargaining and implementation. The costs erode the benefits and increase the costs of transactions (decreasing the net value of trans-boundary collaboration with water transactions), with value often being created by strategies that reduce impediments to collaboration.[6] The second systemic impact concerns rivalry as a mechanism for social adaptation (evolution). In nature, competition drives biological adaptation to changing contexts. In economics, competition is a mechanism for maximizing value. Combining considerations of transaction costs and adaptation through competition, we suggest that organizational rivalries can simultaneously have a negative and a positive impact, increasing some costs but potentially driving beneficial adaptation to changing circumstances.

Boundary conflicts over political authority and resources occur within the nation and between nation states. Within a federation, state-to-state rivalries require coordinating mechanisms (such as the Australian Council of Australian Governments (COAG)).[7] Federations of nation states such as the EEC or the former Soviet Union develop boundary spanning conflict management arrangements, and rivalry over spatial and jurisdictional boundaries is an important consideration in how federations work. When legal mechanisms fail, more destructive conflict can emerge.

There are many relevant forms of firms, boundaries and boundary conflicts within any state. Legal arrangements help manage the role definition of arms of government (e.g. the separation of powers doctrine), and inter-firm rivalry and coordination within government (departmental structures and operations). Rivalry between the State and the citizenry (including private corporations) over the legitimacy of government power is a constitutional concern. This is particularly so in relation to property and civil rights. All these boundary rivalries involve legal arrangements to ensure that the contest for power and resources is non-destructive. That is a core function of the law.

The graphic below suggests dimensions of boundary rivalries. It represents two imaginary nations (C1 and C2) with trans-boundary arrangements. Within each nation there are entities with rivalries between them. There are also rivalries between the firms within one country and those within the other country (e.g. trade competition, where firms contest the marketplace within their own and other countries). These inter-firm rivalries are not the conventional focus of trans-boundary legal scholarship, but are significant in natural resource management, including water governance. In practice, rivalries between firms, government agencies and the like can all impede coordinated governance of water resources.

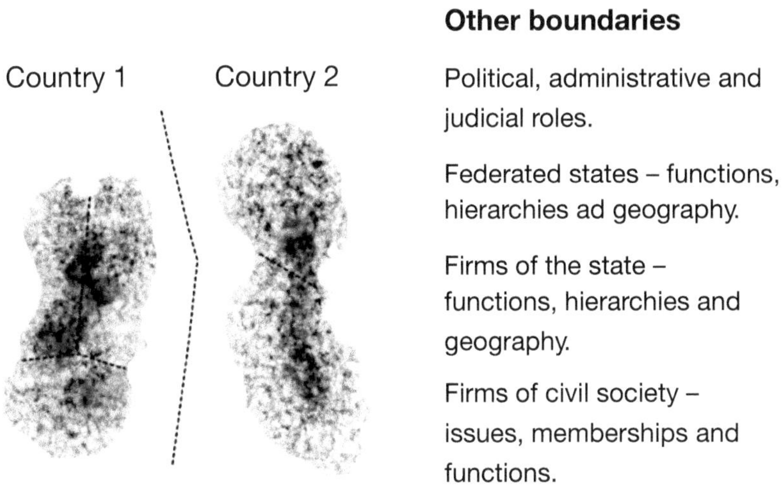

Figure 6.1 Multiple levels and forms of boundary rivalry

The forms of trans-boundary arrangements (and relevant legal aspects) represented in the diagram include:

- Nation state rivalries reflected in political, trade and possibly military mechanisms, with military competition over territory being the alternative to legal resolution. International law and conventions, and negotiated agreements, are the pivotal instruments for boundary definition and spanning.
- Federated state rivalries within the framework of constitutional law, fiscal rules, and coordinating institutions. The rivalries reflect political and economic dynamics, and history. If legal arrangements fail, then counter-productive outcomes can range from inefficient trade or other barriers through to civil war.
- Meta-state federations at the international level, such as the European Union or the USSR. These involve conventions, transnational fiscal and coordinating institutions. Failures of legal boundary spanning can result in diminished effectiveness, and in extreme instances military conflict.
- Government bureaucracies are rivalrous over authority and resources. Political contests at the ministerial level are relevant, as are professional rivalries and stakeholder affiliations. There is rivalry between the arms of government, and levels of government. Legal mechanisms include constitutions and court rulings (e.g. separation of powers), and administrative rules. The likely result of a failure of legal arrangements will be inefficiency. The significance of increased transaction costs within government should not be underestimated, as these can reduce national competitiveness and undermine the confidence of citizens.

- Private firms are rivals over resources and markets. Corporation rivalry is demonstrated by price competition over water. Legal instruments to manage competition include competition laws, laws of contract, civil and sometimes criminal law (as in the instance of water theft or unlawful water infrastructures). There is also rivalry /cooperation between the private and the public sector which is relevant to water governance. The rivalry over boundaries of legitimacy and control is well demonstrated with water law and politics.
- Non-government actors engage in rivalrous behaviour in pursuing their institutional agenda such as industry development or environmental protection. Civil and criminal laws establish the boundaries of legitimate action in pursuit of these various interests.

In the instances we examine below, conflict is shaped by institutional structures generally developed in response to non-water contests, and by normative frameworks shaped by complex social and political histories. The rivalry is often conducted through firms that define themselves by the position that they take on issues which are not typically about water, where the water issue reflects inter-firm competition.

Examples of intra-national water conflicts

Legal and institutional arrangements for water reflect different paradigms for water policy. Jurisdictions that prioritize social justice and human rights typically use regulation to constrain private interest and to protect the less advantaged. Where economic optimization is the focus of policy, social justice concerns are accommodated administratively or by economic adjustment. However in practice water governance seems to be largely about 'satisficing' interests rather than implementing a single paradigm. Contest and compromise across firm boundaries is framed by historically created legal/institutional settings (Martin and Becker, 2011). Changes to the dominant water governance paradigm do shift the institutional structures and value weightings, but resolution also feeds into the structures and value weightings in future water law. Under either paradigm inter-firm conflicts are inevitable, as legal arrangements lead to adjustments to economic and political power and impact upon cultural and social arrangements. This suggests the need for laws that provide a constructive setting for water and water-related conflicts.

We will illustrate the issues using examples of conflicts within intra-national water governance, starting with Australian water law, then examples from other jurisdictions. Australia's national water policy embraces economic optimization, constrained by scientifically determined limits of extraction to protect the system and non-extractive values (Hussey and Dovers, 2006; Dovers and Hussey, 2007; Quiggin, 2008). A focus on economics has resulted in a legal framework that constrains the capacity of the citizen to legally contest the hegemony of science and the market (Martin, 2009).[8] The approach presumes an ordered, united structure (Luhmann, 1988a) within which the 'bureaucratic and professional ways of handling issues' are accepted as 'facts' (Luhmann, 1986, pp. 111–27). The approach prioritizes a few, narrowly specified interests and issues. Conflict is perceived as an interference with efficient market and scientific processes.

Illustrative of a failure to understand the important role of legally mediated conflict is the Australian Water Act 2007 (Water Act) that frames the rules under which sustainable water extraction limits for a major water system are determined (s. 20). Prior to the initial plan being finalized, a Guide to the Draft Plan was released on 8 October 2010. This triggered protests from farmers, rural communities and environmentalists.[9] While the narrative was about water allocation and impacts on human and environmental values, the contests were continuations of positional rivalries over environmentalism, 'city-based' policy making and access to government services, or socio-cultural concerns about rural communities. Political and legal boundaries were defined and defended. The positional clashes were apparent in the media, with violent outbursts in conjunction with meetings or announcements. Bargaining and disputation occurred between states, between states and the federal government, between communities and government, between communities, and between stakeholder firms and everyone else. They were characterized by positional bargaining and limited collaborative engagement.

The way in which conflict issues were (unsatisfactorily) dealt with suggests a failure of the governance approach to harness the (inevitable) conflict, partly reflecting an inadequate consideration of water conflict issues in the legislation. Under the legislation there is no facility for the affected community to contest Authority decisions, or to dispute the model or management decisions (Martin and Kennedy, 2011). The legal instrument reified property holder and scientist concerns. Eventually, complex community concerns about relationships, family futures, cultural integrity and the like were transmuted into the restrictive epistemology of quantitative social economic modelling, and the conflict discourse enabled by the law was limited to the trade-off between the environment and the economy (Murray-Darling Basin Authority, 2010).[10]

Conflicts about plural values and social or cultural interests that do not fit within the preferred paradigm are treated as sources of inefficiency to be excised by cutting off any opportunity for legal contestation. This means dis-enabling the legal system for conflict and adjustment decisions. The 'freezing' of interests suggests potential maladaptation to changing conditions, or to changing values. It is not possible to envisage what water conflicts may arise, but it is possible to predict that they will arise, and that some may not fit easily within the framework established through the legislation.

The Water Act highlights the effects of past boundary competition between the states, and between the states and federal governments (Hussey and Dovers, 2007). Some of these constitutional conflicts were over water, but many were not. The long history of dispute resulted in fractious negotiations, and in complicated legal and financial arrangements for water governance. Because of this boundary conflict history, the Water Act was framed to manage rivalry over authority between the states and the Commonwealth. Arrangements to compensate private water right holders for risk transfer also reflect the public/private property right boundary. This has been the subject of legal rivalry over the power of governments to appropriate private property.

We cannot value the transaction costs of constitutionally linked boundary rivalries, but they would be substantial. The cost impacts of intra-national boundary rivalry on Australian water law are demonstrated by case studies of the implementation of the COAG water reforms, in urban and rural settings (Martin and Shortle, 2010, pp. 705–20). In the urban setting, the implementation of national policy was frustrated by a contest among land-use planning agencies, local government and water agencies. In the rural setting, 'turf' rivalry between state agencies stalled implementation of aspects of the national water reform agenda (Martin *et al.*, 2008).[11] Both cases demonstrate 'us versus them' dynamics and the lack of a boundary-spanning mechanism for collaboration (Carey *et al.*, 2002).

To illustrate this point we consider some water conflicts in other jurisdictions. These demonstrate the varied forms of conflict that can emerge over water and the necessity for flexible conflict management mechanisms within intra-national water laws.[12]

In China, significant conflict over water pollution occurs between citizens and industry, citizens and the regional governments, industry and regional governments, and between regional governments and federal government. There are also issues of the rights of minority peoples, and particularly of industry, in the transfer of water. The contests concern not only water conflict, but also constitutional structure and minority peoples (Du, 2011).

In the United States, conflict over water extraction and contamination is increasing (Benson, 2011; Fort, 2011). Forecasts suggest a growing intensity of that conflict (Resources for the Future, 2013).

In the Andes and Patagonia, conflict is focused around nation state trans-boundary issues and the interests of marginalized communities. Issues of indigenous social justice intersect with nation state questions and a long history of trans-boundary contestation (Tafur, 2011).

In Iceland, the focus of conflict is foreign ownership of hydrological resources for hydroelectricity generation. These are contests over constitutional and political integrity (Haraldsdottir, 2011).

In South Africa, issues of water contamination are interwoven with social well-being. Also interconnected are questions about the control of industrial polluters (Kidd, 2011).

In the Mekong, the proliferation of dams brings together many strands of rivalrous behaviour including international trans-boundary conflict, and contests among local authorities, local businesses and communities over water supply and electricity generation (Cronin, 2009).[13]

It is difficult for policymakers to anticipate what issues will be contested, by whom and for what reasons. It should be expected that resource conflict will be diverse and value laden, rather than narrow and amenable to simple analysis.

The strength of the law is its capacity to cope with axiological challenges and to develop norms that respond to the complexity of society. Using the law to rigidify or exclude conflicts from consideration prejudices aspects of environmental justice or participatory democracy, but also raises concerns about social adaptation.

Implementation of water law presents dilemmas, limitations and conflicts, which are often not well accommodated by institutional arrangements. It is not hard to envisage conflicts in Australia over water nationalism, Indigenous justice, the integrity of water models, linkages between water management and international obligations (Martin and Kennedy, 2011). The adaptive capacity of the legal mechanism is needed to help to adjust governance to changing values. While the specific challenges may be unpredictable, the likelihood that some will arise is not. The international experience suggests that legal conflict mechanisms are fundamental to the efficient and fair adaptation of economic and social systems to unanticipated changes.

As demonstrated above, the issues frequently reflect legal and institutional structures that have evolved through contests over non-water issues (Heinmiller, 2009; Martin and Becker, 2011). These can include constitutional provisions as previously interpreted by courts and negotiated legal/political agreements (such as COAG arrangements and national competition policy, in Australia). Legal resolution helps to create legal/institutional structures for future conflicts over other matters.

Some theoretical frameworks for conflict and the law

Conflict within the boundaries of law is a social adaptation method. Luhmann's social systems theory indicates that a central role of the law is to stimulate and channel conflict in constructive ways (or at least to reduce the potential for destructive conflict). His theory holds that society is constructed through the self-organization of interacting sub-systems – including law, economics, science and politics. The elements of each system self-create and evolve through being 'continually irritated, stimulated, disturbed and faced with changes in the environment' (Luhmann, 1988b). Luhmann's theory holds that the driver of a social system is self-referential communication (Luhmann, 1995, 2004). A contentious part of his approach is the exclusion of humans *per se* as units of analysis within the system's functioning (Banakar and Travers, 2005; Gershon, 2005).[14]

Thus, the legal system constitutes itself through communicative events, developing 'self-referential circles in relation to legal acts, legal norms, legal process, and legal dogmatics, with the result that the law becomes increasingly "autonomous"' (Teubner, 1993, p. 33). It is the legal system that decides what is a legally relevant event, by reference to these communicative events. While the system is self-generating, the external environment influences it (Luhmann, 1988a; Teubner, 1993; Ellsworth, 2011). However, only the legal system can confer a legal norm upon a communication (Luhmann, 1988a). Thus, when a conflict is conducted within the legal subsystem it generates new forms of meaning and interpretation, to guide future conflict. Luhmann suggests that the legal system consists of 'the exploitation of conflict perspectives for the formation and reproduction of congruently (temporally/objectively/socially) generalized behavioural expectations' (Luhmann, 1988a).

Contests that take place outside the legal sphere do not create legally binding norms (other than, perhaps, through statutory intervention). Only if the legal system is obliged to cope with challenging issues can it strengthen its capacity to do so in the future. This suggests the advantage of judicial or other publicly communicative resolution mechanisms over administrative management, which does not establish precedent or provide public communication of norms and has a limited ability to develop social capacity.

We can transpose this view of conflict and the law into trans-boundary water law. Issues of resource competition and national sovereignty are a focus in international trans-boundary water arrangements. Legal norms are created by international law coupled with specific agreements. The rivalries between adjacent states over the Nile, between India and Bangladesh, between states along the Mekong, and many other situations, all speak of the significance of boundary-defined rivalries between nation 'firms'. These contests can result in trade disputes, warfare or other disruptive outcomes, and so legal mechanisms for water conflict are important (Pacific Institute, 2009; Paisley and Grzybowski, 2011). Nation-to-nation conflict management is the focus of many research and intervention projects on trans-boundary water law.[15] Building adaptive capacity through the evolution of legal norms in response to conflicts is well illustrated by international trans-boundary legal arrangements.

With increased scarcity and increasing demand, conflicts will probably increase (Heinmiller, 2009).[16] The examples we have discussed suggest that in the intra-national arena, legal mechanisms for managing inevitable rivalries are under-developed. Legal researchers and policymakers have paid relatively little attention to intra-national water conflict. Non-law experts and governments are attracted to non-court management of social adaptation and conflict (e.g. integrated water resource management, adaptive management and administrative mechanisms). While these are important, often they are not institutionalized as avenues for managing conflicts among citizens, firms and state agencies. There is a need for sophisticated legal governance arrangements that can reduce the adverse impacts of such conflict, while maintaining the adaptive value of legally managed contestation.

Recognition of the centrality of conflict through the courts resonates with long-established legal principles. Two of these are the distinction between the proper roles of the executive, administrative and judiciary, and the long-established aversion of the courts to private arrangements that are intended to oust their jurisdiction. In Australia, the pursuit of 'science-led' and 'economically efficient' water policy arrangements has undervalued the strategic role of the courts and elevated the role of the administration. Defining the social objectives of natural resource governance is important in deciding what is 'efficient' in managing conflict. Too narrow a definition may result in longer-term inefficiency because it limits adaptiveness and potentially undermines social justice goals of society.

There are many theories that could be relevant to understanding these issues. We have referred to social systems theory, as developed by Luhmann and Teubner, or (with a contrasting perspective) Habermas. Community engagement and conflict theories highlight the importance of institutional empowerment of the citizen in the pursuit of public policy goals, and suggest ways of conceptualizing the social challenge of water policy (Innes and Booher, 2004). The jurisprudence of environmental justice emphasizes the significance of not disempowering weaker actors in society in the management of natural resources, including water (Schlosberg, 2007, 2013; Hanak *et al.*, 2011, p. 482). From such sources we propose that legal arrangements which do not adequately contemplate conflicts, or which marginalize actors in these conflicts, are unlikely to be socially optimal or robust over time.

Embracing conflict in water policy

Contests over interests in water will be a permanent and probably increasing feature of the natural resource governance landscape in many countries. It is likely that the types of conflict that will be encountered will be varied and probably often different from those which have previously been dealt with. Various commentators suggest that the future of water will be characterized by significant and potentially catastrophic conflict, in the absence of effective legal norms and institutions (Wolf *et al.*, 2005).[17] It seems on this basis that it would be wise for legal natural resource governance to directly embrace the issue of water conflict, not as an unexpected or undesirable incident of either rights or water use efficiency, but as something which contributes to the evolution of the law.

Poorly managed trans-boundary rivalry exists at many levels, in many forms and in most jurisdictions. There is evidence that poorly managed conflict, or potential conflict, does result in high transaction costs and difficult coordination arrangements in the complex tasks of water governance. From the jurisdictions we have looked at we can observe many types of contestation around interests and values, that these almost inevitably arise during implementation of legal arrangements, and that the form they take is sometimes surprising. International trans-boundary arrangements have long recognized that it is essential to have proper mechanisms for managing boundary-related disputes, but this knowledge does not seem to translate itself into awareness within the intra-national context. If, as we surmise, society is going to be faced with more water conflicts and that those conflicts are likely to be increasingly diverse, then this suggests that we will need new legal norms and new legal institutional arrangements, as well as theories of conflict and its resolution through the law. If many of these conflicts occur within rather than between nation states, then these arrangements will have to be effective at that jurisdictional level.

A first step to more constructive treatment of potential conflict is to recognize that these will arise regardless of the legal regime that is in place. In the case of the Australian Water Act, recognition of the potential for conflict extended only to those interests that had become evident during the preceding political conflict over water. The aspects of conflict that received media (and thus political) attention were principally a contest between production and environmental values, partly viewed through a lens of contests over science (and between scientists). While there was extensive community consultation, this did not convert into awareness of the more diverse interests that were involved, and the potential rivalry between them. Neither did the process recognize the extent to which institutional boundaries and affiliations would become a source of intense conflict, which impacted upon implementation of the 'rational optimization' model that was preferred. The social research conducted for the purpose of that plan did not involve overt conflict prediction or analysis, though the likelihood of conflict was always known. Put in terms that would be recognizable by legal social systems theorists, there was no mechanism for attaching meaning to the signals which existed about likely conflict, and as a result the system was unable to respond appropriately with the development of new norms and mechanisms for managing such issues.

An adjustment to water management will always have the potential to impact adversely on many interests, and these are not limited to economic values. Cultural and political concerns are also relevant. There are (at least) three distinct forms of policy risk that are relevant (Martin and Williams, 2010).[18] The first is political risk to the creation of an effective water governance instrument, and to its implementation. This risk has crystallized in Australian water in the many conflicts between states, between the state and the Commonwealth, between environmental and commercial interests, and between non-government interests and governments. The second is instrumental risk, the likelihood that an instrument will fail to be effective in achieving its primary objectives.

While the use of the market has arguably accelerated adjustment of water use towards more sustainable volumetric boundaries, it is a matter of conjecture whether the social goals of water governance have been achieved. What is, however, clear is that boundary-related conflicts have reduced the cooperation that would have facilitated the achievement of volumetric and broader goals.

The third form of policy risk is of costly spillovers. The conflict that has emerged in Australia's water governance is often around claims of significant spillovers, notably impacts on vulnerable communities.

The autopoetic systems view of law as participant in the evolution of society also suggests that through this process, law itself will evolve new norms and mechanisms for coping with the challenges with which it is faced from its environment. Traditionally, the conflict mechanisms of the law were focused around litigation, the political/legal mechanism of statute and regulation, or administrative ruling. However, faced with new demands, law has evolved new conflict mechanisms, notably forms of alternative dispute resolution. We suggest that, faced with increasing conflict situations, the law will need to further evolve innovative approaches. In particular, we anticipate that conflict over natural resources (including water) will increase and become more complicated. Faced with these demands and the cost of more traditional mechanisms, legal systems will generate innovative norms and institutional arrangements for dealing efficiently with diverse conflicts over resources (Norman *et al.*, 2012).[19] Drawing upon other fields, there are potential innovations that might be imported into law. We are not able in this paper to define precisely how such innovations might translate, but we take the opportunity to highlight some possible directions for future legal institutionalization.

The concept of environmental justice articulates the need to ensure adequate community engagement in land-use decision-making processes and provides a helpful framework for considering the nature of legal and political institutions. The concept initially focused upon distributive notions of justice, such as the inequitable siting of environmental harms in low-socio-economic areas. Contemporary constructions of the concept however express the importance of ensuring procedural justice in land-use decisions. This includes whether those communities burdened by environmental harms are adequately recognized within political and legal processes and whether their meaningful participation in such processes is enabled (Norman *et al.*, 2012; Schlosberg, 2013). The use of environmental justice as a lens to explore regulatory frameworks for the allocation and management of land thus invites consideration of whether factors are present that enable effective participation in land-use decision making. This may include the adequacy of information provision and notice of upcoming decisions, the ability to engage in active consultation as well as genuine and meaningful participation in decisions, whether rights to challenge decisions through an independent arbiter exist, and whether provisions of standing are broad enough to apply to those with an interest) (Hunold and Young, 1998; Millner, 2011).

In order to achieve such procedural conditions for justice, it may be necessary to provide education and capacity-building resources to ensure that the community is aware of its rights and can participate effectively (Millner, 2011). Furthermore, alternative discourses must be recognized and misrecognition must not serve to devalue non-dominant perspectives (Schlosberg, 2007) or suppress them altogether (e.g. via the use of strategic lawsuits against public participation). Financial, information and geographic barriers to participation must also be overcome, which might include the provision of economic support (to obtain legal or other advice), removal of any threats associated with adverse costs orders (which might discourage litigation) and consideration of the physical location of land use decisions (Hunold and Young, 1998; Millner, 2011).

The field of peace studies provides another potentially fruitful avenue, including the adoption of 'peacebuilding' methods of conflict transformation at firm level. With origins in post-WWII efforts by the UN to recover from conflict by cultivating peace, at a general level 'peacebuilding' encompasses activities which aim to overcome conflict by:

> [A]ddressing structural and proximate causes of violence, promoting sustainable peace, delegitimizing violence as a dispute resolution strategy, building capacity within society to peacefully manage disputes, and reducing vulnerability to triggers that may spark violence.[20]

Specific to cases of 'sustained and episodic, non-violent' (Bond, 2012) conflict, actors at the community and firm level may employ a range of peacebuilding strategies in a variety of dispute situations to *transform* conflicts. This is distinct from approaches that seek only to *manage* conflict, which focus only on suppressing or containing or avoiding conflict consequences. Conflict transformation does not seek to stop, or otherwise manipulate or control conflict. It is, as Bond notes, 'less about a prescriptive set of steps that can be applied universally to conflict situations [and] is more about an attitude, a willingness, for parties to honestly enter and engage a relationship space' (Bond, 2012). The parties develop capacities to view conflict as a natural phenomenon through a positive paradigm that invites constructive development (Lederach, 2003).

Even the more mechanistic approaches of alternative dispute resolution within the law have the potential to be useful. The process of communicative framing has shown much promise in cases of intractable land use conflicts, and might be applied to the types of boundary-related conflicts that we have discussed. Frames are communicative devices utilized by both groups and individuals to navigate interactions with one another (Dewulf *et al.*, 2009). The process of framing plays a critical part in the 'creation, evolution, and perpetuation' (Lewicki *et al.*, 2003, p, 15) of conflict through the reflexive definition of issues and proposed resolution. A difference in frames between actors has the potential to create or escalate conflict, as they 'engender divergent interpretations of events, paint parties into negative characters, yield mutually incompatible issues, and focus attention on specific outcomes that impede exploration of alternatives' (Shmueli *et al.*, 2006). Different frames thus have the capacity to advance conflicts towards intractability, because divergent frames aggravate communication problems and become 'self-reinforcing' as they filter new information received (Shmueli *et al.*, 2006).

In the recent water conflicts in Australia, issues of framing are apparent and are directly related to the boundary/firm issues. Thus, rural communities have demonstrated a strong collective affiliation and a sense of being the 'other' in the face of the proposals to alter water flow regimes. The challenge became partly framed as being about the collective identity and interest of rural communities, relative to 'greenies' and 'city slickers'. On the other side, a similar framing and belonging process took place within the environmental community. It is notable that even given substantial consultation and a great deal of information, the political positions of the main stakeholder groups shifted very little if at all. While the topic of the contest was water, the contest was about far more than this. This is evident from the types of argument that were raised both in the media and in the formal submissions to the Murray-Darling Basin draft plan.

The concept of framing has received considerable attention as a tool to examine communication in conflict situations (Shmueli *et al.*, 2006; Brummans *et al.*, 2008; Shmueli, 2008; Dewulf *et al.*, 2009). Most recently, its benefits in broad-scale intractable multi-party conflicts have been explored (Dewulf *et al.*, 2009), where it may be used to 'unravel the complexities of situations in which mutually interdependent parties struggle to cope with their incompatible interests' (Brummans *et al.*, 2008). Shmueli *et al.* posit that frame insights are especially useful in intractable conflict situations, as framing can be 'particularly prominent when conflicts are long lasting and alienation and escalation drive disputants apart' (Shmueli *et al.*, 2006, p. 210). Similarly, Putnam and Wondolleck note that where disputants characterize a conflict as deeply connected to an illegitimate social system, or where the distinct identities of opposing parties are critical to the definition of the conflict, frame insights can open up the possibilities of shared identities or more positive characterizations (Putnam and Wondolleck, 2003).

Conflict theory and practice literature suggest to us that there may be positive directions that can be pursued for the better management of trans-boundary water conflicts, including the non-international versions of trans-boundary conflict which we have discussed. The type of legal approaches that would be required would be more deeply interventionist than those that have traditionally characterized the law. It is possible to envisage an innovative legal regime under which stakeholders would be brought together (rather than merely consultant) in a different type of process in which frames would be explored, and issues discussed.

Conclusions

In writing this chapter our focus has been upon the effectiveness of water law beyond exploring legal instruments or reliance upon doctrinal scholarship. Even from superficial observation, it is clear that Australia's history of water governance shows that conflict is a constant regardless of the type of legal regime that is put in place. Negotiation of institutional interests is always an important element in designing and implementing the law. What this chapter fundamentally suggests is that this must be recognized and that with recognition ought to come conscious management of the conflict potentials. The management of conflict potential involves far more than information provision and rights of legal appeal. Trans-boundary conflict does have a significant cost, and if we are serious about effective water law it is worthwhile investing in managing the potential conflicts in more creative ways. This can only be done systematically if the legal arrangements are in place that institutionalize better processes and help to establish norms based upon accumulating experience in handling diverse conflicts as they emerge. Siphoning off these conflicts into administrative arrangements or cutting off their potential (as has been attempted with the Water Act) comes at a cost of delaying this evolution.

The implications of the recognition of boundary-related conflict as a central component in water law and policy are significant. First, it is very difficult to ignore the diversity of human interest in water policy if you are aiming to identify potential conflicts. Second, a focus upon institutional boundaries as a potential locus of conflict draws attention to the diversity of roles, overlap and institutional interest that are involved in almost any water governance issue. Conflict awareness makes it very difficult to narrow the relevant interests to a subgroup of the broader society. Finally, identifying institutional arrangements which can better manage conflict, rather than attempting to create institutional arrangements which will prevent the emergence of those conflicts, seems likely to require a greater deal of innovation in the design of legal arrangements than is generally evident. Conflict anticipation requires attention to a full range of values and interests, including justice.

A focus on trans-boundary conflict at all levels highlights a number of key aspects of water governance. The first is the link between conflict and transaction costs, interest bargaining over time, and the potential for effective implementation of water laws. It also highlights the diverse political economy of water and power, and the ways in which this diversity affects governance in practice (which is typically far more complex than governments in theory, or as represented in policy analysis or legal instruments). The approach also will bring to the surface the ways in which particular legal arrangements privilege or marginalize particular interests, notably the interests of the less powerful. This in turn should give rise to greater attention to ways in which structural adjustments may be incorporated into legal arrangements to allow for a more equitable treatment of the transitions that are envisaged in water law and policy. This link between water law and the potential for structural adjustments that may take place outside the realm of water highlights the link between water governance and the broader legal and political regimes. This link also extends to the ways in which pre-established institutional arrangements will shape the nature of water conflicts within any legal regime, and the way in which those conflicts will in turn shape the broader institutional structure of the society.

Finally, a focus on potential conflict should help to reduce the significant risks of technical and bureaucratic hubris, which does seem to have been a feature of modern Australian water law. In the context of water law scholarship, emphasis upon conflict and social justice in natural resource governance seems to offer pathways to reintroduce issues of justice and fairness as core questions rather than as peripheral afterthoughts. The approach also places an emphasis upon independent judgement and questioning of power and status, and it does offer the potential for original scholarly work around conflict and interests. Finally, because institutional boundaries and conflicts are such a fundamental source of inefficiencies and difficulties in the implementation of water war, having a clear focus upon these questions may allow legal scholars to move from instrumentalism towards a positive focus on more effective water governance and innovation in the mechanisms for management of conflict that are encompassed by the law.

Notes

1 We refer in this chapter to 'legal governance' as the system of public and private laws and the organizational arrangements to create and implement these, that are used to align (control, regulate or govern) the interests of people and organizations to the public interest.
2 Rio+20 Declaration 'The World We Want', with its clear assertion in Cl 19 of the need to improve implementation. IUCN 'Natural Resources Governance' framework, which is targeting 'scientific' improvement in the effectiveness of natural resource governance including the legal aspects, the UNEP's 'Environmental Governance sub-programme'; and UNEP/WGEA MoU (4/90 implementation – GEO5), through which audit of implementation of a large number of international agreements is to be conducted (though only at the superficial level of verifying their institutionalization in the signatory countries).
3 See *www.ecolex.org*. Ecolex reports 2,070 treaties, of which 670 are multilateral, and the remaining are bilateral; 110,000 national laws and regulations; 33,700 legal and policy literature records; 1,100 court decisions. The records of national laws and court decisions omit many items.
4 An extensive empirical assessment of transboundary water conflict is provided in Wolf (1997). The potential for water conflict to escalate, and approaches for heading this off, are illustrated in Waslekar and Futehally (2013).
5 For a discussion of interfirm rivalry theory see Chen (1996).
6 These issues have been widely researched in studies of corporate rivalry and collaboration, as 'collaborative strategy'.
7 The 2004 Intergovernmental Agreement on a National Water Initiative is illustrative of an inter-state agreement for coordinated water governance within a federated structure.
8 The details of this argument with references are contained in Martin (2011).
9 There are many non-scholarly reports and commentaries on these conflicts. See, e.g., ABC News (2010) 'Tractor convoy as SA irrigators protest' ABC News, 15 October, http://abc.net.au/news/2010–10–15/tractor-convoy-as-sa-irrigators-protest/2299440?section=justin, accessed 24 March 2014; ABC News (2010) 'Water cut treats farmers as 'second-class citizens', ABC News, 4 November, http://abc.net.au/news/2010–11–04/water-cuts-treat-farmers-as-second-class-citizens/2324236?section=justin, accessed 24 March 2014; Crase, L. (2012) 'Resisting the Murray-Darling Basin Plan – in whose interest?' *The Conversation*, 14 August, http://theconversation.com/resisting-the-murray-darling-basin-plan-in-whose-interest-8771, accessed 24 March 2014; Miragliotta, N. (2012) 'Is anyone really listening? The Murray Darling and the limits to community consultation', *The Conversation*, 7 June, http://theconversation.com/is-anyone-really-listening-the-murray-darling-and-the-limits-to-community-consultation-7498, accessed 24 March 2014; Barlow, S. (2011) 'Murray Darling Basin can feed the nation if we just get it right', *The Conversation*, 12 May, http://theconversation.com/murray-darling-basin-can-feed-the-nation-if-we-just-get-it-right-352, accessed 24 March 2014; Burke, T. (2012) 'Solving the century of conflict over riverways', *The Daily Telegraph*, 23 July, http://dailytelegraph.com.au/news/opinion/solving-the-century-of-conflict-over-riverways/story-e6frezz0–1226432292224, accessed 24 March 2014.

10 Detailed in Martin, P. (2010) 'Expert review from a social and economic perspective in Developing the Guide to the Proposed Basin Plan: Peer review reports', Murray–Darling Basin Authority, Canberra, Australia, pp. 194 – 202, http://mdba.gov.au/sites/default/files/pubs/PeerReview280311.pdf, accessed 24 March 2014.

11 For more details of the case studies see Martin, P., Williams, J. and Stone, C. 'Transaction costs and water reform: The devils hiding in the details', Technical Report UNE CRCIF 9.1, Cooperative Research Centre for Irrigation Futures, Australia, http://irrigationfutures.org.au/newsDownload.asp?ID=768&doc=CRCIF-TR0808-web.pdf

12 The majority of these water conflict issues were canvassed in a Special Edition of the *International Journal of Rural Law and Policy*, which arose from an international meeting of water law experts focusing on water law conflict at the University of New England in January 2011: http://epress.lib.uts.edu.au/journals/index.php/ijrlp/issue/view/205

13 Some of the interest group rivalries are documented in Cronin (2009).

14 For many, defining social systems in terms of communications and divorcing human actions and norms as the unit of analysis is a contradictory, dehumanized concept (Banakar and Travers, 2005). However the approach is desirable in order to unpack complexity by '. . . understanding systems in terms of communication and its binds' (Gershon, 2005). For Gershon, the utility of Luhmann's theory comes into play when analysing social systems that participants *see* as systems and which create individual subject positions – for example, legal structures, bureaucracies – but which '. . . do not necessarily engage with subjects' (pp. 99–100).

15 See for example UBC Global Transboundary International Waters Research Initiative; or UNESCO Water Centre at Dundee; Stockholm International Water Institute or the Pacific Institute.

16 For an examination of some aspects of intra-national boundary related conflicts with an Australian flavour, see Heinmiller (2009), which explicitly examines interagency 'turf-war' as elements within water governance of the Murray-Darling Basin of Australia, the Colorado Basin of the United States and the Saskatchewan-Nelson Basin of Canada.

17 For a review of water conflicts at various levels, and indications of the potential for unmanaged conflict to result see Wolf *et al.* (2005).

18 For more details of policy risk see Martin and Williams (2010).

19 For a water law-specific illustration of the potential response of water law to varied demands, see Norman *et al.* (2012).

20 Organisation for Economic Co-operation and Development (2008) 'OECD guidance on evaluating conflict prevention and peacebuilding', Organisation for Economic Co-operation and Development, Paris, http://oecd.org/dac/evaluation/dcdndep/39774573.pdf, accessed 24 March 2014.

References

Banakar, R. and Travers, M. (2005) 'Law, sociology and method', in R. Banakar and M. Travers (eds) *Theory and Method in Socio-Legal Research*, Hart Publishing, Oxford, UK, pp. 1–25.

Benson, R. D. (2011) 'Public on paper: The failure of law to protect public water uses in the western United States', *International Journal of Rural Law and Policy*, 1, 1.

Bond, C. (2012) 'Sustainable Peace in the Mining Context', Paper presented at Eighth International Conference on Environmental, Cultural, Economic and Social Sustainability 2012, Vancouver, Canada, https://csrm.uq.edu.au/publications/sustainable-peace-in-the-mining-context, accessed 24 March 2014.

Brummans, B., Putnam, L., Gray, B., Hanke, R., Lewicki, R. and Wiethoff, C. (2008) 'Making sense of intractable multiparty conflict: a study of framing in four environmental disputes', *Communication Monographs*, 75, 1, 25–51.

Carey, J., Sunding, D. L. and Zilberman, D. (2002) 'Transaction costs and trading behavior in an immature water market', *Environment and Development Economics*, 7, 4, 733–50.

Chen, M. J. (1996) 'Competitor analysis and interfirm rivalry: towards a theoretical integration', *Academy of Management Review*, 21, 1, 100–34.

Coase, R. H. (1937) 'The Nature of the Firm', *Economica*, 4, 16, 386–405.

Cronin, R. (2009) 'Mekong Dams and the Perils of Peace', *Survival*, 51, 6, 147–60.

Dewulf, A., Gray, B., Putnam, L., Lewicki, R., Aarts, N., Bouwen, R. and van Woerkum, C. (2009) 'Disentangling approaches to framing in conflict and negotiation research: A meta-paradigmatic perspective', *Human Relations*, 62, 2, 155–93.

Dovers, S. and Hussey, K. (2007) *Managing Water for Australia: The Social and Institutional Challenges*, CSIRO, Collingwood, Australia.

Du, Q. (2011) 'Developments in water pollution law and policy in China: effective enough to cope with water pollution conflict?', *International Journal of Rural Law and Policy*, art. 2.

Ecolex (2014) 'The Gateway to Environmental Law,'FAO, IUCN and UNEP, www.ecolex.org, accessed 24 March 2014.

Ellsworth, J. (2011) 'Semiotics in legal theory design', in J. M. Broekman and F. J. Mootz (eds) *The Semiotics of Law in Legal Education*, Springer, Netherlands, pp. 129–51.

Fort, D. (2011) 'Too hot to handle: climate change and agricultural water use', *International Journal of Rural Law and Policy*, art 5.

Gershon, I. (2005) 'Seeing like a system: Luhmann for anthropologists', *Anthropological Theory*, 5, 2, 99–100.

Hanak, E., Lund, J., Dinar, A., Gray, B., Howitt, R., Mount, J. and Thompson, B. (2011) 'Managing California's Water: From Conflict to Reconciliation', Public Policy Institute of California, San Francisco, CA, www.ppic.org/content/pubs/report/R_211EHR.pdf, accessed 24 March 2014.

Haraldsdottir, K. (2011) 'Property rights in water and social conflict: an example from Iceland', *International Journal of Rural Law and Policy*, art. 6.

Heinmiller, T. (2009) 'Path dependency and collective action in common pool governance', *International Journal of the Commons*, 3, 1, 131–47.

Holmström, B. and Roberts, J. (1998) 'The boundaries of the firm revisited', *Journal of Economic Perspectives*, 12, 4, 73–94.

Hunold, C. and Young, I. M. (1998) 'Justice, democracy, and hazardous siting', *Political Studies*, 46, 1, 82–95.

Hussey, K. and Dovers, S. (2006) 'Trajectories in Australian water policy', *Journal of Contemporary Water Research and Education*, 135, 1, 36–50.

Hussey, K. and Dovers, S. (2007) *Managing Water for Australia – The Social and Institutional Challenges*, CSIRO Publishing, Collingwood, Australia.

Innes, J. E. and Booher, D. E. (2004) 'Reframing public participation: strategies for the 21st century', *Planning Theory & Practice*, 5, 4, 419–36.

Kidd, M. (2011) 'Poisoning the right to water in South Africa: what can the law do?', *International Journal of Rural Law and Policy*, art. 7.

Lederach, J. P. (2003) *The Little Book of Conflict Transformation*, Good Books, Intercourse, PA.

Lewicki, R. J., Gray, B. and Elliot, M. (2003) *Making Sense of Intractable Environmental Conflict: Concepts and Cases*, Island Press, Washington, DC.

Luhmann, N. (1986) 'The Self-Reproduction of Law and its Limits', in G. Teubner (ed.) *Dilemmas of Law in the Welfare State*, European University Institute – Series A, Walter De Gruyter, Berlin, Germany.

Luhmann, N. (1988a) 'The unity of the legal system', in G. Teubner (ed.) *Autopoietic Law – A New Approach to Law and Society*, Walter De Gruyter, Berlin, pp. 12–35.

Luhmann, N. (1988b) 'Closure and openness: On reality in the world of law', in G. Teubner (ed.) *Autopoietic Law – A New Approach to Law and Society*, Walter De Gruyter, Berlin, pp. 335–48.

Luhmann, N. (1995) *Social Systems*, Stanford University Press, Stanford, CA.

Luhmann, N. (2004) *Law as a Social System* (trans. K. A. Ziegert), Oxford University Press, Oxford, UK.

Martin, P. (2009) *Submission to the Inquiry on the Management of the Murray Darling Basin*, Australian Centre for Agriculture and Law, Armidale, Australia, https://senate.aph.gov.au/submissions/comittees/viewdocument.aspx?id=f0ec73f5-bee3–4832–968a-3c45235dfe7b, accessed 24 March 2014.

Martin, P. and Becker, J. (2011) 'A tale of two systems: conflict, law and the development of water allocation in two common law jurisdictions', *International Journal of Rural Law and Policy*, art. 8.

Martin, P. and Kennedy, A. (2011) 'Water management in rural Australia: the human rights dimension', *Human Rights Defender*, 20, 2, 12–19.

Martin, P. and Shortle, J. (2010) 'Transaction costs, risks and policy failure' in C. Dias Soares, J. Milne, H. Ashiabor, L. Kreiser and K. Deketelaere (eds) *Critical Issues in Environmental Taxation: International and Comparative Perspectives Vol VIII*, Oxford University Press, New York.

Martin, P. and Williams, J. (2010) *Policy Risk Assessment*, Cooperative Research Centre for Irrigation Futures, University of New England, Armidale, Australia, http://irrigationfutures.org.au/news.asp?catID=9&ID=1209, accessed 24 March 2014.

Martin, P., Williams, J. and Stone, C. (2008) *Transaction costs and water reform: the devils hiding in the details*, Cooperative Research Centre for Irrigation Futures, University of New England, Armidale, Australia, http://irrigationfutures.org.au/newsDownload.asp?ID=768&doc=CRCIF-TR0808-web.pdf, accessed 24 March 2014.

Millner, F. (2011) 'Access to environmental justice', *Deakin Law Review*, 16, 1, 189–207.

Murray-Darling Basin Authority (2010) *Developing the Guide to the proposed Basin Plan: Peer Review Reports, Murray–Darling Basin Authority*, Murray-Darling Basin Authority, Canberra, Australia, http://mdba.gov.au/sites/default/files/pubs/PeerReview280311.pdf accessed 24 March 2014.

Norman, E., Cohen, A. and Bakker, K. (2012) *Flashpoints and Collaboration: How Problems Can Inspire Innovative Solutions for Canada, the US, and the Governance of Shared Waters*, Program on Water Governance, Vancouver, Canada.

Oliver, C. (1993) 'Organisational boundaries: definition and functions', *Canadian Journal of Administrative Sciences*, 10, 1, 1–17.

Pacific Institute (2009) *Water Conflict Chronology*, Pacific Institute– The World's Water, Oakland, CA, http://worldwater.org/chronology.html, accessed 24 March 2014.

Paisley, R. K. and Grzybowski, A. (2011) 'Some reflections on the resolution of state-to-state disputes in international waters governance agreements', *International Journal of Rural Law and Policy*, art. 9.

Putnam, L. L. and Wondolleck, J. L. (2003) 'Intractability: Definitions, dimensions and distinctions', in R. J. Lewicki, B. Gray and M. Elliot (eds) *Making Sense of*

Intractable Environmental Conflict: Concepts and Cases, Island Press, Washington, DC, pp. 49–50.

Quiggin, J. (2008) *Key Issues in Australian Water Policy*, University of Queensland, Brisbane.

Resources for the Future (2013) 'The Future of US Water Supplies', Resources for the Future, Washington, DC, http://rff.org/Events/Pages/The-Future-of-US-Water-Supplies.aspx, accessed 24 March 2015.

Santos, F. M. and Eisenhardt, K. M. (2005) 'Organizational boundaries and theories of organization', *Organization Science*, vol 16, no 5, pp. 491–508

Schlosberg, D. (2007) *Defining Environmental Justice: Theories, Movements and Nature*, Oxford University Press, Oxford, UK

Schlosberg, D. (2013) 'Theorising environmental justice: The expanding sphere of a discourse', *Environmental Politics*, vol 22, no 1, pp. 37– 55

Shmueli, D. (2008) 'Framing in geographical analysis of environmental conflicts: Theory, methodology and three case studies', *Geoform*, 39, 2048–61.

Shmueli, D., Elliott, M. and Kaufman, S. (2006) 'Frame changes and the management of intractable conflicts', *Conflict Resolution Quarterly*, 24, 2, 207–18.

Tafur, V. (2011) 'Water law, mining and hydro-energy conflicts in South America: tales from the Andes and Patagonia', *International Journal of Rural Law and Policy*, art. 4.

Teubner, G. (1993) *Law as an Autopoietic System*, Blackwell Publishers, Oxford, UK.

Waslekar, S. and Futehally, L. (2013) *Water Cooperation for a Secure World*, Strategic Foresight Group, Mumbai, India.

Water Act 2007 (Cth).

Wolf, A. T. (1997) 'International water conflict resolution: lessons from comparative analysis', *International Journal of Water Resources Development*, 13, 3, 333–66.

Wolf, A., Kramer, A., Carius, A. and Dabelko, G. (2005) 'Managing water conflict and cooperation', in E. Assadourian *et al.* (eds) *State of the World 2005*, W. W. Norton, New York, http://tbw.geo.orst.edu/publications/abst_docs/wolf_sow_2005.pdf, accessed 24 March 2014, pp. 80–95.

7 Defragging

Overcoming fragmentation in United States water governance

Barton H. (Buzz) Thompson Jr

Introduction

This chapter examines fragmentation in US water management, its deleterious consequences, and potential approaches to overcoming those problems. In the US, water management is often fragmented *geographically* among governments, *hydrologically* between groundwater and surface-water regimes, and *functionally* among different regulatory tasks. Geographic fragmentation fractures waterways and watersheds among multiple governmental units with little, if any, hydrologic logic; it also creates a myriad of local, frequently small water suppliers. Such fragmentation obstructs efficient water management, generates distributional disputes and encourages protectionism. Hydrologic fragmentation undercuts the conjunctive management of groundwater and surface water. Functional fragmentation undercuts efficient, comprehensive management and breeds distributional fights over political power. Although these problems often are discussed separately, they are highly similar – and interrelated – and their potential solutions are also similar.

The US has recently tried to overcome various forms of fragmentation through consensus or multi-stakeholder processes designed to bridge differences and encourage coordination. Such 'soft restructuring' is not the only way to defragment water management (or, to use the term employed in the computer sector, 'defrag' it), nor is it often the best approach. Soft restructuring assumes that fragmented units will engage in sustainable cooperation without either a fundamental change in institutional structure or incentives. Three other methods are also important in defragging US water management: (1) 'hard restructuring' (in which separate governing units merge or create a formal partnership); (2) 'top-down defragmentation' (in which a higher unit of government mandates or incentivizes coordination); and (3) decentralization (in which authority is delegated to smaller entities within the fragmented units that can then reconstitute themselves into more logical orderings; effectively defragging through further fragmentation).

This chapter begins with an overview of fragmentation in the US and its problems. It then provides a case study of interstate fragmentation, the most visible type of fragmentation in the US. Finally, the chapter evaluates several principal means to defrag water management. As part of this evaluation, the chapter also considers the reasons, both generic and particular, why various efforts at defragmentation have failed.

Fragmentation of US water management

Fragmentation of water management is a growing problem in the US. The US has long suffered from both geographic and hydrologic fragmentation. As regulatory systems multiplied in the 20th century, functional fragmentation also arose. The results, not surprisingly, have included distributional disputes (including water protectionism), inefficiency (in both water administration and use) and increased risk of water disruptions.

Geographic fragmentation

With 48 sovereign states sharing the waters of the continental US and a federal system that awards states strong authority, interstate fragmentation was probably inevitable.

Approximately 95% of the freshwater in the continental US comes from interstate sources (Hall, 2010, p. 239). The challenges of geographic fragmentation were not a consideration when state borders were drawn. In the East and Midwest, waterways were often used as borders between states. In the West, the hubristic practice of laying out states with straight lines that ignored natural features sliced and diced waterways.

The US could have managed interstate waterways nationally (Thompson, 1997b). However, the US federal system has strongly favoured state authority. Following the Civil Water, Congress ratified water rights 'recognized and acknowledged by the *local* customs, laws, and the decisions of courts' (14 USC 253, § 9), confirming that states rather than the federal government controlled water rights. In admitting new states, moreover, Congress also approved state constitutions that often explicitly asserted state authority over all waters within the states' borders (Thompson, 1996). At the end of the 19th century, the US Supreme Court held that states, not the national government, enjoyed general authority over their waterways (*United States v. Rio Grande Dam & Irrigation Co.*, 1899). The national government retained authority only as necessary to protect navigability and make use of national property.

Geographic fragmentation is mirrored at the local level. Nationally, there are over 150,000 water agencies. Indeed, there are more water suppliers than elementary, middle and high schools combined. Most service 500 or fewer people (Kearney *et al.*, 2014). Watersheds are often split among scores of different entities. In California, several dozen 'wholesalers' sell water to about 400 large retailers and several thousand smaller, typically rural retailers; hundreds of separate agricultural water districts provide water to the state's farming community (Hanak *et al.*, 2011).

Hydrologic fragmentation

Much of the US also suffers from hydrologic fragmentation. When water managers have authority over both groundwater and surface water, they can increase overall yield and water availability by managing the water on a conjunctive basis (Evans, 2010). Fragmented management, by contrast, can lead to conflicts between users of groundwater and surface water (Thompson, 2011).

Yet for years, most state water agencies enjoyed responsibility only over surface water and, in limited cases, groundwater that constituted the 'subflow' of a surface waterway (a purely legal construct with no scientific meaning). Virtually all western states have now integrated groundwater and hydrologically connected surface water in at least some water basins. However, several major states continue to relegate groundwater management primarily to local agencies with little or no authority over surface water. No state, moreover, has comprehensively integrated groundwater and surface water in all basins and for all uses (Thompson, 2011).

Functional fragmentation

Water management also suffers from functional fragmentation. In the early 20th century, state water agencies enjoyed responsibility over most major water issues. However, as new, frequently environmental, issues arose, governments often created new agencies to manage them.

Consider water quality. While states hold principal responsibility over water quantity, the national government claimed principal authority over water quality in the Clean Water Act, 33 USC. 1972, §§ 1251 *et seq.*). Although quantity and quality are closely related issues, authority is split between different levels of government. Even at the state level, a water agency typically manages quantity, while an environmental agency regulates quality. Indeed, California is the only state to place quantity and quality in the same agency (Thompson *et al.*, 2013).

Authority over instream flow is similarly fragmented. State water agencies generally determine when to permit new diversions. But the national government often sets minimum instream flows under the Clean Water Act, Endangered Species Act, 16 USC. 1973, §§ 1531 *et seq.*), or 'wild and scenic rivers' legislation. Even at the state level, environmental or fish-and-wildlife agencies often establish and hold instream rights (Thompson *et al.*, 2013).

The problems of fragmentation

These forms of fragmentation produce at least three categories of problem. First, they lead to unnecessary and often harmful distributional fights. Geographic fragmentation, for example, has led to competing water claims among states or water-supply organizations, creating costly legal fights. Geographic fragmentation also can generate a synthetic 'we–they' perspective that fuels such competition and undercuts efforts to find managerial options that might benefit all users.

Geographic fragmentation can similarly undercut water markets and the efficient transfer of water. Separate governmental units often protect 'their' water rights for local use and proscribe external transfers. States, for example, often limit out-of-state transfers; interstate agreements sometimes explicitly outlaw them (Thompson, 1991). Local water suppliers often bar or impede transfers outside their borders, while encouraging internal transfers (Thompson, 1993, 1997a).

Functional fragmentation can lead to harmful distributional contests for political power and budgets. For years, the US Army Corps of Engineers and the US Bureau of Reclamation competed to build large reclamation projects in the West. The agencies competed in part by offering large subsidies, leading to sizeable redistributions of taxpayer monies and the construction of economically unjustifiable dams (McCool, 1987; Nicholson-Crotty, 2005). The competition also has generated two separate and only loosely coordinated sets of dams and reservoirs.

Second, fragmentation can lead to inefficient management. For example, states that share interstate rivers often worry more about how to maximize their share of the river's water than about how to holistically manage the river for maximum combined benefit. Lawsuits are more likely than joint management agreements. As in other settings, 'claiming' value impedes efforts to 'create' value (Lax and Sebenius, 2011).

Fragmentation also undercuts coordinated management by adding administrative costs and collective-action challenges. Efforts to improve the management of California's Sacramento-San Joaquin Delta system are illustrative (Gray *et al.*, 2013). The Delta faces a variety of challenges, including salt-water intrusion and endangered fish, stemming from large diversions. Yet management is highly fragmented. The federal and state governments divert water through separate projects. An array of upstream and in-basin water users, including cities, municipal water agencies, and irrigation districts, also divert and use water. The State Water Resources Control Board (SWRCB) manages water rights. Separate national and state fish agencies protect endangered species. The Army Corps regulates wetlands and manages flood risks, the national Environmental Protection Agency (EPA) and the SWRCB share authority over water quality and the Federal Energy Regulatory Commission (FERC) regulates hydropower.

This broad diversity of agencies has made it exceptionally difficult to develop integrated approaches to the multiple Delta stresses. Because the agencies generally focus on their own missions and interests, they frequently miss opportunities for watershed-wide management. Given the cost of coordination, moreover, agencies have less incentive to work together. Conflicting mandates make coordination even more challenging. Efforts to restore degraded parts of the Delta, moreover, require multiple permits and approvals, slowing and undermining those efforts. Finally, because each entity frequently conducts its own science, scientific understanding of the Delta, which is crucial to sustainable long-term management, is fractured and less enlightening (Gray *et al.*, 2013).

Although the Delta is an extreme example of fragmentation, similar problems arise elsewhere. For example, local separation of water suppliers and wastewater agencies has undermined opportunities to recycle and reuse wastewater (Kiparsky *et al.*, 2013). Separation of agricultural and municipal water agencies has impeded opportunities to trade treated municipal wastewater for agricultural freshwater supplies. By reducing economies of scale, fragmentation also has reduced innovation, raised costs and generated unnecessary system failures (Thompson, 1999; Ajami *et al.*, 2014).

Finally, fragmentation of the water industry can generate greater supply uncertainty. Geographically small water suppliers typically have only one principal source of water. Any reduction in that source, whether from drought, pollution or similar factors, can undermine the supplier's ability to meet local demand. While suppliers can reduce this risk through transfer or sharing agreements with other local suppliers (assuming conveyance facilities or wheeling opportunities exist), such agreements require negotiation and add administrative costs.

Interstate rivers: a case study in fragmentation

Interstate fragmentation of major waterways, as discussed in Part I, has long posed a managerial problem in the US, and efforts to deal with the resulting challenges have enjoyed only limited success. Politics and costs have deterred action. Efforts, moreover, have focused more on how to divide rivers among competing states than on effective joint management.

Judicial defragging

The US Supreme Court, which holds exclusive jurisdiction under the US constitution over lawsuits among states (US Constitution, art. III, § 2, cl. 2), has heard interstate water disputes since the early 1900s (Stewart, 1982). The Supreme Court, however, has always felt uncomfortable as an interstate arbitrator. The process for resolving interstate water disputes, many of which involve relatively minor rivers, can be clunky, costly and time consuming, and the disputes put the Court in the middle of often highly inflamed conflicts. Frequently, therefore, the Court has refused to address interstate disputes either on the ground that the dispute is not of such 'serious magnitude' to justify the exercise of its 'extraordinary power' to order state action (e.g. *New York v. New Jersey (1921)*) or for procedural reasons (e.g. *Arizona v. California (1936)*).

More importantly, the US legal system and jurisprudence have limited what the Court has been willing to do. Where states have vied for limited interstate waters, the Court has divided the water through what the Court calls 'equitable apportionment' (Thompson *et al.*, 2013). The Court has never attempted to help states manage interstate waterways on a holistic or prospective basis. The Court also has only recently accepted a case involving environmental needs (*Florida v. Georgia (2014)*).

Interstate compacts

Beginning in the 1920s, states began to negotiate agreements, or 'compacts,' to allocate the waters of interstate rivers, establish ground rules for managing such rivers and create 'compact commissions' to address longer-term issues. To date states have negotiated almost two dozen compacts, beginning with the 1922 Colorado River Compact (US General Accounting Office, 2007).

Most compacts have again proven a limited solution. First, transaction costs have been high, making them difficult to negotiate. For example, California and Nevada took 15 years to hammer out a draft compact for the rivers they share, only to have Congress refuse to confirm the compact (as required by the US Constitution) because of Native American concerns. In many cases, states have reached agreement only because the national government conditioned water storage projects on reaching an agreement. With the decline of national funding, states no longer have the same incentive to negotiate. States have not negotiated a new compact since 1978 (McCormick, 1994).

Because of negotiation challenges, moreover, compacts are often vague, suffer from internal inconsistency and ignore difficult issues. The history of the above-noted California–Nevada compact is illustrative. While California sought a broad compact, Nevada insisted that the negotiations address only the division of unused water. To make agreement possible, the negotiators ignored water-quality issues and groundwater. Even with issues narrowed, the negotiators ultimately left 'many thorny issues' up to a compact-created commission (Pisani, 1978).

Like the Supreme Court, compacts also have focused more on dividing available water supplies than on promoting joint management. Although most compacts create commissions to oversee administration, the commissions often have limited managerial authority and require unanimity (US General Accounting Office, 2007). Many compacts, moreover, are relatively inflexible, preventing states from using the compacts adaptively to address emerging or changed conditions. Efforts to amend an existing compact generally run into the same challenges as the negotiation of a new compact; to date, only three have ever been amended (McCormick, 1994).

Exceptions do exist. Unlike most interstate compacts, for example, the Delaware River Basin Compact has promoted more comprehensive, basin-wide planning, rather than simply dividing up interstate waters and setting basic ground rules for river management. Dissatisfied with US Supreme Court rulings that merely apportioned the waters of the Delaware River, local states negotiated a compact designed to manage the Delaware as a regional resource (Muys, 1971). The Compact also created the Delaware River Basin Commission (DRBC), with one representative from each state and from the federal government, to develop and implement a comprehensive regional plan for both current use and long-term development. Importantly, the Compact gives the DRBC extensive authority over all aspects of river management, including water allocation, water quality, hydroelectric generation, recreation, flood control and watershed protection, and provides for simple majority vote on most issues other than interstate allocation of the waters (Dellapenna, 2011; Thompson *et al.*, 2013).

Most observers believe that the DRBC has been effective in delivering more holistic management of the Delaware River, despite significant issues and changing demands and conditions. In particular, the DRBC has played a major role in:

> (1) developing a unified drought management plan; (2) adopting a joint reservoir operations program; (3) resolving interstate conflicts over water allocations and transfers; (4) implementing comprehensive water quality standards; (5) promulgating water conservation standards; and (6) providing a vehicle for joint investment in new water management facilities.
>
> (Weston, 1989)

National legislative intervention

Congress could also help to overcome interstate fragmentation. John Wesley Powell, the second director of the US Geologic Survey who wrote a major report in the late 19th century on the 'arid regions' of the West, urged Congress to divide the region into 150 separate drainage basins (Powell, 1890a, b). Powell's vision, however, clashed with political realities. Many political borders were already established, creating political allegiances and vested economic interests; local leaders had no interest in ceding control over 'their' waters (Pisani, 1982).

In 1963, however, the US Supreme Court held that Congress had given the US Department of the Interior broad authority over the development and use of the waters of the lower basin of the Colorado River (*Arizona v. California (1963)*). Although scholars have questioned whether Congress actually intended to give Interior such authority (Patashnik, 2014), the decision effectively made Interior water master for the lower Colorado basin, reducing the problems of fragmentation. Interior can nudge the lower-basin states to take actions needed for effective management and to manage the lower stretch of the river, with its multiple reservoirs, on a comprehensive basis.

Whether or not Congress intended to assert such authority over the Colorado River, Congress has otherwise proven reticent to directly intervene in the management of interstate waterways. Congress generally sees little advantage in wading into often politically heated interstate disputes. Senators from outside a region, moreover, are hesitant to vote to impose any type of national solution on an interstate river over the opposition of local states, for fear that Congress might subsequently impose a national regime on rivers in their own region.

Congress took a step toward more active national involvement in the management of interstate rivers in the Water Resources Planning Act of 1965. That Act authorized the creation of regional river basin commissions to (1) coordinate federal, state and local water development plans, and (2) plan and continually update comprehensive, coordinated joint plans. Congress provided funding both for the commissions and for state water planning. To help coordinate actions of US agencies, the Act also created a national Water Resources Council. Seven regional commissions ultimately formed under the Act, covering all or part of 32 states (Mandarano *et al.*, 2008).

Neither the regional commissions nor the Water Resources Council were ever successful in delivering effective, comprehensive and integrated management of regional river basins, and President Ronald Reagan abolished them by executive order in 1981 (Goldfarb, 1994). Their failure has been attributed to several factors. First, the regional commissions did not have the authority to directly manage, develop or regulate basins. Instead, they had to depend on the fragmented authority of the states and national government to implement the plans that they developed. Second, states viewed the national involvement suspiciously, particularly since most commissions were dominated by federal representatives. Compounding this problem, Congress failed to provide states with sufficient incentive to actively participate. For example, states did not need to comply with basin plans to obtain and spend federal funds for water projects. Third, the National Water Council never succeeded in coordinating federal water planning, and the federal representatives to the commissions often failed to speak with one voice. Functional fragmentation thus undermined efforts to overcome geographic fragmentation. Finally, Congress never adequately funded the commissions, yet coordinated planning can be very expensive (Mandarano, 2008).

Lessons

US efforts to manage interstate rivers suggest several lessons. First, solving fragmentation challenges is seldom easy. Second, geographic fragmentation often leads to allocative disputes that can make it more difficult to develop integrated solutions. Third, new institutions like the Delaware River Basin Commission can be crucial to integrated planning and management, but they must have adequate authority and funding. Finally, top-down resolution of interstate fragmentation only works where integrated management is either mandated or states are provided with adequate incentives to participate.

Defragging options

Governments have several options in attacking fragmentation. Recent emphasis has been on coordination and consensus – what I will call 'soft restructuring', because the structure and authority of governmental units remain the same. Under 'hard restructuring', fragmented units either merge, geographically or functionally, into new 'uberunits' or create formal partnerships embodied in separate organizations with their own authority. The DRBC is an example of hard restructuring through an interstate compact. A higher level of government (e.g. the national government) can force or encourage lower units to work together in effective collaborations through 'top-down defragmentation'. The Water Resources Planning Act of 1965 is an example of (ultimately unsuccessful) top-down defragmentation. Finally, in limited situations, governments might delegate authority down to actors within the separate units who can work together, freed of the rivalries or other pathologies of their units. Markets are perhaps the best example of this approach, which cures fragmentation, counterintuitively, through further fragmentation.

Soft restructuring

In recent years, governments have often tried to cure fragmentation through collaboration and other informal partnerships (Thompson, 2012). Collaboration has held out the promise of bridging trans-jurisdictional divides without top-down mandates or organizational restructuring. Analysts also have argued that key characteristics of collaboration – e.g. unanimity, lengthy discussions and joint analyses – lead to better, more acceptable and enduring solutions. According to Schlager and Heikkila (2009, p. 370), 'participants tend to be very satisfied with decisions, they view the decision process as fair, they build trust in one another, and they are more willing to work together with their fellow participants in future endeavors'.

Some collaborations have proven successful. As discussed in Part II, interstate compacts often create commissions that have no significant independent authority but can serve as fora for consensus decision-making.[1] Although many have concluded that such compact commissions are doomed to be ineffective (e.g. Muys, 2004), Schlager and Heikkila (2009) find that they have often helped resolve disputes. Of 23 conflicts studied (involving 14 interstate rivers), commissions helped resolve 8. A majority of the disputes, surprisingly, involved zero-sum distributional disputes.

Collaborative processes, however, have clear limitations. First, they do not work well where there is fundamental disagreement over participants' rights. Second, they work best when issues are simple and discrete rather than complex and comprehensive. Although compact commissions often successfully address distributional disputes, 'compared to other venues, they do not address the most complex conflicts' (Schlager and Heikkila, 2009). As discussed, various efforts to manage interstate waters through collaboration have succeeded only by narrowing and simplifying issues. Third, many collaborations prove unsustainable in the long run as interests flag or fundamental disagreements come to the foreground. Finally, collaborative processes that require long-term implementation work best only when coupled with hard restructuring (i.e. a new entity).

California's Sacramento-San Joaquin River Delta presents an illustration of these weaknesses. As described earlier, the Delta has a highly fragmented governance structure. Partly as a result, its aquatic ecosystem faces multiple stresses and is in decline. Altered water flows, degraded water quality, wetland destruction, modification of other habitat and the invasion of exotic species have harmed both biodiversity and local water users. In response, California courts have called for a more 'global perspective' (*United States v. State Water Resources Control Board* (1986)).

Many of the governmental units agreed in the 1990s to join together in the Bay Delta Program (CALFED) to undertake a more comprehensive analysis of Delta challenges, coordinate short-term actions and collaboratively develop a long-term managerial plan (Little Hoover Commission, 2005; Layzer, 2008; Hanak *et al.*, 2011). Despite a decade of effort, however, CALFED ultimately failed. The highly charged nature of the challenges was one problem. As CALFED moved from discussion to action, support for it waned. Structural weaknesses were also problematic. Solving the challenges required agencies to give up existing control in favour of shared power and often to make substantive concessions. Yet agency interests remained largely the same, and CALFED's momentum remained dependent on the political leaders who had initiated and sustained the process. When leadership changed, governmental support and funding evaporated. CALFED also never generated any type of formal governance structure through which agencies could coordinate actions and scientific studies (Gray *et al.*, 2013; Little Hoover Commission, 2005).

Watershed planning has also demonstrated some of the weaknesses of soft restructuring. Over several decades, hundreds of informal watershed groups, inter-agency working groups and watershed planning councils have formed (Blomquist and Schlager, 2005). The most effective efforts, however, have had limited focus. Most watershed planning has addressed water quality in small watersheds rather than more comprehensive, regional management. Because few groups have enjoyed the authority to implement plans, moreover, most efforts have involved sharing information and eliciting ideas, rather than long-term trans-jurisdictional management (Nakamura and Born, 2007). 'On a spectrum from "strong" to "weak", most watershed planning has been at the "weaker" end' (Blomquist and Schlager, 2005, p. 103).

Efforts to promote more comprehensive trans-jurisdictional planning have been a hard sell. Despite growing international interest in integrated water resource management (IWRM), few regions in the US have attempted this (Thompson, 2012). The complexity, difficulty, time needs and costs of IWRM have deterred agencies. IWRM also has sometimes threatened vested political interests.

Hard restructuring

A major problem with soft restructuring is that it does not cure fragmentation itself. It addresses a manifestation of the disease, rather than the disease. With rivalrous organizations still in place (often with vested interest groups pushing them apart) or with authority divided among unrelated organizations, long-term sustainability is often difficult. Defections from agreements, moreover, remain a serious risk.

In the private marketplace, companies often solve fragmentation by restructuring. One business might acquire or merge with another; or the two businesses might create a partnership. Whether businesses restructure or attempt to coordinate through contracts or informal agreements depends on comparative transaction costs, opportunities and incentives for defection, and legal restraints (e.g. anti-trust laws) (Williamson, 1975).

Hard restructuring is far less common in governmental water management, but it does take place. One common form of hard restructuring is the creation of a new entity (often, but not always, a partnership) to enable and coordinate joint action. The DRBC is an example of an effective interstate partnership. When Southern California cities recognized the value of joint water imports, they formed the Metropolitan Water District of Southern California (MWD), an independent regional special district, with its own authority, leadership team and staff, but with board members from the participating cities. Although technically a water wholesaler, MWD promotes coordinated management among its members.

To jointly manage groundwater and surface water in California, which suffers from a dual system, local stakeholders have sometimes created new regional agencies with authority to engage in conjunctive management (Thompson, 2011). For example, the Santa Clara Valley Water District (SCVWD), which has authority to both import surface water and manage groundwater in California's Santa Clara Valley, 'integrates surface water and groundwater to provide a flexible and reliable water system' (Micko, 1991, p. 1). Conjunctive management by such entities has proven both successful and enduring (Blomquist *et al.*, 2001).

Small water suppliers also sometimes merge into larger entities to ensure a safer and more sustainable supply, garner economies of scale and engage in broader watershed planning and management (US Environmental Protection Agency, 2002). Large suppliers can annex a smaller, neighbouring system (Feiock and Carr, 2001) or they can create a special district, such as MWD, to manage activities. Less common, but not exceptional, is the merger of multiple suppliers.

Governmental procedures can influence whether hard restructuring occurs. Most states authorize some forms of annexations, mergers, or special districts (e.g. Kan. Stat. §§ 19–3545 et seq.). However, procedural requirements can vary dramatically. Some states, for example, require voter approval through petitions, referenda and the like (Feiock and Carr, 2001). Not surprisingly, such processes undercut hard restructuring by increasing transaction costs (Carr and Feiock, 2000). State intervention also can promote hard restructuring. New York copied a successful programme for school mergers to promote similar mergers in the water sector. Vermont actively seeks out opportunities for consolidation.

The private sector also can help promote hard restructuring of water management. Where inefficiently small water systems have been either unwilling or unable to merge, a single private water supplier has sometimes contracted to run the units, providing economies of scale and coordination without formally merging the systems (Thompson, 1999).

Top-down restructuring

Top-down mandates and incentives can be critical in overcoming fragmentation. As noted, defragmentation often confronts collective action problems and can threaten agencies and stakeholders with vested interests in the fragmented structure. Legislatures and courts at the national and state level can play a critical role in overcoming these obstacles. Five examples illustrate.

Arizona v. California

The US Supreme Court's decision in *Arizona v. California* (1963) provides an illustration of judicial restructuring. Prior to the decision, Arizona and California were at loggerheads over the allocation and management of the Colorado River. As noted in Part II, the Supreme Court effectively defragmented management of the lower basin by awarding broad authority to the US Department of the Interior, which has led and encouraged greater coordination.

The Bay-Delta Conservation Plan

The US and California endangered species acts can also encourage coordinated planning on waterways with endangered fish species. Both laws constrain water exports and other actions that jeopardize listed species. Water agencies, however, can escape the prescriptions by developing conservation plans. Although each agency could develop its own plans, both California and the US have encouraged agencies and stakeholders to jointly develop comprehensive regional plans (known as 'regional habitat conservation plans' or 'natural community conservation plans').

As the CALFED process described above unraveled, agencies and other Delta stakeholders began to negotiate a Bay Delta Conservation Plan (BDCP) to meet the requirements of the endangered species laws through a 'comprehensive conservation strategy for the Delta designed to restore and protect ecosystem health, water supply, and water quality' (Bay Delta Conservation Plan Steering Committee, 2012, p. 1-1). Progress initially was promising. A BDCP Implementation Office would have coordinated with and among key agencies. A separate Permit Oversight Group, comprising the national and state fish agencies responsible for administering the endangered species laws, was proposed to 'coordinate agency review of the actions being implemented under the Plan' (Bay Delta Conservation Plan, 2013, pp. 7–15). Ultimately, however, California chose to take a different approach to comply with the endangered species laws.

The Delta Reform Act

Even more promising is California's Delta Stewardship Council (DSC), a new coordinating body mandated by the legislature in the Delta Reform Act (Gray *et al.*, 2013). The DSC, which consists of gubernatorial and legislative appointees, is responsible for developing and implementing a comprehensive Delta Plan that ensures a reliable water supply while protecting and enhancing the ecosystem. The Delta Plan will coordinate the work of dozens of water entities on such issues as estuarine health, water quality, water conservation, and storage and conveyance facilities. All state and local agencies will need to certify that their actions are consistent with the plan. Anyone who disagrees with the certification can appeal to the DSC.

The Delta Reform Act also requires the DSC to establish performance standards to 'enable the council to track progress in meeting the objectives of the Delta Plan'. These must include measurable assessments of both the:

> health of the Delta's estuary and wetland ecosystem for supporting viable populations of aquatic and terrestrial species, habitats, and processes, including viable populations of Delta fisheries and other aquatic organisms . . . [and the] reliability of California water supply imported from the Sacramento and San Joaquin River watershed.
>
> (California Water Code 2015, § 85211)

The Delta Reform Act is very different from earlier efforts to bridge the fragmentation in Delta management. The Act recognizes the importance of (1) a top-down approach, (2) clear performance objectives and, perhaps most importantly, (3) a new entity with responsibility for ensuring a consistent and comprehensive approach among agencies working in the Delta. As the DSC has explained, the Act is a 'distinct departure from CALFED and the *status quo* of disparate agencies struggling to tackle complex modern resource problems'. According to the DSC, its roles are to set a 'comprehensive, legally enforceable direction' for Delta management and to 'ensure coherent and integrated implementation of that direction through coordination and oversight of State and local agencies proposing to fund, carry out, and approve Delta-related activities' (Delta Stewardship Council, 2012, p. 39).

State incentives

Some states also encourage or require local water agencies to restructure in the face of problematic fragmentation. At the weaker end of the spectrum, some states require local water suppliers that plan to invest in new infrastructure to undergo a state review to see whether the project requires consolidation among the participating agencies. At the stronger end of the spectrum, some states authorize a central governmental agency to evaluate fragmentation and either require local agencies to consolidate or take over and run the local agencies. Other policies occupy a middle ground. Some states, for example, authorize state agencies to recommend consolidation. Other states give water suppliers that consolidate either preferences in governmental funding or regulatory relief (US Environmental Protection Agency, 2002).

Promotion of IRWM

California promotes IWRM, although in a confusing twist calls it 'integrated regional water management' (IRWM).[2] Finding increased coordination 'necessary to maximize the quality and quantity of water available to meet the state's agricultural, domestic, industrial and environmental needs' (California Water Code §§ 10531(b)), California's Integrated Regional Water Management Planning Act of 2002 (the IRWM Act) authorizes regional water management agencies to jointly prepare and adopt integrated plans addressing any matters over which they have authority, including urban and agricultural water management, recycling, groundwater, conservation, water quality, ecosystem health and flood control. California has provided more than $350 million in matching grants to both reduce the cost of IRWM and thus one of the principal hurdles, and encourage participation by paying for part of the cost of projects adopted in an IRWM.

The IWRM Act has been highly successful in encouraging agencies to engage in IRWM (Hanak *et al.*, 2011). A successful example of IRWM is the Santa Ana Watershed Project Authority's development of a 'one water, one watershed' plan (Santa Ana Watershed Project Authority, 2010; Thompson, 2011). The Santa Ana watershed, which covers 2650 square miles of Southern California and is home to over five million residents, was once a poster child for fragmentation; more than 100 local, regional, state and federal agencies, including five large water districts and a number of smaller suppliers, shared jurisdiction. The agencies formed the Santa Ana Watershed Project Authority (SAWPA) to help overcome this fragmentation. SAWPA has broad authority to address not only traditional water supply and quality issues, but also habitat restoration, invasive species and flood control. According to SAWPA, its 'one water, one watershed' plan is a 'living document' that permits ongoing integration and coordination among the multiple agencies.

While the IWRM Act has encouraged multiple planning efforts, its likely success in dealing with the problems of fragmentation is still an open question. Many observers have questioned local agencies' commitment to IWRM, as well as IWRM's effectiveness as a defragging tool. In their view, most agencies have participated in IWRM to obtain funding for infrastructure projects they already planned to pursue, not for the integrated planning. Indeed, some have joked that IRWM stands not for 'integrated regional water management', but for 'I really want mine' (Hanak *et al.*, 2011).

IWRM, furthermore, does not appear to have led in most cases to systematic integration of decision making at a watershed or comparable scale. One study of IWRM in the San Francisco Bay region found that those who had participated in IWRM believed that it had increased 'collaborative implementation' and integration, but only slightly (Lubell and Lippert, 2011). Less than a quarter of the participants reported that IWRM had promoted functional or geographic integration. Participants, moreover, did not believe that any increase in collaboration improved the ability to solve problems.

The results to date raise questions on whether soft restructuring can effectively overcome fragmentation, even when the state provides incentives to participate. Because IWRM is still new in California, however, the process and its results might improve over time.

Defragging by fragging

Another, counterintuitive, option for solving the problems of fragmentation in some, perhaps limited, contexts is to further decentralize authority. Fragmentation poses a challenge not simply because it divides management, but also because it erects walls that can make little policy sense and create unnecessary rivalries. Where the problems of fragmentation stem from the walls that are created, rather than from the lack of comprehensive management, further decentralization of authority might be a successful solution.

The best example of defragging by fragging is decentralized water markets. As noted earlier, fragmentation of water supplies has often led to protectionism in which water suppliers refuse to allow transfers of water outside their boundaries. In many cases, these transfers would promote both efficiency and overall state policy. Opposition to such transfers often stems from the vested interests of local water users who benefit from a protected market (Thompson, 1993).

In these settings, the protectionism can be overcome by giving individual water users the power to transfer their water. In 1992, the US Congress did exactly this by allowing individual users of water in the US's massive California Central Valley Project to transfer water to other California water users or water agencies, even over the opposition of their water districts. In the early 1990s, California considered similar legislation that would have allowed individuals to transfer water even in the face of local district opposition (Thompson, 1993).

A key question is whether private water marketing is sui generis, or whether decentralization of decision making could also help overcome other problems of fragmented management. In theory, this strategy might help in other settings where fragmentation gives one agency the ability to block a broader perspective. For example, fish agencies often control instream flows even though other agencies might also have an interest in flow levels. To overcome this challenge, some states have authorized other agencies and stakeholders to acquire water rights for environmental use. This has allowed the other agencies to address instream flow even when fish agencies have limited interest or resources.

Conclusion

As discussed, there are multiple ways to try to solve current water fragmentation challenges. Among these approaches, hard restructuring and top-down defragmentation have proven particularly effective. Collective action, however, is a major barrier to any solution. Hard restructuring, for example, requires often rivalrous agencies to discuss, develop and form a new entity. Top-down defragmentation requires proponents of integration to convince national or state governments of the importance of defragmentation. There are thus no easy solutions to fragmentation once it arises, to which the many failed efforts at defragmentation testify. Critical to future defragmentation will be educating both policymakers and the public of the problems of fragmentation and encouraging continual collaboration and discussion toward solutions.

Notes

1 The Delaware River Basin Commission is an exception, since it has significant managerial authority of its own.

2 California is the first state in the United States to promote IWRM.

References

Ajami, N. K., Thompson, B. H. and Victor, D.G. (2014) *The Path to Water Innovation*, The Hamilton Project, Washington, DC.

Arizona v. California (1936) 298 US 558.

Arizona v. California (1963) 373 US 546.

Bay Delta Conservation Plan (Draft) (2013) 'Chapter 7: Implementation Structure' in *2013 Public Review Draft Bay Delta Conservation Plan*, State of California, Sacramento, CA.

Bay Delta Conservation Plan Steering Committee (2012) *Progress Report on the Bay Delta Conservation Plan*, State of California, Sacramento, CA.

Blomquist, W. and Schlager, E. (2005) 'Political pitfalls of integrated watershed management', *Society and Natural Resources*, 1, 1, 101–17.

Blomquist, W., Heikkila, T. and Schlager, E. (2001) 'Institutions and conjunctive water management among three western states', *Natural Resources Journal*, 47, 3, 653–83.

California Water Code 2015.

Carr, J. B. and Feiock, R. C. (2000) 'Boundaries entrepreneurs & city/county consolidation', unpublished manuscript, on file with the author.

Clean Water Act, 33 USC 1972.

Dellapenna, J. W. (2011) 'The Delaware and Susquehanna river basins', in A. J. Kelley (ed.) *Waters and Water Rights*, 4th edn., LexisNexis, New Providence, NJ.

Delta Stewardship Council (2012) *Final Draft Delta Plan (Working Draft)*, State of California, Sacramento, CA.

Endangered Species Act, 16 USC 1973.

Evans, A. (2010) 'The groundwater/surface water dilemma in Arizona: a look back and a look ahead toward conjunctive management reform', *Phoenix Law Review*, 3, 1, 269–92.

Feiock, R. C. and Carr, J. B. (2001) 'Incentives, entrepreneurs, and boundary change: a collective action framework', *Urban Affairs Review*, 36, 3, 382–405.

Florida v. Georgia, Original, order granting leave to file bill of complaint (US, No 22O142 ORG, 3 November 2014).

Goldfarb, W. (1994) 'Watershed management: slogan or solution?', *Boston College Environmental Affairs Law Review*, 21, 3, 483–510.

Gray, B., Thompson, B., Hanak, E., Lund, J. and Mount, J. (2013) *Integrated Management of Delta Stressors: Institutional and Legal Options*, Public Policy Institute of California, San Francisco, CA.

Hall, N. (2010) 'Interstate water compacts and climate change adaptation', *Environmental & Energy Law & Policy Journal*, 5, 2, 237–324.

Hanak, E., Lund, J., Dinar, A., Gray, B., Howitt, R., Mount, J., Moyle, P. and Thompson, B. (2011) *Managing California's Water: From Conflict to Reconciliation*, Public Policy Institute of California, San Francisco, CA.

Integrated Regional Water Management Planning Act of 2002 (California).

Kansas v. Colorado (1907) 206 US 46.

Kansas Statutes 2012.

Kearney, M. S., Harris, B. H., Hershbein, B., Jácome, E. and Nantz, G. (2014) *In Times of Drought: Nine Economic Facts about Water in the United States*, The Hamilton Project, Washington, DC.

Kiparsky, M., Sedlak, D., Thompson, B. H. and Tuffer, B. (2013) 'The innovation deficit in urban water: the need for an integrated perspective on institutions, organizations, and technologies', *Environmental Engineering Science*, 30, 8, 395–408.

Lax, D. A. and Sebenius, J. S. (2011) *The Manager as Negotiator*, The Free Press, New York.

Layzer, J. A. (2008) *Natural Experiments: Ecosystem-Based Management and the Environment*, MIT Press, Cambridge, MA.

Little Hoover Commission (2005) *Still Imperiled, Still Important: The Little Hoover Commission's Review of the CALFED Bay-Delta Program*, Little Hoover Commission, Sacramento, CA.

Lubell, M. and Lippert, L. (2011) 'Integrated regional water management: a study of collaboration or water politics-as-usual in California, USA', *International Review of Administrative Science*, 77, 1, 76–100.

McCool, D. (1987) *Command of the Waters: Iron Triangles, Federal Water Development, and Indian Water*, University of Arizona Press, Tucson, AZ.

McCormick, Z. L. (1994) 'Interstate water allocation compacts in the western United States – some suggestions', *Water Resources Bulletin*, 30, 3, 385–95.

Mandarano, L. A., Featherstone, J. P. and Paulsen, K. (2008) 'Institutions for interstate water resources management', *Journal of the American Water Resources Association*, 44, 1, 136–47.

Meyers, C. J. (1966) 'The Colorado River', *Stanford Law Review*, 19, 1, 1–75.

Micko, J. L. (1991) 'Surface water and groundwater conjunctive use management: Santa Clara County, California', Paper presented at Innovation in Western Water law and Management (Summer Conference), Boulder, CO, http://scholar.law.colorado.edu/innovation-in-western-water-law-and-management/13, accessed 5 January 2015.

Muys, J. (1971) *Interstate Water Compacts: The Interstate Compact and the Federal-Interstate Compact*, National Water Commission Legal Study No. 14, Washington, DC.

Muys, J. (2004) 'The western states' experience with interstate water issues', paper presented at the American Bar Association Conference on Eastern Water Resources: Law, Policy and Technology, San Diego, CA.

Nakamura, L. and Born, S. M. (2007) 'Substate institutional innovation for managing lakes and watersheds: a Wisconsin case study', *Water Resources Bulletin*, 29, 5, 807–21.

New York v. New Jersey (1921) 459 US 296.

Nicholson-Crotty, S. (2005) 'Bureaucratic competition in the policy process', *Policy Studies Journal*, 33, 3, 341–61.

Patashnik, J. (2014) '*Arizona v. California* and the equitable apportionment of interstate waterways', *Arizona Law Review*, 56, 1, 1–51.

Pisani, D. J. (1978) 'The strange death of the California-Nevada compact: a study in interstate water negotiations', *Pacific Historical Review*, 47, 4, 637–58.

Pisani, D. J. (1982) 'State vs. nation: federal reclamation and water rights in the progressive era', *Pacific History Review*, 71, 4, 833–49.
Powell, J. W. (1890a) 'The irrigable lands of the arid region', *Century Magazine*, 38, 5, 766–76.
Powell, J. W. (1890b) 'Institutions for the arid lands', *Century Magazine*, 40, 1, 111–16.
Sacramento-San Joaquin Delta Reform Act 2009.
Santa Ana Watershed Project Authority (2010) *One Water One Watershed: 2010 Integrated Regional Water Management Plan*, Santa Ana Watershed Project Authority, Riverside, CA.
Schlager, E. and Heikkila, T. (2009) 'Resolving water conflicts: a comparative analysis of interstate river compacts', *The Policy Studies Journal*, 37, 3, 367–92.
Stewart, R. B. (1982) 'Interstate resource conflicts: the role of the federal courts', *Harvard Environmental Law Review*, 6, 2, 241–64.
Thompson, B.H., Jr. (1991) 'Interstate transfers: *Sporhase*, compacts, and free markets', in *Western Water Law in the Age of Reallocation*, American Law Institute, Philadelphia, PA.
Thompson, B.H., Jr. (1993) 'Institutional perspectives on water policy and markets', *California Law Review*, 81, 3, 671–764.
Thompson, B.H., Jr. (1996) 'Environmental policy and state constitutions: the potential role of substantive guidance', *Rutgers Law Journal*, 27, 4, 863–925.
Thompson, B.H., Jr. (1997a) 'Water markets and the problem of shifting paradigms', in T. L. Anderson and P. J. Hill (eds) *Water Marketing: The Next Generation*, Rowman & Littlefield, Lanham, MD.
Thompson, B.H., Jr. (1997b) 'Water federalism: governmental competition and conflict over western waters', in T. L. Anderson and P. J. Hill (eds) *Environmental Federalism*, Rowman & Littlefield, Lanham, MD.
Thompson, B.H., Jr. (1999) 'Privatization of the municipal water industry', *Looking Ahead*, 30, 5, 1–4.
Thompson, B.H., Jr. (2011) 'Beyond connections: pursuing multidimensional conjunctive management', *Idaho Law Review*, 47, 2, 273–323.
Thompson, B.H., Jr. (2012) 'A federal act to promote integrated water management: is the CZMA a useful model?', *Environmental Law*, 42, 1, 201–40.
Thompson, B.H., Jr., Leshy, J. D. and Abrams, R. H. (2013) *Legal Control of Water Resources*, Thomson-West Publishers, St Paul, MN.
United States Constitution.
United States v. Rio Grande Dam & Irrigation Co. (1899) 174 US 690.
United States v. State Water Resources Control Bd. (1986) 227 Cal. Rptr. 161.
US Environmental Protection Agency (2002) *Restructuring and Consolidation of Small Drinking Water Systems: A Compendium of State Authorities, Statutes, and Regulations*, US Environmental Protection Agency, Washington, DC.
US General Accounting Office (2007) *Interstate Compacts: An Overview of the Structure and Governance of Environmental and Natural Resource Compacts*, US General Accounting Office, Washington, DC.
Water Resources Planning Act, Pub L No 89–80, 1965 Stat. 244.
Weston, R. T. (1989) 'The Delaware River basin: courts, compacts and commissions', in *Boundaries and Water: Allocation and Use of Shared Resources*, University of Colorado Natural Resources Law Center, Boulder, CO.
Williamson, O. E. (1975) *Markets and Hierarchies: Analysis and Antitrust Implications*, Free Press, New York.

8 Beyond the traditional governance of trans-jurisdictional groundwater

Unconventional approaches to cross-boundary aquifer management in the United States

Rebecca Nelson and Meg Casey

> The answer [addressing how interstate groundwater resources should be managed] is not self-evident. Sensible management of groundwater is different enough [between states] without considering state lines.
>
> (Leshy, 2008, p. 1491)

> The beginning steps would be informal and ad hoc; more formal arrangements would be made as time passed. Thus one might have cooperative data-gathering, leading to administrative agreements, leading to interstate compacts creating new government agencies.
>
> (Agnew and Busch, 1971, p. 81)

Introduction

Groundwater bodies and groundwater-connected rivers often span jurisdictional boundaries in the United States (US), but groundwater rarely appears in traditional legal arrangements for apportioning and managing water across these boundaries.

Managing trans-jurisdictional groundwater is an important issue in the US: the nation holds significant trans-jurisdictional aquifers, growing populations continue to seek new, out-of-state water sources and biodiversity protection legislation collides with trans-boundary water allocation arrangements across the country. This chapter explores trans-jurisdictional groundwater governance in the US, with a focus on the architecture, challenges and innovations that arise in managing groundwater across state boundaries. It extends the usual focus of discussion beyond interstate compacts and other traditional legal mechanisms to unconventional modes of governing trans-jurisdictional groundwater, including approaches based on collaborative science, planning and centralized regulatory authorities.

This chapter commences with a brief overview of groundwater resources in the US and a short survey of the range of traditional, legally formal approaches used to manage trans-jurisdictional water across the US: interstate compacts and agreements, or apportionment by federal legislation or court decree. It finds that these apply relatively rarely to groundwater. Using key scholarly criticisms of these governance approaches as a springboard, it then suggests a series of five key functions and characteristics desirable for trans-jurisdictional groundwater governance. These are: (1) facilitating the building of high-quality, agreed technical information on which to base management decisions; (2) viewing a groundwater system holistically, recognizing connections with surface waters and ecosystems; (3) involving a broad range of participants and (4) management tools; (5) making management arrangements flexible enough to deal with new knowledge and changing circumstances; and (6) minimizing political barriers to adopting the governance arrangements.

The chapter then looks beyond traditional approaches to trans-jurisdictional groundwater governance. Political and institutional cultures in the western US provide a fertile base for creating unconventional, typically non-regulatory groundwater governance arrangements (Nelson and Casey, 2013): this is one end of the spectrum of unconventional approaches to trans-boundary governance. At the other end appear uncommonly centralized regulatory approaches. We find three key manifestations of innovative, unconventional governance at either end of the regulatory spectrum. These mirror the categories that Agnew and Busch hypothesized in 1971: (1) collaborative science, (2) administrative planning and (3) centralized regulatory authorities. We briefly draw out and use examples of these innovative arrangements to explore their capacity to address the key functions and characteristics of trans-boundary groundwater management referenced above. The examples were chosen based on situations raised as important by the staff of western US water agencies in interviews and at two workshops held at Stanford University and the University of Sydney, both in 2011–2012 (Nelson, 2012a, 2014; Nelson and Casey, 2013). The chapter concludes with a reference table summarizing the ways in which unconventional approaches address the key functions we have outlined, a summary of the lessons offered by the examples presented and a working agenda for further research.

Importance of managing interstate groundwater in the US, and traditional mechanisms for doing so

Groundwater is an important source of water supply for both human and environmental water uses across the US. An estimated 130 million people depend on groundwater for drinking and 65% of groundwater is used for irrigation (Hall, 2010). Even states that rely heavily on surface water for consumptive purposes may, in fact, be relying indirectly on groundwater, given that groundwater discharges to rivers as baseflow (Santhi *et al.*, 2008; Barlow and Leake, 2012). Around the US, a wide variety of species and ecosystems, from salmon in baseflow-dependent rivers to riparian vegetation communities, to spring communities, to prairie wetlands, also depend on groundwater (Blevins and Aldous, 2011).

With this degree of groundwater dependence, it can be surprising to note that some states lack important components in the armoury of data desirable to manage groundwater. In states where extraction is particularly high, as in California, groundwater use is rarely metered and data describing groundwater levels are sometimes remarkably sparse (Nelson, 2012b). Sometimes the boundaries of aquifers are not known with confidence (Theesfeld, 2010). No state routinely monitors groundwater-dependent ecosystems, though there are the beginnings of a federal approach to the issue (e.g. US Department of Agriculture, Forest Service, 2012). As with many other parts of the world, therefore, dealing with data paucity is a key element of trans-jurisdictional groundwater management in the US.

Groundwater in the US is strikingly trans-boundary by nature. Of the 62 named principal aquifers of the US recognized by the United States Geological Survey (USGS), a conservative estimate is that 36 of them (58%) cross a state border (Maupin and Barber, 2005). In the absence of coordinated action, management efforts on one side can be undermined by other actions, or lack of action, across the border.

Much US legal scholarship on trans-jurisdictional aquifers has focused on formal, traditional legal arrangements for managing groundwater across states, and particularly the limitations of these arrangements (e.g. Hall, 2006; Davenport, 2008; Leshy, 2008; Hathaway, 2011; Hall and Catavaro, 2013). A brief tour through these arrangements is warranted as a prelude to the main focus of this chapter: expanding the discussion of US trans-jurisdictional groundwater governance to less conventional management arrangements. Conventionally, three methods have been used to apportion water and secure water rights in trans-jurisdictional water systems in the US: (1) Congressional allocation (that is, Congress directly allocating trans-boundary water); (2) equitable apportionment (where the Supreme Court allocates water in a litigation setting); and (3) interstate compacts (a form of agreement between states) (Hall, 2006). These arrangements are designed to pre-empt or, more commonly, respond to, disputes between states, and involve a narrow range of actors: state governments as the main actors and the federal government as the mediator. Neither of the first two traditional mechanisms has ever been applied to groundwater (Sax *et al.*, 2006). Furthermore, few interstate compacts expressly refer to groundwater (Clemons, 2004) and none has been created to govern an entire aquifer, as opposed to groundwater existing within the boundaries of a surface water basin. Those that do expressly mention groundwater (usually found in the eastern US) tend not to deal with water allocation, focusing instead on water quality (Dellapenna, 2005).

Even if these traditional mechanisms were to be applied more frequently to groundwater, scholars and practitioners have found key flaws with these mechanisms, at least in their usual form. Turning these criticisms around, we suggest six key functions and characteristics as desirable for trans-boundary groundwater governance mechanisms: a first key function is facilitating the building of agreed technical information on which to base management decisions. Traditional legal approaches are often the scene for extensive and expensive 'combat science', where experts from each side 'duel' over technical assumptions favourable to their state or local jurisdiction (Hanak *et al.*, 2011, pp. 131–2; Nelson and Casey, 2013, pp. 4, 6 and 26).

Recognizing the virtues of systems thinking to complex problems, a second key function is viewing a groundwater system holistically, recognizing connections with surface waters and ecosystems. Traditional legal approaches have tended to take a fairly narrow view of the water system and goals for that system, tending to focus on allocating water for consumptive purposes and viewing the system as isolated, for example, from other water sources and ecosystems. They address goals for ecological protection relatively rarely and, in important cases, do not recognize connections between groundwater bodies and surface water (Sax *et al.*, 2006; Thompson, 2011).

Third, involving a comprehensive range of participants in governance encourages this holistic view of a water system. It also generates trust and acceptance of decisions and can produce innovative management solutions (Bergkamp *et al.*, 2009). Traditional legal approaches, particularly those based on litigation, tend not to involve a comprehensive range of non-state participants.

Fourth, adopting a broad view of the tools available to manage a water system and providing for their use is equally desirable to optimize the set of solutions chosen. By contrast, traditional legal approaches tend to focus on controlling allocations of water or weak requirements to exchange information, rather than a broader suite of management tools addressing water supply or demand measures.

Adapting to changing circumstances is also vital, yet traditional legal approaches can 'become frozen in place, difficult to modify with changing circumstances' (Grant, 2003, p. 150), where those changes relate to values or biophysical conditions. This can be because allocations of water to different states or individual users (in the case of intrastate groundwater law) are typically quantified as volumes, and interstate compacts and (particularly in the western US) state groundwater allocation systems tend to lack provisions requiring that these volumes be reviewed in future to account for changing climatic conditions, for example (Adler, 2010).

Being politically palatable is a final, pragmatically desirable element of trans-boundary water governance. There can be significant political barriers to using traditional mechanisms, since compacts, for example, require state legislative and federal Congressional approval (Sax *et al.*, 2006).

We return to these functions and characteristics as a framework for analysing the unconventional mechanisms for interstate groundwater governance, to which we now turn. These approaches could prove valuable both in inspiring components to be implemented within existing traditional legal approaches in the future, as well as potentially acting as stand-alone governance mechanisms, in the US or further afield.

Unconventional architecture for governing interstate groundwater

As they contemplated potential interstate groundwater governance mechanisms in the Pacific Northwest over four decades ago, Agnew and Busch (1971) mused that governance could evolve from cooperative data-gathering to administrative agreements to new government agencies. In the intervening years, we have indeed witnessed several new, unconventional approaches to governing trans-jurisdictional groundwater, though whether these initial forms are merely steps towards new agencies is an open question (see section on Research Agenda). We turn now to examine these approaches in more detail, titling them collaborative science, administrative planning and centralized regulatory authorities. In important ways, these unconventional approaches to governing trans-jurisdictional groundwater are showing themselves capable of addressing the key governance functions and characteristics noted above. Nonetheless, they are giving rise to their own challenges, which stem chiefly from their generally voluntary nature. In this section, we provide notable examples of each unconventional approach and point to key features that seem to address key governance functions, as well as those that present new challenges.

Collaborative sciences as governance

As foreshadowed above in the section on traditional mechanisms, one of the most challenging and universally relevant aspects of groundwater management is paucity in available groundwater data, and therefore, understanding of the resource. Yet scientific data have a clear and direct role in groundwater law, policy and management, since the task of allocating resources relies fundamentally on accurately understanding the resources available and the impacts of using them. Trans-boundary aquifers are particularly difficult to characterize, because hydrologic information describing the entire aquifer must be gathered and considered in a cohesive manner (Milman and Ray, 2011).

Technical differences between the scientific understandings of a shared source have proven a fertile ground for litigation, as well as a logical place to initiate trans-jurisdictional cooperation. Recognizing the necessity of comprehensive studies and data exchange, some jurisdictions in the US have agreed to collaborate on technical studies of trans-boundary aquifers. This form of governance aims to create a common understanding of the source and related issues rather than address allocation questions or limit access to the resource. If well executed, collaborative science can both build a common scientific understanding of the trans-jurisdictional problems at hand among a broad range of state and non-state actors, and form the cross-institutional relationships for later cooperative management. These effects of joint technical studies have also been noted by US groundwater managers more broadly, as we have described elsewhere (Nelson and Casey, 2013). These apparent benefits, as well as the general intention of the studies to facilitate later management actions, justify their characterization as a type of 'governance'.

This collaborative science approach is well illustrated by the case of Washington and Idaho, which have collaborated with the USGS since 2004 to produce a groundwater budget and numeric groundwater flow model for the cross-border Spokane Valley-Rathdrum Prairie (SVRP) Aquifer, with the intention of enabling the states to evaluate management tools for the water system (Kahle and Bartolino, 2007). These neighbouring states have purposefully sought to work through technical differences collaboratively to form the foundation of trans-jurisdictional governance of aquifers and avoid 'duelling models' or 'combat science' and litigation (Idaho Department of Water Resources, undated(a)). The effort was triggered by concerns about the trans-boundary implications of proposals by Idaho power companies to withdraw substantial quantities of groundwater. However, the water system was also generally experiencing increased competition and multiple sources of stress, including concerns that the achievement of water quality standards in Washington might be affected by groundwater pumping in Idaho depleting stream flows, and Idaho issuing pollution discharge permits under different standards (*Arkansas v. Oklahoma* (1992); Patton, 2011).

The SVRP aquifer is the sole source of drinking water for over 500,000 people in growing communities on either side of the border, and an important source of agricultural irrigation water in the sub-humid to semi-arid area (Kahle and Bartolino, 2007). It is connected to a number of regional lakes and the Spokane River, which flows from Idaho into Washington, losing and gaining water from the aquifer along different reaches (Kahle and Bartolino, 2007). Prior to the collaborative study, major differences in hydrogeological interpretation existed within a substantial body of scientific work (Agnew and Busch, 1971) and individuals' groundwater rights had not yet been (and are still not) quantified (Idaho Department of Water Resources, undated(b)). The inconsistencies led to an early conclusion that without a major water study of the SVRP area, 'the two states will continue to have an interstate ground-water problem of considerable magnitude, and insufficient conclusive information for reaching sound water-management decisions' (Agnew and Busch, 1971, p. 57).

The SVRP technical study proceeded collaboratively, under ground rules set forth in a relatively detailed, formal memorandum of understanding (US Geological Survey *et al.*, 2003). Personnel of the water agencies of both states, and the USGS formed a modelling team, which made technical decisions by consensus (Hsieh *et al.*, 2007). That team was managed by a Project Technical Leadership Team with representatives from each of the three agencies, advised by a community-based Technical Advisory Committee and reviewed by a Peer Review Team (US Geological Survey *et al.*, 2003). A much wider set of stakeholders was involved through participation on a Policy Advisory Committee and, less formally, through public meetings (Idaho Department of Water Resources, undated(c)). Unusually, initial federal funding for the project was sought by a coalition of business and environmental groups (Haynes and Stoeffel, 2007). The final numerical model is intended to be used not just in dealing with water quantity and water rights decision making, but also in planning and zoning at local levels (US Geological Survey *et al.*, 2003), and is available to the public (Hsieh *et al.*, 2007). At the conclusion of the study, the parties formally established an ongoing, bi-state technical modelling committee to establish protocols for the use of the model and to develop and distribute enhancements to it (Idaho Department of Water Resources and Washington Department of Ecology, 2007). As of November 2013, the committee continues to meet at least annually, though funding shortages over the last five years have hindered their ability to undertake significant projects (Gregory, 2013).

Though the SVRP technical work was not specifically directed to developing a joint management plan, an approach discussed further in the next section, it may well lay the foundations for doing so. The technical exercise has created strong working relationships between agency staff in each state (Gregory, 2013). It also created a possible template for collaboration, particularly elements such as establishing process rules, and has identified sources of funding (Washington Department of Ecology, 2008). In addition, the model is clearly informing management efforts, such as Idaho's Comprehensive Aquifer Management Plan, which adopts groundwater consumption goals in seeking to minimize 'jurisdictional conflicts with Washington' and includes a bi-state advisory committee (Idaho Water Resource Board, 2011, p. 14).

The SVRP example shows how overlying states can collaborate to conduct a consistent, cohesive study in order to create a shared scientific depiction of a groundwater source – a crucial foundation for effective groundwater governance. It is difficult to reach this shared vision where jurisdictions base their understanding of the source on studies that only assess part of an aquifer. The processes involved a much broader swathe of stakeholders than would likely have been substantively involved, had the technical work occurred as part of a traditional approach to trans-jurisdictional governance that focused solely on allocation, and the resulting project is intended to be more useful to that broad range of groups. The joint studies inquired broadly into matters beyond groundwater quantity, including groundwater quality and connections with surface water (though there is no evidence that they inquired into groundwater-dependent ecosystems in a general sense). Despite their relative legal informality, these kinds of studies appear to have benefited from concrete ground rules, for example, in structuring committees and meetings. Though there is no guarantee that such studies will lead to actual management action, at minimum, they equip participants with the knowledge and relationships to undertake such action in a collaborative way.

In addition to securing ongoing funding, an important, though less evident, challenge of collaborative science stems from the kind of information learned in the study. Given the high degree of hydrogeological uncertainty that is likely at issue, particularly where the study seeks to examine an aquifer in its entirety for the first time, scientific studies can open a significant and complex can of worms. In the case of the SVRP, the committees initially set out to determine how much water was stored in the aquifer; by the end of the study, the key question had become, 'How do we manage the water differently so that we can keep doing the things we want to do?' (Gregory, 2013). The study exposed the problem as unsustainable use – a culture of unsustainable practices – rather than merely an issue of conflicting uses. While committee members recognize that money alone cannot effect the kind of cultural change needed meaningfully to reshape water-use practices, there is strong consensus that the funding put toward the joint studies has been money well spent, as would be additional investment in continued education and outreach (Gregory, 2013).

Administrative planning as governance

Planning-based approaches to water management are increasingly common across the globe (Nelson, 2012b). They differ substantially, however, in structure and objective, ranging from those that resemble investment plans for monitoring and infrastructure, to regulatory plans that focus on water allocation. Planning-based approaches also feature in trans-boundary groundwater governance in the US. In many ways, they are particularly suited to this application. Unlike conventional legal approaches to trans-boundary groundwater management, planning approaches are inherently forward-looking, including into the very long term. They do not, at least in theory, rely on a party having suffered damage as a pre-condition to action. They can use a suite of management approaches tailored to local circumstances, some of which – particularly infrastructure projects such as aquifer storage and recovery – would be less likely to occur under a conventional legal approach. They can take a holistic view of the basin setting, bringing in a wide range of actors that have a wide range of impacts on groundwater, and deal with a broad range of concerns, from water availability, to quality, to ecological impacts.

The example of the Platte River in the mid-West US illustrates a planning-based approach to managing trans-boundary groundwater. A particularly outstanding feature is that it is driven by ecological considerations – an important omission from much western US groundwater law (Thompson, 2011; Nelson, 2012a). The Platte River Recovery Implementation Program (PRRIP) is a large-scale partnership between Colorado, Wyoming, Nebraska and the US Government, which has significantly affected interstate groundwater management – albeit as a side effect of other concerns. The PRRIP focuses on ecological restoration of the Platte River, which flows through these states, with a particular emphasis on protecting habitat for four endangered fish and bird species protected under the federal Endangered Species Act (ESA). The PRRIP was born in 1997 from a settlement agreement for the re-licensing of a Platte River hydropower and irrigation dam facility. The re-licensing process led to findings that diversions for the dam posed potential jeopardy to ESA-listed species. The resolution of this issue led to an agreement that established the PRRIP to cover multiple water projects along the Platte River. This removed the need for many individual water projects in the basin to undergo lengthy processes as a result of separate reviews of their federal licences (Platte River Recovery Implementation Program, 2006; Hoffman and Zellmer, 2013). The PRRIP is being developed in an incremental way and is now undergoing implementation of its first 13-year 'increment' (2007–2019).

Vital components of the PRRIP relate to reducing groundwater extraction or increasing recharge. These actions reduce depletions and increase inflows of groundwater to the Platte River, respectively, helping to meet target streamflows for endangered species. The Water Plan component of the PRRIP (Platte River Recovery Implementation Program, 2006) includes a wide range of implementation tools. The parties, including the federal government, must offset new depletions to the river caused by pumping groundwater (Platte River Recovery Implementation Program, 2006; Freeman, 2010). 'Incentive-based water conservation and water supply activities' also relate to groundwater (Platte River Recovery Implementation Program, 2006, p. 1). These projects include paying groundwater-using farmers to reduce water use by undertaking deficit irrigation, fallowing or on-farm efficiency improvements; replacing surface water use with groundwater artificially recharged through modified irrigation canals when flows exceed ecological requirements; and undertaking artificial recharge designed to increase flows in the Platte River at critical times, which incidentally achieves enhanced wildlife values at the recharge sites (Platte River Recovery Implementation Program, 2006). Though the ultimate goals of the PRRIP have been described as ecologically modest, it represents an impressive collection of obligations and commitments to projects that recognize the place of groundwater management within a larger, connected trans-boundary water system. These things are particularly impressive when one considers that at the start of the negotiations for the PRRIP, the water quantity laws of two of the states, Nebraska and Wyoming, did not treat groundwater and surface water as connected. In addition, the Platte negotiations triggered substantial new legislation in Nebraska to link groundwater and surface water law and management to support Nebraska's ability to comply with its PRRIP obligations (Freeman, 2010).

The fundamental challenge of a planning approach to trans-boundary water governance lies in achieving collaboration and commitment – making sure the parties come to the table and stay there until a deal is done – despite the voluntary nature of many planning approaches. In the absence of a regulatory mandate, agreements form only where the parties can 'transcend their self-interests and cooperate' and overcome what may be 'a history of contention, [boundaries] dividing their geographic areas, collisions of interests within and among their units and levels of government, incompatible water interests, and conflicting visions' (Freeman, 2010, location 950). The states involved in the PRRIP were motivated to collaborate by a desire for regulatory certainty and simplicity relative to the alternative arrangements that would apply under the ESA (Freeman, 2010). A further challenge stemming from the voluntary nature of planning approaches is the potential for within-state opposition to the agreed measures, which presents as a key risk for negotiating agencies. Controls on groundwater were highly contentious within certain states in the PRRIP (Freeman, 2010).

Whereas the parties to the planning approach evidenced in the PRRIP implemented the plan as separate entities, the final unconventional approach to trans-jurisdictional groundwater governance presented here goes several steps further, by creating a separate, supra-jurisdictional entity to carry out the governance role, including regulating within-state groundwater pumping projects.

Centralized regulatory authority as governance

Though these are not common, some states have managed to develop centralized regulatory and operational management of the water sources within a basin. The Susquehanna Compact is an example of such an arrangement, under which a commission (consisting of one governor per state and a federal representative) was created and authorized to promulgate laws and develop comprehensive programmes related to water resource management within the basin. This Compact provided the basis for an expansion of groundwater governance in an unconventional, centralized way.

The Susquehanna Compact is rare, and itself unconventional, in that it expressly addresses groundwater and allows for supra-state direct regulation of important groundwater uses, that is, imposing mandatory controls on groundwater uses (noting that we have conceived of governance as much broader than imposing legal mandates). Its commission has also taken additional, non-traditional steps to regulate groundwater use and enhance sustainability efforts since the Compact's inception (Grant, 2004). In the Susquehanna, groundwater management has been steadily evolving since the adoption of the Susquehanna Compact in 1970. In 1978, the Susquehanna River Basin Commission (SRBC) promulgated regulations that limited large groundwater withdrawals within the participating jurisdictions (New York, Maryland and Pennsylvania) – specifically, those exceeding 100,000 gallons per day (gpd), or 20,000 gpd used consumptively within 30 consecutive days. These large extractions were identified as a common gap in the participating states' regulation, and the SRBC limited these withdrawals to the quantity and rate needed to meet the 'reasonably foreseeable needs of the [proposed] project' without causing adverse impacts on listed matters, which include streamflow, water quality and groundwater levels (SRBC Regulation § 803.43). In assessing potential adverse impacts, the SRBC reviews 'a constant-rate pumping test, metering, monitoring and reporting, mitigation of adverse impacts, water conservation, and a docket reopener provision', the last of which can be used to impose additional requirements on already approved provisions (SRBC Regulation § 803.32). With flow models and scientific data gathered in the 1980s, the SRBC went on to draft the basin's first Groundwater Management Plan in 1993, modified in 2005 (SRBC, 2005). The Plan instructs the SRBC to regulate groundwater withdrawals and consumptive use of water, coordinate groundwater quality issues, conduct a variety of planning studies (watershed-based, special studies, and water budget analyses) and provide education and public outreach services (SRBC Regulation § 803.4). The SRBC reviews proposed groundwater projects against considerations that relate to, among other things, protection of fisheries and aquatic habitat, preventing undue salinity and diluting and abating pollution (SRBC Regulation § 803.4). Importantly, it has the power to modify and approve as modified, or may disapprove, a project if it determines that the project is not in the best interest of the conservation, development, management or control of the basin's water resources, or is in conflict with the comprehensive plan (SRBC Regulation § 803.41(b)).

The SRBC has continued to expand regulation of the trans-boundary water source. In 2012, it adopted a Low Flow Protection Policy (SRBC, 2012) based on earlier recommendations of The Nature Conservancy (DePhilip and Moberg, 2010). The Policy allows the SRBC to limit, condition and even deny new groundwater uses (based on cumulative impact) to avoid significant impacts in low-flow seasons. A project must compensate for its consumptive use of groundwater when stream flow is less than the low flow criterion (SRBC Regulation, § 803.42(a)(ii)).

Centralized regulation-based approaches are not the typical vehicle for trans-jurisdictional groundwater management in the US. These require an enormous amount of cooperation and commitment on the part of participating members. The example discussed here stemmed from, and expanded out from, an interstate compact that recognizes trans-boundary groundwater management principles early. We can look to this example for direction on how states might recognize a higher-level perspective on trans-boundary groundwater depletion and initiate collaborative action to address gaps in state regulatory systems. That approach can adopt a broad view of the water system, use a variety of management tools and can involve stakeholders in innovative ways.

Lessons

The aim of this paper has been to shed light on unconventional approaches to governing trans-jurisdictional groundwater in the US, and how these can address key desirable governance functions and characteristics with which traditional approaches may struggle. We recap these methods in Table 8.1, below, and summarize them here. Each cell of the table displays one key example of a method of addressing each function or characteristic, which corresponds to the unconventional approach noted in the column heading. The table suggests that in many cases, unconventional approaches appear better equipped to deliver these functions than traditional mechanisms, at least in the forms in which they have traditionally appeared.

Collaborative science, administrative planning and centralized regulatory authorities can all use structured collaborative approaches, such as joint technical committees and funding for scientific studies, to reduce the potential for

Table 8.1 Ways in which unconventional approaches to trans-jurisdictional groundwater governance can address key governance functions, drawing on examples discussed in this section

Governance function	*Collaborative science*	*Administrative planning*	*Centralized regulatory authority*
Structure to build and agree technical information	SVRP used a joint technical team under a formal Memorandum of Agreement	Collaborative science was part of the PRRIP planning process	Collaborative scientific studies were part of the Susquehanna groundwater management plan
Comprehensive view of the water system	SVRP studies addressed groundwater–surface water interactions	The PRRIP considered groundwater in the context of ecological restoration and surface water connections	Controls on groundwater pumping in the Susquehanna were intended to benefit fisheries, streamflow and the bay
Comprehensive range of participants	SVRP arrangements included a Policy Advisory Committee of stakeholders	The PRRIP included NGO representation on a statutory committee and consultation requirements	NGO advice became formal policy in the Susquehanna
Comprehensive range of management tools	May not directly contemplate or lead to management action	The PRRIP involved supply-side manipulations (e.g. managed aquifer recharge) and demand-side tools (e.g. offset requirements)	Supra-state permit requirements for individual users apply in the Susquehanna, as well as demand-side measures, such as conservation and public education
Ongoing adaptability	SVRP arrangements included a permanent ongoing technical committee	The PRRIP is developed in 13-year increments	Susquehanna River Basin Commission can impose additional requirements on already permitted extractions
Potential for political barriers to use	Low	Low	High due to reduction in state authority/ autonomy

agreements over trans-boundary resources to be caught up in expensive disputes about the underlying scientific characterization of the resource (row 2 of table). Unconventional approaches also ‘see’ groundwater systems, groundwater stakeholders and groundwater management tools in a comparatively broad way. They can embrace a broader view of groundwater as a resource connected to other water sources and ecosystems by collecting data on these connections, or explicitly requiring them to be considered or protected in planning or centralized permitting functions (row 3 of table). They can embrace a broader range of stakeholders through stakeholder representation on technical, planning or other advisory committees, and the resulting work products can be useful to a broader audience (row 4 of table). They can use both supply-side management tools – for example, infrastructure options such as managed aquifer recharge, as well as demand-side management tools – for example, water conservation requirements and public education to reduce water use (row 5 of table). They can also include simple mechanisms to enable adaptation to changing circumstances, such as automatic time-based review requirements and ongoing leadership groups (row 6 of table). These elements suggest that these tools of unconventional approaches could effectively address the key desirable functions and characteristics of trans-boundary governance proposed in the section of this chapter dealing with the importance of managing interstate groundwater.

This is not to say, however, that unconventional modes of governance are without challenge. Collaborative science and administrative planning do not feature the legal formality of traditional mechanisms, such as interstate compacts. The lack of a formal regulatory 'stick' in these approaches can raise challenges associated with motivating parties to come to the table and ensuring they have the resources to stay there. Funding 'carrots' to motivate parties to come to the table and continue to participate may not necessarily be secure. On the other hand, a powerful motivator is to offer an unconventional mode of governance, say a planning process, as an alternative to a more formal, bureaucratic and inconvenient way to fulfil a regulatory requirement. This is apparent in relation to ESA requirements in the PRRIP.

Another way to encourage participation (or at least keep discouragement to a minimum) is to adopt a 'less is more' principle, particularly in the face of hydrogeologic complexity. Under this approach, parties could 'develop limited agreements between themselves premised on the known characteristics of groundwater systems, rather than on their unbounded unknowns' (Davenport, 2008, p. 173). The incentive of minimized initial commitment 'may in fact produce more results in terms of managing interstate groundwater resources than would reaching an agreement splitting those resources' (Davenport, 2008, p. 173).

Once at the table, the structure of the SVRP project suggests that concrete ground rules in the form of formal committee structures and decision-making processes can help even legally informal approaches to create durable leadership groups. However, in the absence of a regulatory stick, there is no guarantee that commitment to joint science or planning approaches will turn into real, on-ground management action. Nonetheless, joint science and planning processes that are well regarded by multiple constituencies can facilitate management by creating shared concepts and 'shared logics' when it comes to management decisions (Freeman, 2010).

As opposed to the challenges posed by the relative informality of joint science and planning approaches, it is the heightened formality and power of a centralized regulatory institution that acts as the key challenge of this latter approach. Participating entities may discern too high a risk in giving up their autonomy in the service of centralized decision making. The example presented here may be particularly unsettling in this context, in that the SRBC has continually strengthened its powers to regulate groundwater-using projects in a way that supersedes state law.

Strikingly, federal participants are a common presence among both traditional and unconventional approaches. They take on a veritable 'hodgepodge' of roles (Leshy, 2008, p. 1485). The federal government can drive trans-boundary water governance that includes groundwater, as in the PRRIP. Through the USGS, it plays a key role in technical studies, as in the SVRP studies. In this context, federal participants may be welcome for their perceived impartiality (Nelson, 2012a). The federal government may also play a role by agreeing to restrain its own groundwater pumping where this has inter-jurisdictional effects, as in its PRRIP offset commitments (PRRIP, 2006). In other circumstances, it might also contribute to trans-boundary groundwater governance through federal water project planning and funding, offering other incentives, exercising powers over the use of federal lands and asserting water rights in wildlife refuges that rely on interstate flows of water compromised by excessive groundwater pumping (Leshy, 2008).

The broad-ranging lessons and observations drawn from the examples of unconventional groundwater governance presented here raise a rich set of possible research questions to which we now turn.

Research agenda

In line with the preliminary, exploratory nature of this chapter, we have presented three unconventional approaches as independent from each other and described them through examples described at a high level, at a point in time, largely based on documentary research (albeit research spurred by approaches to governance indicated by interviews). A clear line of future work could seek to confirm these observations through empirical study – for example, by surveying or interviewing participants of each process outlined in this paper as an example and extending the range of examples used to draw conclusions. In this section, we raise some additional questions for future research, which consider the links between different groundwater governance approaches, including the order, sequentially or other, in which different approaches are adopted and tend to develop from and into one another; opportunities for traditional approaches to employ key features of governance that we have associated with unconventional approaches; potential for conflict between approaches and available means of resolution; and funding issues and financial structures associated with the various approaches. Future research might also further investigate approaches and actors within groundwater governance particularly at the inter-local level and various implications of the natural environment for the approach adopted.

Further examination of the links between different approaches to governance may reveal important themes and considerations for trans-jurisdictional management. A first type of link might consider the temporal character of groundwater governance and whether Agnew and Busch were correct in assuming that these approaches do in fact develop sequentially. Key research questions could be: Under what circumstances does a collaborative science approach mature into planning arrangements, which then mature into centralized regulatory institutions? Or, conversely, to what extent do centralized regulatory institutions tend to be founded on previous joint technical studies or administrative planning approaches?

A second type of link between approaches relates to the potential for particular mechanisms that we have found associated with unconventional approaches to be used in some instances of a traditional approach. Key research questions could be: To what extent have recent manifestations of traditional mechanisms adopted – or could they adopt – 'unconventional' mechanisms, for example, court processes that mandate an initial collaborative scientific process (as in the settlement of litigation over the Republican River interstate compact (Barfield, 2009))? Future compact drafters may wish to consider using joint scientific studies and explicit provisions to address the issue. Related work could investigate the extent to which traditional trans-jurisdictional water governance mechanisms trigger or are succeeded by an unconventional approach, for example, a joint scientific body established after litigation over an interstate compact.

A third type of link between approaches might investigate costs and funding structures – a key challenge for unconventional approaches. Research could look to other Congressionally funded projects for structures providing greater financial security. Key research questions regarding other forms of governance might ask: How are financial responsibilities structured within a centralized regulatory structure? For example, where a commission designates a groundwater protection area within a single state, who bears the cost of implementing management? How are impacts to water-dependent economies of participating jurisdictions considered? Empirically, how do the costs of unconventional approaches compare against the cost of traditional approaches?

A final area of inquiry relates to the types of parties involved in trans-jurisdictional groundwater governance. We have highlighted the fact that unconventional governance mechanisms include non-state as well as state actors. The SVRP even suggests that non-state actors can catalyse the formation of a trans-boundary groundwater governance mechanism. This raises a question, which, to our knowledge, has not been answered: Going beyond government-driven approaches to trans-boundary groundwater governance, to what extent do NGOs and businesses drive unconventional governance solutions independently or jointly with governments? In what types of frameworks does this occur? Through partnerships with government agencies? If the organization or business has no enforcement power, how do they get participants to the table?

Conclusion

The US presents an extreme case of groundwater boundaries, with numerous state boundaries spanned by important aquifers, to say nothing of international and local boundaries. Legal mechanisms which are traditionally used to manage trans-boundary water in the US and state-based groundwater laws may struggle to facilitate several key desirable functions and characteristics of trans-boundary groundwater management. These relate to building agreed scientific information; adopting a broad view of the water system, a broad range of participants and a broad range of management tools; ensuring that arrangements are adaptable; and minimizing political barriers to adopting the arrangements. In practice, traditional legal mechanisms are rarely used to manage trans-boundary groundwater in the US. Instead, unconventional approaches to governance have arisen, and seem to offer relatively simple mechanisms for delivering many key desirable functions of trans-boundary water governance. The examples discussed in this chapter fall into three major categories of approaches: collaborative science, administrative planning and centralized regulatory authorities. They present not only possible stand-alone approaches for managing trans-boundary aquifers that are currently not managed in a coordinated way, but the mechanisms that they use to address key governance functions might lend themselves to being grafted onto traditional legal approaches. The examples also raise many questions for further research. Within these questions might lie at least part of the answer to addressing better trans-jurisdictional groundwater problems, whether trans-boundary governance mechanisms are unknown or well established and whether they are found within the US or further afield.

References

Adler, R. W. (2010) 'Climate Change and the Hegemony of State Water Law', *Stanford Journal of Environmental Law*, 29, 1–61.

Agnew, A. F. and Busch, R. W. (1971) *Interstate Ground-Water Aquifers of the State of Washington, Physical and Legal Problems – A Preliminary Assessment*, State of Washington Water Research Center, Pullman, Washington, DC.

Arkansas v. Oklahoma (1992) 503 US 91.

Barfield, D. W. (2009) 'Collaborative Groundwater Model Development', Paper presented at World Environmental and Water Resources Congress 2009, Great Rivers, Missouri, http://ascelibrary.org/doi/pdf/10.1061/41036(342)477.

Barlow, P. M. and Leake, S. A. (2012) 'Streamflow Depletion by Wells: Understanding and Managing the Effects of Groundwater Pumping on Streamflow', Circular 1376, US Geological Survey, Washington, DC.

Bergkamp Jr, G., Hobbs, G. J. and Iza, A. (2009) 'Linking Policies to Realities', in A. Iza and R. Stein (eds) *Rule: Reforming Water Governance*, IUCN, Gland, Switzerland.

Blevins, E. and Aldous, A. (2011) 'Biodiversity value of groundwater-dependent ecosystems', *Wetland Science and Practice*, 28, 2, 18–24.

Clemons, J. (2004) 'Interstate water disputes: A road map for states', *Southeastern Environmental Law Journal*, 12, 2, 115–42.

Davenport, J. H. (2008) 'Less is more: A limited approach to multi-state management of interstate groundwater basins', *University of Denver Water Law Review*, 2, 1, 139–80.

Dellapenna, J. W. (2005) 'Interstate struggles over rivers: The southeastern states and the struggle over the "Hooch"', *New York University Environmental Law Journal*, 12, 828–900.

DePhilip, M. and Moberg, T. (2010) *Ecosystem Flow Recommendations for the Susquehanna River Basin: Report to the Susquehanna River Basin Commission and United States Army Corps of Engineers*, The Nature Conservancy, Harrisburg, PA.

Freeman, D. M. (2010) *Implementing the Endangered Species Act on the Platte Basin Water Commons*, Kindle edn, University Press of Colorado, Boulder, CO.

Grant, D. L. (2003) 'Interstate water allocation compacts: When the virtue of permanence becomes the vice of inflexibility', *University of Colorado Law Review*, 74, 1, 105–80.

Grant, D. L. (2004), 'Water apportionment compacts between states', in R. Beck (ed.) *Waters and Water Rights*, The Michie Co, Charlottesville, VA.

Gregory, G. (Washington Department of Ecology) (2013) Phone interview, 13 November.

Hall, N. D. (2006) 'Toward a new horizontal federalism: Interstate water management in the Great Lakes region', *University of Colorado Law Review*, 77, 405–56.

Hall, N. D. (2010) 'Interstate water compacts and climate change adaptation', *Environmental and Energy Law and Policy Journal*, 5, 237–324.

Hall, N. D. and Cavataro, B. L. (2013) *Interstate Groundwater Law in the Snake Valley: Equitable Apportionment and a New Model for Trans-boundary Aquifer Management*, Wayne State University Law School Research Paper mo. 2013–24, Wayne State University, Detroit, MI.

Hanak, E., Lund, J., Dinar, A., Gray, B., Howitt, R., Mount, J., Moyle, P. and Thompson Jr., B. (2011) *Managing California's Water: From Conflict to Reconciliation*, Public Policy Institute of California, San Francisco, CA.

Hathaway, D. L. (2011) 'Trans-boundary groundwater policy: developing approaches in the western and southwestern United States', *Journal of the American Water Resources Association*, 47, 1, 103–13.

Haynes, B. and Stoeffel, K. (2007) 'SVRP bi-state study', Paper presented at Spokane Valley-Rathdrum Prairie Aquifer Study Results and Policy Summit, Spokane Valley, Washington, DC, https://idwr.idaho.gov/news/multimedia/spokane_rathdrum_aquifer_study_5_8_2007/spokane_rathdrum_main.htm>, accessed 30 March 2015.

Hoffman, C. and Zellmer, S. (2013) 'Assessing institutional ability to support adaptive, integrated water resources management', *Nebraska Law Review*, 91, 4, 302–60.

Hsieh, P. A., Barber, M. E., Contor, B. A., Hossain, M. A., Johnson, G. S., Jones, J. L. and Wylie, A. H. (2007) 'Ground-water Flow Model for the Spokane Valley-Rathdrum Prairie Aquifer, Spokane County, Washington, and Bonner and Kootenai Counties, Idaho', *Scientific Investigations Report* 2007–5044, US Geological Survey, Reston, VA.

Idaho Department of Water Resources (IDWR) (undated(a)), 'Frequently Asked Questions about Northern Idaho Adjudication', Idaho Department of Water Resources, Boise, ID, http://idwr.idaho.gov/WaterManagement/NorthIdAdju/PDFs/NIA_FAQ_Card.pdf, accessed 30 March 2015.

Idaho Department of Water Resources (undated(b)), 'Northern Idaho Adjudication', Idaho Department of Water Resources, Boise, ID, http://idwr.idaho.gov/WaterManagement/NorthIdAdju/, accessed 30 March 2015.

Idaho Department of Water Resources (undated(c)), 'Spokane Valley-Rathdrum Prairie Hydrologic Project', Idaho Department of Water Resources, Boise, ID, https://idwr.idaho.gov/WaterInformation/projects/svrp/, accessed 30 March 2015.

Idaho Department of Water Resources and Washington Department of Ecology (2007) 'Memorandum of Agreement (MOU) for Maintenance and Utilization of the Numerical Model of the Spokane Valley-Rathdrum Prairie Aquifer', https://idwr.idaho.gov/WaterInformation/projects/svrp/PDFs/svrp_MOA_10–26–07.pdf, accessed 30 March 2015.

Idaho Water Resource Board (2011) *Rathdrum Prairie Aquifer Comprehensive Aquifer Management Plan*, Idaho Water Resource Board, Boise, ID.

Kahle, S. C. and Bartolino, J. R. (2007) *Hydrogeologic Framework and Ground-water Budget of the Spokane Valley Rathdrum Prairie Aquifer Spokane County, Washington, and Bonner and Kootenai Counties, Idaho*, Scientific Investigations Report 2007–5041, US Geological Survey, Reston, VA.

Leshy, J. D. (2008) 'Interstate groundwater resources: The federal role', *Hastings West-Northwest Journal of Environmental Law and Policy*, 14, 1475–98.

Maupin, M. A. and Barber, N. L. (2005) 'Estimated withdrawals from principal aquifers in the United States, 2000', Circular 1279, US Geological Survey, Reston, VA.

Milman, A. and Ray, I. (2011) 'Interpreting the unknown: uncertainty and the management of trans-boundary groundwater', *Water International*, 36, 5, 631–45.

Nelson, R. (2012a) 'Instituting Integration: Findings of the Comparative Groundwater Law and Policy Program's Workshop 1', Stanford University, Stanford, CA, http://waterinthewest.stanford.edu/sites/default/files/RNelsonWorkingPaper3.pdf, accessed 30 March 2015.

Nelson, R. (2012b) 'Assessing local planning to control groundwater depletion: California as a microcosm of global issues', *Water Resources Research*, 48, 1, W01502.

Nelson, R. (2014) 'Groundwater wells versus surface water and ecosystems: An empirical approach to law and policy challenges and solutions', JSD dissertation, Stanford University, Stanford, CA.

Nelson, R. and Casey, M. (2013) 'Taking policy from paper to the pump: Lessons on effective and flexible groundwater policy and management from the Western US and Australia', Stanford Woods Institute for the Environment, Stanford, CA, http://waterinthewest.stanford.edu/sites/default/files/CGLPP_Wkshp_2_Report%20final_0.pdf, accessed 30 March 2015.

Patton, B. (Bureau Chief, Planning Bureau, Idaho Department of Water Resources) (2011) Interview, 31 October, Boise, ID

Platte River Recovery Implementation Program (PRRIP) (2006) 'Attachment 5, Water Plan', Platte River Recovering Implementation Program, Denver, CO and Kearney, NE, https://platteriverprogram.org/PubsAndData/ProgramLibrary/PRRIP%20Attachment%205%20-%20water_plan_final.pdf, accessed 30 March 2015.

Santhi, C., Allen, P. M., Muttiah, R. S., Arnold, J. G. and Tuppad, P. (2008) 'Regional estimation of base flow for the conterminous United States by hydrologic landscape regions', *Journal of Hydrology*, 351, 1–2, 139–53.

Sax, J. L., Thompson Jr, B. H., Leshy, J. D., and Abrams, R. H. (2006) *Legal Control of Water Resources: Cases and Materials*, 4th edn, Thomson/West, St Paul, MN.

Susquehanna River Basin Commission (SRBC) (2005) 'Groundwater Management Plan for the Susquehanna River Basin', Susquehanna River Basin Commission, Harrisburg, PA, http://srbc.net/programs/docs/GW_Mngt_Plan_June2005/GWMP%20Final.pdf, accessed 30 March 2015.

Susquehanna River Basin Commission (SRBC) (2008) 'Standard for Water Withdrawals, 18 CFR § 806.23', Susquehanna River Basin Commission, Harrisburg, PA, http://gpo.gov/fdsys/pkg/CFR-2008-title18-vol2/pdf/CFR-2008-title18-vol2-sec806–23.pdf, accessed 30 March 2015.

Susquehanna River Basin Commission (SRBC) (2012) 'Low Flow Protection Policy Related to Withdrawal Approvals (Policy No 2012–01)', Susquehanna River Basin Commission, Harrisburg, PA, http://srbc.net/policies/docs/2012–01_Low_Flow_Protection_Policy_Related_to_Withdrawal_Approvals_12–14–12_fs170475.PDF, accessed 30 March 2015. An Act Consenting to the Susquehanna River Basin Compact 1970, 84 Stat. 1509.

Theesfeld, I. (2010) 'Institutional challenges for national groundwater governance: policies and issues', *Ground Water*, 48, 1, 131–42.

Thompson Jr, B. H. (2011) 'Beyond Connections: Pursuing Multidimensional Conjunctive Management', *Idaho Law Review*, 47, 273–323.

US Department of Agriculture, Forest Service (USFS) (2012) 'Groundwater-Dependent Ecosystems: Level 1 Inventory Field Guide: Inventory Methods for Assessment and Planning', USFS, Washington, DC, http://fs.fed.us/geology/GDE_Level_I_FG_final_March2012_rev1_printing.pdf, accessed 30 March 2015.

US Geological Survey, Idaho Department of Water Resources and Washington Department of Ecology (2003) 'Memorandum of Understanding (MOU) for Research of the Spokane Valley-Rathdrum Prairie Aquifer', https://idwr.idaho.gov/WaterInformation/projects/svrp/PDFs/svrp_Final-MOU.pdf, accessed 30 March 2015.

Washington Department of Ecology (2008) 'Idaho and Washington Agree on Water Management Protocols', Press release 08–018, Washington Department of Ecology, Washington, DC, http://ecy.wa.gov/programs/wr/ero/svrp_summit.html, accessed 30 March 2015.

9 Muddied water

(Un)cooperative governance and water management in South Africa

Michael Kidd

Introduction

Being a developing country that is water scarce, it is imperative that South Africa manages its water resources effectively (South Africa, 2004, 2010a). There are currently several serious problems in relation to the country's water resources – both quality and quantity issues – that will not be easy to solve. These problems have the potential to undermine South Africa's efforts to provide its people their constitutionally protected right to access water, and to hamper the provision of water to economic sectors requiring it. This chapter focuses on the complex institutional governance structure applicable to water in South Africa, which is an apparent impediment, although not the only one, to effective engagement with South Africa's water problems.

Water governance in South Africa is a classic example of trans-jurisdictional governance in several senses. First, there is the straddling of jurisdictions in relation to what the Constitution of the Republic of South Africa, 1996 refers to as 'spheres of government': national, provincial and local. The local sphere of government, often the least resourced, is responsible for several critical aspects of water governance. As the discussion below will show, local government is often not coping with water governance and the other governmental spheres have failed to account for this deficiency. The second example relates to the horizontal jurisdictional issues caused by the 'silo effect' of different (primarily national government) departments' involvement in water governance matters. The third instance (and there may be more) is a temporal one: pre-1994 water governance legacy issues constitute a contributing factor to many of the problems faced by the country today.

This analysis will primarily focus on the first scenario outlined above; however, the others will be touched upon as they each interact. Although addressing governance issues will in no way totally solve the water problems faced by South Africa, doing so will constitute an important first step to solving them.

This chapter will outline the water context in South Africa, including national challenges to water quality and quantity. This outline will be followed by a discussion of the applicable legal framework, including analysis of cases that have dealt with water governance issues. Analysing gaps in the existing legal framework and its implementation in relation to water governance will highlight that the water situation in South Africa has elements of being a 'wicked problem' (one that defies 'resolution because of the enormous interdependencies, uncertainties, circularities, and conflicting stakeholders implicated by any effort to develop a solution' (Lazarus, 2009, p. 1159)). Nevertheless, efforts to address this problem will be considerably strengthened by effectively resolving trans-jurisdictional governance issues.

The water situation in South Africa

Despite regional differences in rainfall, South Africa is a dry country overall (South Africa, 2004, p. 15). Currently, most catchment areas in the country are either already water-stressed (demand exceeds supply) or are likely to be water-stressed within a few years (South Africa, 2006). In 1994, at the end of apartheid, many South Africans (overwhelmingly black people) did not have adequate access to water. In 1997, between 12 and 14 million South Africans (out of a total population of about 40 million) were without access to safe water and over 20 million without adequate sanitation (South Africa, 1997, p. 15). The situation has improved significantly, with 89.9% of the population now having access to piped water (not necessarily in homes – the norm is within 200 m of a dwelling), and 77.9% having access to 'improved sanitation' (Statistics South Africa, 2014). Although impressive at first glance, these statistics indicate that approximately 5 million South Africans are still without access to piped water and about 10 million without access to sanitation. The statistics are also somewhat misleading because they do not take account of the fact that many people's access to water is very unreliable.

The facts outlined above indicate that South Africa is faced with the challenge of providing its citizens with water from increasingly dwindling resources. Other demands on water resources should also be kept in mind, the biggest of which is agriculture (using approximately 60% of national water resources) (South Africa, 2013, p.10). These dynamics are reflected in the primary aims of the National Water Act of 1998, which highlights both conservation (and proper management) of water resources and improving access to water for those historically deprived of such access (see preamble and s. 2, National Water Act).

This sketches the broad context within which the problems mentioned in the introduction present themselves. One of the problems mentioned above is increasing water scarcity, particularly in more populous areas of the country. This issue will not be considered in detail in this chapter. Rather, this chapter will focus on the pervasive problem of water pollution by untreated sewage. This pollution has adverse impacts on the provision of potable water to people (among other effects).

Sewage treatment and the Green Drop Reports

In the last five years, The Department of Water Affairs has released two reports called the Green Drop Reports (South Africa, 2010b, 2012a), which report on South Africa's wastewater quality management performance. According to the more recent report:

> [The] Green Drop process measures and compares the results of the performance of Water Service Authorities and their Providers [these terms defined below], and subsequently rewards (or penalizes) the municipality upon receiving evidence of their excellence (or failures) according to the minimum standards or requirements that have been defined.
>
> (Department of Water Affairs, 2012a, p. 1)

The results of the 2010 report provided cause for alarm. Only 449 (53%) of the 852 municipal wastewater systems were assessed (South Africa, 2010b, p. 4). Of these 449 systems, 203 (45%) scored higher than 50% in the assessment (scores being allocated on a range of functions and competences relating to wastewater treatment) (South Africa, 2010b, p. 4). Those systems that were not assessed were unlikely to have been sufficiently well run to have scored passing grades. This observation is based on the reasons provided in the report for non-assessment, which usually indicated chaotic 'management' of the treatment systems concerned (South Africa, 2010b, p. 4).

The 2010 Report provides more detailed aggregated data as to the various aspects contributing to the overall scores. The ratings for wastewater compliance with licence conditions/general authorizations or special limits are a particular cause for concern. This assessment examined whether wastewater released into water resources from the treatment works complied with the relevant legal standards for emission of wastewater or water containing waste into water resources. According to s. 1 of the National Water Act, a 'water resource' is defined as including 'a watercourse, surface water, estuary, or aquifer' and a 'watercourse' is defined as '(a) a river or spring; (b) a natural channel in which water flows regularly or intermittently; (c) a wetland, lake or dam into which, or from which, water flows; and (d) any collection of water which the Minister may . . . declare to be a watercourse, and a reference to a watercourse includes, where relevant, its bed and banks'.

Only 65 of 449 wastewater systems (14.5%) were in full compliance (South Africa, 2010b). Ignoring non-compliance that appeared (on my understanding of the assessment criteria) to be formal rather than substantive, there were still 299 treatment works (66.5%) that were not in substantive compliance with the legal requirements (South Africa, 2010b). 'Formal non-compliance' refers to treatment works that reported incorrectly or made some similar error in providing data, rather than failing to meet the applicable standards, which would be a matter of 'substantive compliance'.

The 2011 report did not contain the same detailed aggregated data, but only the overall scores (as well as individual details for the different water treatment works). In this report, there was almost blanket coverage: all 156 municipalities operating treatment plants were assessed, comprising 821 water treatment plants (South Africa, 2012a, p. 11). The wider coverage is the likely cause of the percentage of plants that scored over 50% declining from 49 to 44%, even though the average score improved from 37 to 45% (South Africa 2012a, p.11). The report also indicates that results vary considerably between the provinces. More than 65% of the treatment plants in the strongest economic provinces (Western Cape, Gauteng and KwaZulu-Natal) scored above 50% in the assessment. In comparison, less than 20% of the plants in each of North West, Free State, Limpopo and Northern Cape scored above 50%. Eastern Cape barely performed better, with 26% of its plants scoring above 50% (South Africa 2012a, p. 14).

These results indicate that a considerable number of sewage treatment plants in the country (the vast majority in certain parts of the country) are pumping untreated (or inadequately treated) sewage into the country's rivers and streams. This has adverse impacts on agriculture, and people who rely on river water for drinking are being put at risk (Newmarch, 2010). Moreover, excess sewage loads leads to an overabundance of nutrients, which causes eutrophication. According to Harding (2010, p. 6), eutrophication is 'probably the single greatest threat to water resources' in the country. Approximately 35% of impounded water resources in South Africa are seriously impaired (eutrophic to hypertrophic), while a further 20 to 30% are 'incipiently problematical' (Harding, 2010, p. 2). All of the major impoundments in Gauteng, 'the economic heartland of the country', are 'grossly impaired' (Harding, 2010, p. 2). In order to combat eutrophication, nutrient loads should be reduced by between 25 to 85%, 'with reductions in excess of 70% typically required in order to meet desired management targets' (Harding, 2010, p. 3). In the four years since this observation, the situation is unlikely to have improved.

What are the reasons for the problem? As will be explained further below, municipalities are responsible for sewage treatment. Whereas municipalities were almost exclusively urban entities in the apartheid Constitutional regime, the post-1994 position is that South Africa consists of more than 180 'wall-to-wall' municipalities. Many of these municipalities are short of finances and many are very poorly managed. The Auditor-General's 'Consolidated general report on the audit outcomes of local government' for 2012–13 found that only 30 out of 319 audited institutions (278 municipalities and 42 'municipal entities') received a 'clean audit' (South Africa, 2012–13a, p. 16). At the other end of the scale, a total of 151 municipal entities were rated as having either: (1) a qualified audit ('unable to adequately and accurately account for all the financial effects of the transactions and activities they conducted' – i.e. in some aspects there was inadequate accounting); (2) an 'adverse opinion' ('conditions regarding to unreliable financial statements were common in most areas of the financial statements' – auditees in this category are described as having 'demonstrated extreme levels of lack of accountability for financial statements'); or (3) a 'disclaimer of opinion' ('unable to provide the required evidence to enable the auditors to perform tests to satisfy themselves in the fair presentation of the financial statements') (South Africa, 2012–13a, pp. 17–18). During this review period, the Auditor-General found that a total of 11.6 billion Rand

(R) (871 million USD) in irregular expenditure had been incurred by municipalities (South Africa, 2012–13a, p. 19).

Many of the problems relating to municipal performance in sewage treatment are finance related. There is often inadequate provision made for maintenance and operations, and anecdotal evidence suggests that the budget allocation for sewage treatment is sometimes seen as an emergency 'slush fund', causing already inadequate funding to diminish further. Inadequate budgeting also has negative impacts on the recruitment of adequately skilled and adequate numbers of personnel. According to the Centre for Development and Enterprise (CDE) (2010, p. 9), between the 1990s and 2010 the number of engineers working in South Africa's water sector dwindled from between 3,000–4,000 to 1,400. In some cases, it is unlikely that the municipal budget is big enough for proper budgeting for sewage treatment, and in others it is not seen as a priority.

Although it may appear that this is a new problem – providing further ammunition for critics of the post-1994 ruling party – there is a significant legacy issue at play in relation to sewage treatment. Facilities that were designed in the past for servicing primarily the white population in towns and cities are now required to service more people. This is not only because of equitable extension of the services to all inhabitants, but also because of the influx of people into urban areas (South Africa, 2006a; Turok, 2012). Tempelhoff characterizes this legacy issue as follows:

> Previously the country's water infrastructure had focused primarily on a population of about 6 million (comparatively affluent) whites, some people of colour in segregated urban townships, and on meeting the needs of the highly productive (and lucrative) industrial and agricultural sectors. The same infrastructure, after 1994, had to make provision for a population of more than 42 million people, of whom many were now living in informal settlements on the fringes of the country's urban areas. Needless to say the existing infrastructure was simply inadequate.
>
> (Tempelhoff, 2009, p. 165)

The inherited problem has been exacerbated by new housing developments and urban influx, and the water infrastructure has not kept pace with this development (Tempelhoff, 2009, p. 165). Inadequate maintenance of existing infrastructure has further compounded the problem. In 2010, estimates suggested that it would cost over R500 billion (37.5 billion USD) to address backlogs in sanitation services and to repair and increase the capacity of infrastructure, 30 to 40% of which was not working (Centre for Development and Enterprise, 2010, p. 31). According to Donnelly (2012), the Minister of Environmental and Water Affairs was reported as stating that 'over R570 billion will be needed for investment across South Africa's water value chain, in the coming 10 years'.

This demonstrates that poor performance in the Green Drop Report assessments is not solely due to municipal incompetence (although this often plays a role), but there are other, almost intractable, contributing factors. The scale of the problem and the cost of addressing it are such that it is clearly beyond all but the best-resourced municipalities (which are generally performing wastewater treatment adequately). Municipalities cannot be expected to address untreated sewage alone with their own, often meagre, resources. It is a national problem, and as such, it requires national solutions. This issue will be discussed later, after considering problems relating to the provision of drinking water and analysis of South Africa's legal framework.

Provision of water services

The Department of Water Affairs publishes a companion assessment to the Green Drop Report, which deals with provision of (drinking) water services. This report, the Blue Drop Report (South Africa, 2012b), assesses the entire municipal drinking water quality management business and assesses risk involved in the process for all municipalities. The 2012 Blue Drop Report identifies municipalities that were issued with warnings in relation to drinking water. A total of 18 municipalities, some of them covering large areas and each including several separate human settlements, had warnings issued with respect to the quality of water they provided (two in the Eastern Cape, seven in the Free State, five in Mpumalanga, one each in the Northern Cape, Western Cape, North West and Limpopo). Most of the warnings advise against drinking water provided by these municipalities without some kind of home disinfection treatment.

In addition to drinking water quality issues, there are many areas where water supply is unreliable (in some cases these municipalities are the same ones mentioned in the Blue Drop Report). To be clear, this refers to places where there is an existing water supply infrastructure, not places where access to water is still awaited. In other words, many people who have previously had sufficient access to water are now suffering from unreliable (and in some cases little, if any) access to water services (see, e.g. De Waal, 2012; E News Channel Africa (ENCA), 2014; South Africa Broadcasting Company (SABC), 2014). In Carolina (Mpumalanga Province), residents are alleged to have been without a piped water supply for at least five months in 2012, leading to the Federation for Sustainable Environment case (2012) discussed below. Despite litigation, which appears on paper to have been successful, residents in October 2014 indicated that the water supply had still not been restored.

In mid-2014, the media reported that three infants had died in May in the Bloemhof area (in the North West Province) allegedly as a result of the town's water supply having been contaminated by a sewage spillage that had not been properly addressed (Fourie, 2014). In June 2014, in another part of the same province, Biesiesvlei, at least 15 children died from diarrhoea allegedly caused by water contaminated by sewage (Moore, 2014). It is interesting to note that Biesiesvlei is part of the Ditsobotla Local Municipality, which is in turn part of the Ngaka Modiri Molema District Municipality. The latter is one of the municipalities reported in the 2012 Blue Drop Report as having water of dubious quality. This municipality scored 28% in the 2012 Green Drop assessment. The Green Drop Report indicates that 'the situation in Ngaka Modiri Molema is considered critical from a regulatory view and holds high risk to public health and the environment' (South Africa, 2012a, p. 315). The Report indicated (p. 316) that '10 of the 11 wastewater treatment plants [in the municipality] do not have appropriate managerial or technical skills in place to conduct wastewater services management, operation and maintenance'; and '11 of 11 plants do not meet effluent quality standards, which is underscored by the lack of monitoring in 7 of the 11 plants'. In the Auditor-General's audit report for 2012–13, this municipality was rated as the 'worst performing district' (South Africa, 2012–13b, p. 14), incurring irregular expenditure in that period of R953 million (71.5 million USD) (South Africa, 2012–13b, p. 108). The same report indicates that only 51% of inhabitants of this municipality have access to piped water in their homes, 35% have communal access and 8% no access to water. Only 33% have flush or chemical toilets, 60% rely on pit or bucket toilets, while 8% have no access.

In November 2014, the Constitutional Court dismissed an appeal by the municipality arising from the dissolution of the council and replacement by an Administrator under section 139 of the Constitution of the Republic of South Africa, 1996 (the Constitution), in the case of *Ngaka Modiri Molema District Municipality v. Chairperson, North West Provincial Executive Committee and Others* (2014). The municipality had sought to bring an urgent, interdict application against the dissolution of the municipality and installation of the Administrator. The Constitutional Court agreed with the High Court's dismissal of the application, observing that:

> It needs to be stressed that the potential prejudice and urgency lie not in the harm suffered by the Municipality or the municipal councillors, but in the continued disruption of basic essential services to the people and communities the Municipality is supposed to serve. The people who may suffer the real harm are not party to these proceedings. It is because of the alleged failure in its executive obligation to them that the Municipality was dissolved.
>
> (*Ngaka Modiri Molema District Municipality v. Chairperson, North West Provincial Executive Committee and Others*, 2014, para. 9)

The above details regarding the Ngaka Modiri Molema District Municipality are intended as an example of the woes, particularly in relation to water and sanitation service delivery, currently experienced at local government level in South Africa. Although this example is one of the worst, similar problems are experienced throughout the country to varying degrees, particularly in poorer regions. We must endeavour to answer the question: What role is the law able to play in addressing this sort of situation and what, if any, are the trans-jurisdictional governance impediments to possible solutions?

The applicable legal framework

The Constitution of the Republic of South Africa (the Constitution)

At the apex of South Africa's legal system is the Constitution of the Republic of South Africa, 1996. Section 27 provides that everyone has the right to have access to, inter alia, water (s. 27(1)(b)). Where this right has not already been realized, s. 27(2) enjoins the state to take reasonable legislative and other measures, within its available resources, to achieve the progressive realization of the rights set out in s. 27.

The Constitution also sets out the different competences for the three spheres of government. Legislative competence is divided between each level of government by means of two Schedules to the Constitution, schs. 4 and 5. Schedule 4 contains the 'functional areas of concurrent national and provincial legislative competence' and sch. 5 'functional areas of exclusive provincial legislative competence'. 'Water' (in the sense of water management generally) is on neither of these lists. This means that legislative competence over water is exclusively national. That, however, is just the start – the situation becomes more complex. Part B of sch. 4 contains matters that are referred to as local government matters. In terms of section 156, a municipality has executive authority in respect of, and has the right to administer, the local government matters listed in Part B of both schs. 4 and 5 (s. 156(1)(a)). A municipality may make and administer by-laws for the effective administration of these matters (s. 156(2)). This is significant for present purposes, because listed in Part B of sch. 4 is 'Water and sanitation services limited to potable water supply systems and domestic waste-water and sewage disposal systems'. This means that municipalities have executive authority over these matters. This is reflected in the more specific powers and duties of municipalities provided for in the Water Services Act 1997 discussed below.

The municipal executive powers referred to in s. 156 are subject to s. 155(6)(a) and (7). Section 155(6)(a) provides for the monitoring and support of local government by the relevant provincial government in the province. Section 155(7) provides that the national government and the provincial governments have the legislative and executive authority to oversee the effective performance by municipalities of their functions, by regulating municipalities' exercise of their executive authority. This oversight function is bolstered by s. 139, headed 'Provincial supervision of local government'. This section provides that, when a municipality cannot or does not fulfil an executive obligation in terms of legislation, the relevant provincial executive may intervene by taking any appropriate steps to ensure fulfilment of that obligation. The province may issue a directive to the Municipal Council, describing the extent of the failure to fulfil its obligations and stating any steps required to meet its obligations. The province may also assume responsibility for the relevant obligation in that municipality to the extent necessary.

There are other significant aspects of the Constitution. The objects of local government set out in s. 152 include ensuring the provision of services to communities in a sustainable manner, to promote social and economic development and to promote a safe and healthy environment. Section 153 requires a municipality to structure and manage its administration, to give priority to the basic needs of the community in budgeting and planning processes and to promote the social and economic development of the community. Section 154 provides that the national government and provincial governments, by legislative and other measures, must support and strengthen the capacity of municipalities to manage their own affairs, to exercise their powers and to perform their functions.

Finally, the Constitution envisages that the different spheres of government must operate in a cooperative manner. Although the three spheres are, in terms of s. 40, 'distinctive, interdependent and interrelated', they must observe and adhere to the principles set out in s. 41 and must conduct their activities within the parameters provided in the Chapter (i.e. ss. 40 and 41). Section 41(1) sets out the principles of cooperative government and intergovernmental relations, which include that all spheres of government and all organs of state within each sphere must exercise their powers and perform their functions in a manner that does not encroach on the geographical, functional or institutional integrity of government in another sphere, and cooperate with one another in mutual trust and good faith. Such cooperation entails that they must avoid legal proceedings against one another.

An organ of state involved in an intergovernmental dispute must make every reasonable effort to settle the dispute by means of mechanisms and procedures provided for that purpose, and must exhaust all other remedies before it approaches a court to resolve the dispute (s. 41(3)). The Constitution requires the enactment of an Act of Parliament to establish or provide for structures and institutions to promote and facilitate intergovernmental relations; and provide for appropriate mechanisms and procedures to facilitate settlement of intergovernmental disputes (s. 41(2)). This Act is called the Intergovernmental Relations Framework Act 2005, and it provides in s. 40 – that all organs of state must make every reasonable effort to avoid intergovernmental disputes when exercising their statutory powers or performing their statutory functions; and to settle intergovernmental disputes without resorting to judicial proceedings.

There have been several cases involving disputes relating to Constitutional competences, particularly in the sphere of municipal planning (which is also in pt. B of sch. 4). In the *City of Johannesburg Metropolitan Municipality v Gauteng Development Tribunal* case (2010), parts of the Development Facilitation Act 1995 were declared unconstitutional because they provided for an unacceptable interference by a provincial body (the development tribunal) into a matter of exclusive municipal competence, namely municipal planning. The court observed that:

> It is . . . true that the functional areas allocated to the various spheres of government are not contained in hermetically sealed compartments. But

> that notwithstanding, they remain distinct from one another. This is the position, even in respect of functional areas that share the same wording, like roads, planning, sport and others. The distinctiveness lies in the level at which a particular power is exercised.
>
> (*City of Johannesburg Metropolitan Municipality v Gauteng Development Tribunal* case, 2010, para. 55)

and:

> The constitutional scheme propels one ineluctably to the conclusion that, barring functional areas of concurrent competence, each sphere of government is allocated separate and distinct powers which it alone is entitled to exercise.
>
> (*City of Johannesburg Metropolitan Municipality v Gauteng Development Tribunal* case, 2010, para. 56)

This decision was followed, albeit in different circumstances (the question was whether mining rights granted in terms of national minerals legislation would entitle the holder to exercise the rights notwithstanding the failure to obtain the necessary municipal planning permission to re-zone the land for mining purposes), in the *Maccsand (Pty) Ltd v City Of Cape Town And Others* case (2012). The court held that, due to exclusive municipal planning competence, the requisite zoning authorization was required prior to commencing mining.

In the *Minister of Local Government, Development Planning and Environmental Affairs, Western Cape v The Habitat Council and Others* case (2014), the Constitutional Court provided some clarity on the extent of the oversight function that can be played by provincial (and national) government over municipalities in terms of s. 155(7), when it confirmed the declaration of a section in provincial legislation as unconstitutional and invalid because it allowed the province to intervene into municipal planning matters. The court characterized the oversight function in para. 21 as 'hands-off' and continued to observe:

> 'regulating' in section 155(7) means creating norms and guidelines for the exercise of a power or the performance of a function. It does not mean the usurpation of the power or the performance of the function itself. This is because the power of regulation is afforded to national and provincial government in order 'to see to the effective performance by municipalities of their functions'.
>
> (*Minister of Local Government, Development Planning and Environmental Affairs, Western Cape v The Habitat Council and Othersl* case, 2014, para. 22)

If the Constitution itself is not sufficiently complex in relation to the provision of water and sanitation services, the other applicable legislation makes the situation even more so.

The National Water Act 36 of 1998 (NWA)

Section 3 of the NWA is particularly relevant to this chapter:

> (1) As the public trustee of the nation's water resources the National Government, acting through the Minister, must ensure that water is protected, used, developed, conserved, managed and controlled in a sustainable and equitable manner, for the benefit of all persons and in accordance with its constitutional mandate.
>
> (2) Without limiting subsection (1), the Minister is ultimately responsible to ensure that water is allocated equitably and used beneficially in the public interest, while promoting environmental values.
>
> (3) The National Government, acting through the Minister, has the power to regulate the use, flow and control of all water in the Republic.

This section has not yet been considered in the courts.

The NWA is also relevant in relation to setting standards that water treatment works have to meet. If they are working properly, water treatment works release purified effluent into water resources. The release of wastewater is a 'water use' in terms of s. 21 of the NWA (the relevant 'water use' is 'discharging waste or water containing waste into a water resource through a pipe, canal, sewer, sea outfall or other conduit'). This water use requires a licence under s. 22. The licence specifies the standards that the released treated water must meet. If the treatment works discharges less than 200 m^3 of water per day, it will be subject to a general authorization as per s. 39 of the NWA, which discharges the user's obligation to obtain a licence. Operating in terms of a general authorization requires compliance with standards in any event. In many cases, as indicated in the Green Drop Report, these standards are not being met.

Water Services Act 108 of 1997 (Water Service Act)

The Water Services Act regulates the provision of water and sanitation services. One of the Act's main objects is to provide for the right of access to basic water supply and the right to basic sanitation necessary to secure sufficient water and an environment not harmful to human health or well-being (s. 2(a)). In terms of s. 1, 'basic sanitation' means the 'prescribed minimum standard of services necessary for the safe, hygienic and adequate collection, removal, disposal or purification of human excreta, domestic wastewater and sewage from households, including informal households'. A 'sanitation service' means 'the collection, removal, disposal or purification of human excreta, domestic wastewater, sewage and effluent resulting from the use of water for commercial purposes'. Two important institutions required to carry out functions in terms of this Act are Water Services Authorities and Water Services Providers. The former is 'any municipality . . . responsible for ensuring access to water services'.

A 'water services provider' is 'any person who provides water services to consumers or to another water services institution [in turn defined as 'a water services authority, a water services provider, a water board and a water services committee'], but does not include a water services intermediary' ('any person who is obliged to provide water services to another in terms of a contract where the obligation to provide water services is incidental to the main object of that contract'). In practice, water services authorities may also be (and often are) water services providers, although the former may contract with other bodies, including private bodies, to carry out the provision of water services (see s. 19). Water services authorities are under a duty to all consumers or potential consumers in their areas of jurisdiction to progressively ensure efficient, affordable, economical and sustainable access to water services (which includes sanitation services) (s. 11).

Section 9 of the Act empowers the Minister to prescribe compulsory national standards relating to the provision of water services, among other related matters. Every water services institution is required to comply with these standards. The Minister has prescribed standards in a notice entitled 'Regulations relating to compulsory national standards and measures to conserve water' (GN R509 *GG* 22355, 8 June 2001). Regulation 3 provides that the minimum standard for basic water supply is:

- (a) the provision of appropriate education in respect of effective water use; and
- (b) a minimum quantity of potable water of 25 litres per person per day or 6 kiloliters per household per month:
  - (i) at a minimum flow rate of not less than 10 litres per minute;
  - (ii) within 200 metres of a household; and
  - (iii) with an effectiveness such that no consumer is without a supply for more than seven full days in any year.

If the water supply is interrupted, reg. 4 provides that the water services institution must take steps to ensure that, where water services are interrupted for a period of more than 24 hours for reasons other than those contemplated in s. 4 of the Water Service Act (which deals with conditions for provision of water services, and includes matters dealing with discontinuation or limitation of services), a consumer has access to alternative water services comprising at least 10 litres of potable water per person per day and sanitation services sufficient to protect health. The provision of water is governed not only by the Water Service Act, but also by municipal by-laws.

Since s. 27 of the Constitution contains an internal modifier providing for the progressive realization of the right to water; it is not envisaged that everyone will immediately have a claim to have the right realized. The Constitution recognizes that it is not feasible to meet such socio-economic rights overnight and that the government will need to roll out the services in a reasonable way and within a reasonable timeframe. Many problems arise, however, in situations where people have had existing water services that have been interrupted, discontinued or are provided with impure water. This was the factual basis of the *Federation for Sustainable Environment* case (2012).

This case concerned the water supply to consumers in Carolina and its surrounding residential areas. The town's water supply was contaminated with acid mine water and residents had not received piped water supplies since February 2012. The respondents included government bodies at all three levels: the national Department of Water Affairs (DWA), provincial government (Mpumalanga) and various local government authorities responsible for the position in the town. The applicants sought an order declaring, first, that the respondents' failure to provide access to potable water for more than seven full days in the area, as required by the relevant regulations mentioned above, was unlawful.

They also sought an order directing respondents to provide temporary water in accordance with the regulations; and that they (respondents) engage meaningfully with the applicants regarding 'the steps being taken to ensure that potable water can once again be supplied through the water supply services in . . . Carolina, Mpumalanga' and as to 'where, when, what volume, and how regularly temporary water will be made available in the interim'. They also asked the court to direct the respondents to report back to the court within one month of the order of court to indicate what measures in this regard had been taken.

The court observed that the DWA was responsible for the implementation of both the Water Services Act and the National Water Act, and that all the other respondents (provincial and local authorities) were responsible for 'providing water resources' in Carolina. The court highlighted the national and provincial government's responsibility to support municipalities in terms of s. 154 of the Constitution. It also observed that the municipality has responsibilities in terms of s. 152 of the Constitution, which it summarized as entailing striving 'towards improving the quality of life of the community' (*Federation for Sustainable Environment* case, 2012, para. 14).

Yet it was argued that it was inappropriate to order the national and provincial government authorities cited in the case (who were opposing the application only on the basis of potential costs orders against them) to meet the applicant's prayers, as the matter in question was a municipal responsibility. Although national and provincial government are 'committed' to providing financial assistance (which had been provided), 'they are debarred from interfering and imposing their will on the local government' (*Federation for Sustainable Environment* case, 2012, para. 19). The court agreed with this argument and decided, in effect, to absolve them of any responsibility in the case (although not in so many words, this seems to be the effect of this aspect of the judgement).

The court ordered that the relevant district municipality officials were to provide temporary potable water in accordance with the relevant regulations within 72 hours of the order. In addition, the court required these officials to engage meaningfully with the community concerning measures to be taken and to report back to the court within one month.

Analysis

Examining the Blue Drop and Green Drop Reports, together with the legal provisions outlined above, it is clear that numerous municipalities are breaking South African law. What is the government's response to this? Despite the observation in the Green Drop Report that the object of the report is to provide both bouquets and brickbats, the emphasis is very much on the former (as is the case of the Blue Drop Report). Other than adverse publicity, there have been no enforcement efforts aimed at the delinquent municipalities. The Carolina case was brought by a non-governmental body and is still embroiled in legal-technical squabbling. Is it possible to ascertain why these municipalities have been (apparently) handled with kid gloves?

At first glance, it might be tempting to lay the blame for this lack of response at the door of cooperative government, which is aimed at avoiding confrontation between organs of state. Ultimately, it is the national government (and, in particular the Department of Water Affairs acting through the Minister) that is responsible for the water situation in the country, and it is clear from s. 3 of the National Water Act that the Minister has the statutory authority to respond to the problems outlined here. Although cooperative government requires dialogue before litigation, there is not even any evidence of such dialogue (unless the Green and Blue Drop Reports can be characterized as such). The power within s. 155(7), for national government to regulate (in addition to those legal requirements discussed above) the exercise by municipalities of their executive authority, has been largely unused in regard to these problems.

Provinces have the more immediate power of supervision of municipalities under s. 139 of the Constitution. This has been exercised in many instances (such as in the Ngaka Modiri Molema District Municipality), but the impetus for such moves seems to be a pervasive failure of governance by a municipality (particularly in relation to financial management) rather than a 'mere' failure to carry out water services functions.

Despite the presence of these powers, and the obvious fact that the law is not being followed, as evidenced by published government assessments, these problems appear to be persisting year after year. The Department of Water Affairs published, for comment, Draft Regulations relating to Compulsory National Standards for Process Controllers and Water Services Works (GN R813 in GG 36958 of 23 October 2013). These have however not yet been enacted and this is only a small step in the right direction. The norms and standards would provide for minimum qualifications for persons operating water works, depending on the classification of the process in question.

National government may set norms and standards or regulate local municipalities by legislating accordingly, but national government is somewhat removed from the local level. The intervention measures with the most 'bite' are those in s. 139 of the Constitution, but these are provincial government powers. A lack of action against water services failures is possibly due to the provincial governments having no water management functions themselves. Provincial governments are not identifying the problems, but national government is.

Another possible contributing factor is the scale of the problem. It is clear from the details set out by this chapter that the problems are widespread and the cost of addressing the shortcomings is enormous. This is not a problem that can be addressed simply by calling in the relevant municipality, laying down the law and then expecting it to toe the line. In most of the cases, huge amounts of money will have to be spent on upgrading or replacing infrastructure and hiring appropriately qualified technical personnel to address the problems. This is money which most municipalities do not have, nor are likely to be able to raise. The funds will have to come from national government and it has even been mooted that the private sector may have to contribute (see, e.g. Donnelly, 2012). It may be that provincial government and the Minister feel that more direct engagement with the municipalities will be futile because the latter will not be able to do anything about it and the Minister will then have to find the money to address the problem (as was the case in Carolina).

While this may be an explanation, it is not an excuse for doing nothing. Section 3 of the NWA provides for the Minister of Water Affairs to manage South Africa's water as 'public trustee', but it makes it clear that the Minister is acting on behalf of national government as a whole. Water is a critical resource, yet one often gains the impression that it is regarded as somewhat peripheral by government instead of a crucial policy issue. While official policy suggests that it is vital, this 'talk' is not being evidenced in the 'walk'. Just as water conservation and sustainable management of resources is being championed verbally, there are hundreds of mining applications in the most important catchment areas in the country. This might explain the Minister's inaction – there is a recognition that fixing the problems will cost a large amount of money, and the priorities for government expenditure lie elsewhere.

This is somewhat speculative, but one cannot expect the Minister to take the National Water Act's s. 3 duties seriously when she is essentially let off the hook by judgements such as the *Federation for Sustainable Environment* case. Even though the Constitution does not provide for national government interference in the affairs of municipalities, the Minister, as custodian of the nation's water resources, has a statutory duty to regulate municipal water management. In addition, s. 154 of the Constitution provides that the national government, by both legislative and other measures, must support and strengthen the capacity of municipalities to manage their own affairs, to exercise their powers and to perform their functions. This does not mean that the Minister must usurp the municipality's functions, but the very least that can be expected is that the Minister insist that the municipalities obey the law and back this up with remedial action if necessary. This may well require the national executive to commit funds to this purpose.

Conclusion

Far more could be written on this topic in the South African context, but space prevents more than just touching on the situation. Trans-jurisdictional water governance is certainly not the cause of the problems outlined by this chapter, but its inherent issues seem to be a certain impediment to resolving South Africa's water problems. The three spheres of government appear to be leaving the ball to each other, with the result that the ball is dropped. Current efforts at 'naming and shaming' organs of state that are quite brazenly not complying with the law are clearly not effective, and the current 'hands-off' approach to cooperative government in the water services sector is not assisting. One hopes that more people will not die from inadequate water service delivery before positive steps are taken, but that may be too much to hope for. For the most part, there is adequate law to address these problems. However, the failure to use it effectively (and in some cases, a failure to use it at all) seems to rest upon political will.

References

Centre for Development and Enterprise (CDE) (2010) 'Water: A Looming Crisis?', http://cde.org.za/south-africa-s-future/85–4-south-africa-s-future-and-issues-of-national-importance/128-water-a-looming-crisis, accessed 20 March 2015.

City of Johannesburg Metropolitan Municipality v Gauteng Development Tribunal 2010 (6) SDA 182 (CC).

Constitution of the Republic of South Africa, 1996.

De Waal, M. (2012) 'South Africa's ultimate problem: water delivery failure', *Daily Maverick*, 17 August, http://dailymaverick.co.za/article/2012-08-17-south-africas-ultimate-problem-water-delivery-failure/#.VIAoL8kcODo, accessed 29 November 2014.

Donnelly, L. (2012) 'South Africa's water crisis just got expensive', *Mail & Guardian*, 16 April, http://mg.co.za/article/2012–04–16-sas-water-crisis-just-got-expensive, accessed 20 November 2014.

Draft Regulations relating to Compulsory National Standards for Process Controllers and Water Services Works, South African Government Notice 813/36958, 23 October 2013.

E News Channel Africa (ENCA) (2014) 'Water crisis looms – municipal benchmarking kicks in', 24 June, http://enca.com/municipal-benchmarking-avoid-looming-water-crisis, accessed 29 November 2014.

Federation for Sustainable Environment and Others v Minister of Water Affairs and Others Unreported case 35672/12 (GNP), www.saflii.org/cgi-bin/disp.pl?file=za/cases/ZAGPPHC/2012/128.html&query=federation%20for%20sustainable%20environment>, accessed 15 November.

Fourie, M. (2014) 'Three infants die from polluted water in Northwest: Justice must be done', *Daily Maverick*, 9 June, http://dailymaverick.co.za/opinionista/2014-06-09-three-infants-die-from-polluted-water-in-northwest-justice-must-be-done/#.VIBCu8kcODo, accessed 15 November 2014.

Harding, W. R. (2010) 'Eutrophication of impounded water resources in South Africa: Descent into crisis', Paper presented at WISA Conference, Durban, South Africa, http://anthonyturton.com/admin/my_documents/my_files/WISA_2010_Presentation.pdf, accessed 20 November 2014.

Intergovernmental Relations Framework Act, Act 13 of 2005.

Lazarus, R. J. (2009) 'Super wicked problems and climate change: Restraining the present to liberate the future', *Cornell Law Review*, 94, 5, 1153–233.

Maccsand (Pty) Ltd v City Of Cape Town And Others 2012 (4) SA 181 (CC).

Minister of Local Government, Development Planning and Environmental Affairs, Western Cape v The Habitat Council and Others [2014] ZACC 9.

Moore, N. (2014) 'North West's water contamination: A dark story of lost and ruined lives', *Daily Maverick*, 9 July, http://dailymaverick.co.za/article/2014-07-09-north-wests-water-contamination-a-dark-story-of-lost-and-ruined-lives/, accessed 15 November 2014.

National Water Act, Act 36 of 1998.

Newmarch, J. (2010) 'Sewage treatment "unacceptable"', *Business Day*, 28 April, http://allafrica.com/stories/201004280360.html, accessed 20 March 2015.

Ngaka Modiri Molema District Municipality v Chairperson, North West Provincial Executive Committee and Others [2014] ZACC 31.

Regulations Relating To Compulsory National Standards and Measures to Conserve Water, South African Government Notice 509/22355, 8 June 2001.

South Africa (1997) 'White Paper on a National Water Policy for South Africa', Department of Water Affairs and Forestry, Pretoria, South Africa, https://dwaf.gov.za/Documents/Policies/nwpwp.pdf, accessed 20 March 2015.

South Africa (2004) 'National Water Resource Strategy (NWRS)', Department of Water Affairs and Forestry, Pretoria, South Africa, https://dwaf.gov.za/Documents/Policies/NWRS/Default.htm, accessed 20 March 2015.

South Africa (2006) 'Water available for allocation per water management area', Directorate: National Water Resource Planning, Department of Water Affairs and Forestry, Pretoria, South Africa, https://dwa.gov.za/WAR/documents/WMAWaterAvailibility23Feb06.pdf, accessed 20 March 2015.

South Africa (2010a) 'Integrated water resource planning for South Africa', Department of Water Affairs, Pretoria, South Africa, https://dwaf.gov.za/Documents/Other/Water%20Resources/IntegratedWRPlanningSA-2010.PDF, accessed 20 March 2015.

South Africa (2010b) 'Green Drop Report 2009', Department of Water Affairs, Pretoria, South Africa, https://dwa.gov.za/Documents/GreenDropReport2009_ver1_web.pdf, accessed 20 March 2015.

South Africa (2012a) 'Green Drop Report 2011', Department of Water Affairs, Department of Water Affairs, Pretoria, South Africa, https://dwaf.gov.za/dir_ws/GDS/Docs/UserControls/DownloadSiteFiles.aspx?id=32, accessed 20 March 2015.

South Africa (2012b) '2012 Blue Drop Report', Department of Water Affairs, Pretoria, South Africa, http://indicators.hst.org.za/indicators/SocioEconomic/BlueDrop_2012.pdf, accessed 20 March 2015.

South Africa (2012–13a) 'Consolidated general report on the audit outcomes of local government', Auditor-General, New Muckleneuk, South Africa, http://agsa.co.za/Portals/0/MFMA%202012-13/2012_13_MFMA_Consolidated_general_report.pdf, accessed 20 March 2015.

South Africa (2012–13b) 'North West General Report on the audit outcomes for local government 2012–13', Auditor-General, New Muckleneuk, South Africa, http://agsa.co.za/Portals/0/MFMA%202012-13/2012_13_MFMA_North_West_general_report.pdf, accessed 20 March 2015.

South Africa (2013) 'National Water Resources Strategy', Department of Water Affairs, Pretoria, South Africa, https://dwaf.gov.za/nwrs/LinkClick.aspx?fileticket=u_qFQycClbI%3d&tabid=91&mid=496, accessed 20 March 2015.

South Africa Broadcasting Company (SABC) (2014) 'Grahamstown without water again', 22 September, http://sabc.co.za/news/a/7e4428804591506b9bcf9b90ca3f4715/Grahamstown-without-water-20142209, accessed 29 November 2014.

Statistics South Africa (2006) 'Migration and urbanization in South Africa', Report no. 03–04–02, Statistics South Africa, Pretoria, South Africa, www.hsrc.ac.za/en/research-outputs/ktree-doc/10986, accessed 20 March 2015.

Statistics South Africa (2014) 'Household delivery statistics', Statistics South Africa, Pretoria, South Africa, http://beta2.statssa.gov.za/?page_id=739&id=2=1, accessed 15 November 2014.

Tempelhoff, J. W. N. (2009) 'Civil society and sanitation hydropolitics: A case study of South Africa's Vaal River barrage', *Physics and Chemistry of the Earth*, 34, 3, 164–75.

Turok, I. (2012) 'Urbanisation and Development in South Africa: Economic Imperatives, Spatial Distortions and Strategic Responses', International Institute for Environment and Development, London, http://pubs.iied.org/10621IIED.html, accessed 20 March 2015.

Water Services Act, Act 108 of 1997.

10 The flow of laws

The trans-jurisdictional laws of the longest river in Aotearoa New Zealand

Jacinta Ruru

Introduction

Flowing from the slopes of Mount Ruapehu in the centre of the North Island of Aotearoa New Zealand, through Lake Taupo and the Waikato plains to the Tasman Sea south of Auckland, is the country's longest river, the Waikato River (425 km). While the waters in this river do not traverse country borders (New Zealand is an island country) or state lines (New Zealand is a unitary state), the river does cross Indigenous Maori tribal boundaries and New Zealand local government lines. This river navigates its journey from the mountains to the sea through a myriad of overlapped contemporary laws – Indigenous Maori tribal laws and the Western New Zealand laws of legislation and common law. The purpose of this chapter is to demonstrate the applicability of the concept of trans-jurisdiction in a local place, rather than a typical nation-to-nation or federal context, in order to illustrate that different laws operate across one another even in local places, including the Waikato River. Legal norms can coexist culturally, structurally, horizontally and vertically (see Friedman, 1996, p. 29; Nelken and Feest, 2001). This chapter therefore highlights a common tension in many colonized countries between first Indigenous and imposed colonial laws, and it explores how to navigate contemporary, respectful and possible legal solutions to hard questions about the governance of water.

The Waikato River is an apt river to provide an insight into *local* trans-jurisdiction because of the high-profile legislation and case law arising in the context of this river. Notably, in 2010, the New Zealand Parliament enacted groundbreaking legislation that has introduced a co-management vision for this river which gives voice to several Maori tribes and Government. However, other laws also slice and dice across this river. There is national legislation that governs the sustainable management of water (Resource Management Act 1991) and there have been relevant appeals from judicial decisions to the Environment Court and beyond, some of which have specifically involved the Waikato River. There is the national state-owned enterprise (SOE) legislation that demonstrated its relevance in late 2013 when the Government partially sold shares in Mighty River Power Ltd, a corporation which had built a hydroelectricity dam in the Waikato River. (Maori had made an urgent claim in 2012 to the Waitangi Tribunal and had taken judicial review proceedings to the Supreme Court in early 2013 in an attempt to halt the share sale.) And then there is the common law and, interestingly, a crucial 2014 Supreme Court judgement grounded in common law and equity arguments about continuing Maori interests in a specific stretch of the Waikato riverbed. This chapter thus takes a three-pronged view of the matrix of law, structured first with an insight into the first laws of the River – Maori law; then the overlaid state national law including resource management law, Treaty of Waitangi claim settlement legislation and state-owned enterprise law; and finally the common law. These three forms of law – all operative – create the basis to explore trans-jurisdictional issues in the waters of Aotearoa New Zealand.

Indigenous laws

The legal history of Aotearoa New Zealand began when the first peoples – the Maori tribes – settled on these shores sometime about AD 1100. Grouped into distinct peoples, Maori became, literally, *tangata whenua* (the people of the land). Of the about forty distinct *iwi* (tribes) and hundreds of *hapu* (sub-tribes), they all derive their identity from the mountains, rivers and lakes (see Calman and Reed, 2004). For instance, '*Ko wai au?*' asks who am I, but more literally this translates as 'Who are my waters?'. The answer will depend on which tribe and sub-tribe to which the person belongs. For example, for me, the Waikato River is one of my ancestral waters with which I identify through my *whakapapa* (genealogy) on my paternal grandfather's side and whose tribal affiliation is Raukawa. All tribes have these geographical identity markers linked to water. The link between land and water and humans is a common feature of the Maori language. For instance, *iwi* means both 'tribe' and 'bone', *hapu* means both 'sub-tribe' and 'to be pregnant', *whanau* means both 'extended family' and 'to give birth', *whenua* means 'land' and 'afterbirth', and *wai* means 'water' but also 'memory' and 'who' (Williams, 1971; Maori Language Act 1987; Mead, 2003).

In many ways, the Maori legal system reflects this personified worldview. It is predominantly values, not rules, based. It encapsulates a certain way of life that depends on the relationships between all things, including between people and gods, different groups of people, and people and everything in the surrounding world. Integral values thus include *whakapapa* (genealogy), *whanaungatanga* (family relationships), *mana* (authority), *mauri* (spiritual life-principles/life force), *tapu* (sacredness), *rahui* (prohibition or conservation) and *manaaki* (hospitality) (see Jackson, 1992; Durie, 1996, 2000; Tohe, 1998; Kawharu, 2000; New Zealand Law Commission, 2001; Erueti, 2004; Frame, 2010–11; Benton *et al.*, 2013; Stephens and Boyce, 2013). Another key concept, *utu*, plays a regulatory role because everything given or taken requires a return of some kind in order to maintain harmony and balance ensuring reciprocal acts of continued generosity. *Kaitiakitanga* (guardianship), as a further example, is about 'more than managing relations between environmental resources and humans; it also involves managing relationships between people in the past, present and future' (Kawharu, 2000, p. 352).

The Maori phrase for law – *tikanga Maori* – involves an 'obligation to do things in the "right" way' or 'way(s) of doing and thinking held by Maori to be just and correct' (New Zealand Law Commission, 2001, p. 16). *Tikanga* are said to be 'tools of thought and understandings . . . They help us to differentiate between right and wrong' (New Zealand Law Commission, 2001, p. 16). To illustrate in a specific context, contracts concern not the transfer of rights for a prescribed consideration or immediate return as is typical in Western transactions, but the establishment of a permanent and personal relationship with reciprocal obligations where the main benefit to both sides would come in the course of time (see Durie, 1996).

The Maori legal system has developed strong rules for protecting the *mauri* of water. For instance, according to Maori law it is abhorrent to force artificially the mixing of waters from different catchments and to mix waters with human sewage (the preference being for sewage to be treated on land). Moreover, recently some Maori tribes have put forward a 12-point 'indicia of ownership' framework for establishing contemporary customary proof of ownership of water (see Waitangi Tribunal, 2012, p. 38). It states that:

- The water resource has been relied upon as a source of food.
- The water resource has been relied upon as a source of textiles or other materials.
- The water resource has been relied upon for travel or trade.
- The water resource has been used in the rituals central to the spiritual life of the *hapu*.
- The water resource has a *mauri*.
- The water resource is celebrated or referred to in *waiata* (song).
- The water resource is celebrated or referred to in *whakatauki* (proverb).
- The people have identified *taniwha* (water spirit/monster) as residing in the water resource.
- The people have exercised *kaitiakitanga* over the water resource.
- The people have exercised *mana* or *rangatiratanga* (sovereignty/self-determination) over the water resource.
- *Whakapapa* identifies a cosmological connection with the water resource.
- There is a continuing recognized claim to land or territory in which the resource is situated, and title has been maintained to 'some, if not all, of the land on (or below) which the water resource sits'.

Many of these Maori laws remain alive and practised by the tribes who have a whakapapa connection with the Waikato River, namely Waikato-Tainui, Raukawa, Te Arawa and Ngati Tuwharetoa (and Ngati Maniapoto via the Wairoa River, which is a major tributary to the Waikato River). Does state law and common law recognize these Maori laws as part of the trans-jurisdiction of this place?

National state laws

In 1840, with the signing of Te Tiriti o Waitangi (the Treaty of Waitangi) between many Maori chiefs and the British Queen, and specific British proclamations, Aotearoa New Zealand became a British colony. Today, Aotearoa New Zealand is an independent constitutional monarchy with a legal system that adheres strongly to the constitutional principle of parliamentary supremacy. Several national state laws are applicable to the Waikato River.

The Resource Management Act 1991

The central statute that manages the use of all water in New Zealand is the Resource Management Act 1991 (RMA) (see Grinlinton, 2002; Nolan, 2014). The New Zealand Ministry for the Environment retains some responsibility to influence the management of water, primarily through setting national environmental standards and policy statements (RMA, ss. 24–29A and 43–58A). The National Policy Statement for Freshwater Management 2014 (NPSFM) is an example. It articulates strong objectives to safeguard and improve the quality of water. A key objective is to:

> . . . provide for the involvement of iwi and hap_, and to ensure that tangata whenua values and interests are identified and reflected in the management of freshwater including associated ecosystems, and decision-making regarding freshwater planning, including on how all other objectives of this national policy statement are given effect to.
>
> (Ministry for the Environment, 2014, p. 18)

Day-to-day control of water is vested in regional governments (see RMA, ss. 30–31 and 9–23; Scott, 2006). These governments prepare plans that contain rules for the sustainable management of land, air and water and stipulate when and where proposed activities may require resource consents permitting use. The RMA gives regional councils the specific power to assert rules and guidelines for the take, use, damming, and diversion of freshwater (RMA, s. 14). The Waikato Regional Council's geographic boundaries encompass the Waikato River.

Pursuant to the RMA, the Waikato Regional Council can control the use of land within its region for the purpose of the maintenance and enhancement of the quality and quantity of water in water bodies (s. 30(1)(c) (ii) and (iii)). The Council can set maximum and minimum water levels and flows (s. 30(1)(e)); control discharges of contaminants into water, and discharges of water into water; and, if appropriate, establish rules in a regional plan to allocate the taking or use of water (s. 14(3)(b)-(e)). Moreover, the RMA directs that all decision makers, including the Waikato Regional Council, in formulating plan rules and issuing resource consents must recognize the Māori relationship with water. Specifically, s. 6(e) mandates that all persons exercising functions and powers in relation to managing the use, development and protection of natural and physical resources must recognize and provide for matters of national importance, including the relationship of Maori and their culture and traditions with water. Other matters of national importance such as preserving the natural character of lakes and rivers from inappropriate development, protecting outstanding natural landscapes and significant habitats of indigenous fauna, and maintaining public access to lakes and rivers are also contained in this list and thus must be balanced together.

Additionally, s. 7(a) of the RMA directs that all decision makers shall have particular regard to *kaitiakitanga* (defined in s. 2 to mean 'the exercise of guardianship by the tangata whenua of an area in accordance with tikanga Maori in relation to natural and physical resources; and includes the ethic of stewardship'). Again, it is one of several factors that must be considered, like the efficient use and development of natural and physical resources, climate change and the benefits to be derived from the use and development of renewable energy. Furthermore, s. 8 requires 'all persons exercising functions and powers under [the RMA], in relation to managing the use, development, and protection of natural and physical resources, shall take into account the principles of the Treaty of Waitangi (Te Tiriti o Waitangi)'.

In addition, several other sections in the RMA create mandatory requirements on local authorities to be inclusive of Maori. For example, regional councils, when preparing or changing a regional policy statement, must '[t]ake into account any relevant planning document recognized by an iwi authority, and lodged with the council, to the extent that its content has a bearing on resource management issues of the region' (s. 61(2A)(a)). Section 62(1)(b) directs that a regional policy statement must state the resource management issues of significance to tribal authorities in the region.

Moreover, since 2005, all local authorities must keep and maintain, for each tribe and sub-tribe within its region or district, a record of who they are and how they can be contacted (s. 35A1). The RMA provides for some substantial possibilities for Maori to be more actively involved in the governance of natural resources, including water. For example, the RMA empowers a local authority to transfer any one or more of its functions, powers, or duties to any tribal authority (although note no local authority has yet done this) (s. 32(2)). Additionally the RMA enables a local authority to make a joint management agreement with a tribal authority and group that represents sub-tribes for the purposes of the RMA. (Note that the Treaty settlement legislation discussed below has motivated the Waikato Regional Council to enter several joint management agreements for the Waikato River) (s. 36B.)

Hence, the RMA provides a legal base for Maori interests to be considered in making decisions about the use of water. Yet, up until recently, this base has done little to significantly protect Maori interests. Since the enactment of the RMA in 1991, there have been more than twenty instances where Maori, as objectors, have appealed council decisions that approved resource consents to take water, discharge wastewater into water or dam water (see Ruru, 2013).

In all of these cases Maori speak of the importance of the water to them culturally, including the belief that water has its own *mauri*, and the importance of these water sites for food gathering, namely fishing. In most of these cases Maori have lost – sometimes outright, sometimes partially–where, for example, the duration of the resource consent has been reduced in accordance with some of the concerns raised by Maori but the volume of water take has remained the same (see Ruru, 2010). These cases all begin in the Environment Court, which hears appeals relating to the issuance or non-issuance of resource consents and, thereafter, appeals to the higher courts are restricted to points of law (see Palmer, 2010). Some cases have involved the Waikato River and/or its tributaries.

For instance, Tainui argued that the resource consents issued to build a 86 ha engineered landfill would have potential adverse effects on the tributaries of the Waikato River, particularly the Clune Stream. The Environment Court concluded that the existing conditions to which the design is subject 'will adequately protect the Maori interests' (*Land Air Water Association v. Waikato Regional Council*, para. 494). In another set of cases where a tribe opposed the issuing of consents to Genesis Power to enable the Tongariro hydroelectric power development scheme to continue operating, the Maori tribe lost. Ngati Rangi Trust opposed the consents, primarily because it involved diversion of water from the Whangaehu, Whanganui and Moawhango rivers into Lake Taupo and then into the Waikato River. Ngati Rangi Trust argued that these diversions would inhibit their cultural traditions due to reduced flow of water, reduced water levels, degraded water quality and a change to the ecological system that affects the food chain in the water. While the Environment Court restricted the consents from 35 years to 10, the High Court overruled that decision and the Court of Appeal endorsed the High Court's judgement (see the *Ngati Rangi Trust v. Genesis* cases).

The RMA provisions provide some basis for recognizing some Maori laws such as *kaitiakitanga* values, but only as one of many factors to be considered. It is the Treaty claim settlement statutes that have been the catalyst for stronger recognition and provision of Maori relationships with water, and thus the means for increasing the trans-jurisdictional nature of this place.

Treaty of Waitangi Settlement legislation

Since the mid-1980s, New Zealand has been seriously committed to reconciling with Maori. More than 20 settlement statutes have now been enacted with Maori tribes throughout the country, providing financial, commercial and cultural redress for Crown actions or inactions that breached the principles of the Treaty of Waitangi – a document signed between the British Crown and Maori in 1840. In both the English and Maori language versions of the Treaty, it guaranteed Maori continued ownership of their lands and treasures. However, many instances occurred in the subsequent colonial settlement of New Zealand where the Crown breached this guarantee. Since the mid-1990s, the government has been entering Treaty of Waitangi claim settlements with Maori tribes to rectify these breaches. All of these settlements include extensive Crown apologies and financial, commercial and cultural redress. Cultural redress seeks to recognize the claimant group's spiritual, cultural, historical or traditional associations with the natural environment, often through creating opportunities for Maori to be involved in management decision making of the natural resource in question. The settlement process itself is conducted through the Office of Treaty Settlements as a separate unit within the Ministry of Justice. The settlements aim to provide the foundation for a new and continuing relationship between the Crown and the claimant group based on the Treaty of Waitangi principles.

Many of these settlement statutes recognize the specific importance of water to Maori identity, health and well-being; some have been particularly revolutionary in developing cultural redress options that give tribes co- or joint environmental management responsibilities for lakes and rivers (see, for example, Te Arawa Lakes Settlement Act 2006). The settlement statutes provide an additional avenue for Maori to advance their interests and connection to water other than that offered through the RMA. The water redress settlements for the Waikato River are heralded as revolutionary. There are two settlement statutes specific to the Waikato River:

- Waikato-Tainui Raupatu Claims (Waikato River) Settlement Act 2010 (often referred to as the Waikato River Act);
- Ngati Tuwharetoa, Raukawa and Te Arawa River Iwi Waikato River Act 2010 (often referred to as the Upper Waikato River Act).

These two statutes – referred to here as the Waikato River Acts – present an entirely different water redress package than that offered in any other water settlements (note that there is another, further, linked statute for the Waipa River which is the significant contributor to the Waikato River: Nga Wai o Maniapoto (Waipa River) Act 2012). These Waikato River Acts are premised on a vision that seeks to restore the health and well-being of the Waikato River through providing for cooperative management of the river with those tribes that genealogically link to it. These are the first statutes in New Zealand to elevate Maori to co-management roles with the Crown in regard to freshwater, and therefore demonstrate strongly the value of considering the trans-jurisdiction notion as encompassing local place-based scenarios. This refashioning of traditional colonial power structures to provide better for Maori laws, values and customs illustrates a new contemporary understanding and appreciation for the layers of operative laws, customs and values in one place. Trans-jurisdictions thus become local when national and regional governments, along with Maori tribal authorities, are recognized as valued entities.

To explain this point, a key component of the visioning in these Acts is to restore and protect the relevant tribal relationships with the Waikato River and its tributaries in accordance with a Maori worldview including their economic, social, cultural and spiritual relationships (see the Vision and Strategy for the Waikato River in the schedules of the Waikato River Acts). These statutes rely on co-management tools to recognize Crown breaches of the Treaty, the importance of tribal relationships with the environment including water and, in doing so, provide an example of trans-jurisdictional law in practice.

All the settlement statutes provide extensive and respectful Crown apologies and form the central basis for reconciliation and, in doing so, reach across to the jurisdiction of Maori law. The Waikato River Acts record the cultural tribal relationships with the river, including the Waikato-Tainui tribal belief that the Waikato River is a *tupuna* (ancestor). For example, reproduced here in English is the statement recognizing this belief (Waikato-Tainui Raupatu Claims (Waikato River) Settlement Act 2010, s. 8(3)):

> The Waikato River is our tupuna (ancestor) which has mana (spiritual authority and power) and in turn represents the mana and mauri (life force) of Waikato-Tainui. The Waikato River is a single indivisible being that flows from Te Taheke Hukahuka to Te Puuaha o Waikato (the mouth) and includes its waters, banks and beds (and all minerals under them) and its streams, waterways, tributaries, lakes, aquatic fisheries, vegetation, flood plains, wetlands, islands, springs, water column, airspace, and substratum as well as its metaphysical being. Our relationship with the Waikato River, and our respect for it, gives rise to our responsibilities to protect te mana o te Awa and to exercise our mana whakahaere in accordance with long established tikanga to ensure the well-being of the river. Our relationship with the river and our respect for it lies at the heart of our spiritual and physical well-being, and our tribal identity and culture.

The core part of the Waikato River Acts is that they endorse a new vision and strategy that 'is intended by Parliament to be the primary direction-setting document for the Waikato River and activities within its catchment affecting the Waikato River' (see s. 5(1) in both Acts). Key components of the Vision and Strategy include, for example, restoration and protection of the health and well-being of the Waikato River and the tribal relationships with the river, including their economic, social, cultural and spiritual relationships (see Waikato-Tainui Raupatu Claims (Waikato River) Settlement Act sch. 2, cl. 3.).

As explained in the Waikato-Tainui Raupatu Claims (Waikato River) Settlement Act, the Waikato River Authority is the new statutory body responsible for setting the primary direction through the vision and strategy for the river (s22(2)). The Authority consists of 10 members, including one member appointed from each of the tribes which link with the river (Te Arawa, Tuwharetoa, Raukawa, Maniapoto, a member appointed by the Waikato River Clean-up Trust, and five members appointed by the Minister for the Environment in consultation with other Ministers such as Finance, Local Government, Maori Affairs) (sch. 6, cl. 2(1)). This Act gives power to the new

Waikato River Clean-up Trust. The Trust's primary objective is 'the restoration and protection of the health and well-being of the Waikato River for future generations' (s. 32(3)). The Trust must also prepare a new integrated river management plan, along with relevant central departments, local authorities and other appropriate agencies (s. 36(1)). The integrated river management plan must include consideration of conservation, fisheries and regional council components (s. 35(3)). Moreover, the legislation requires the development of a joint management agreement to be in force between each local authority and the Trust (s. 41(1)). The Waikato Raupatu River Trust and Waikato District Council *Joint Management Agreement* was entered into in 2010 and provides the principles for 'an enduring relationship between the parties through the shared exercise of functions, duties and powers' under the relevant legislation (Waikato Raupatu River Trust and Waikato District Council, 2012, p. 4). Maori law – as expressed using the term *tikanga* – is a pivotal component of the guiding principles. It requires 'respect [of] the mana whakahaere rights and responsibilities of Waikato-Tainui in accordance with tikanga to ensure the balance and mauri of the Waikato River are maintained' (Waikato Raupatu River Trust and Waikato District Council, 2012, p. 5).

The Waikato River Acts embrace reconciliation and partial restitution. The settlements illustrate one way in which co-management can be achieved within the confines of existing resource management legislation, and illustrate a meeting of a kind between Maori law and state law. The settlement statutes recognize the value of *tikanga* Maori and go further than the RMA by committing to some power sharing of governance roles. But the Waikato River Acts do more than merely create this meeting point of two legal and cultural systems. The Acts require positive action: governance across the different legal systems. This is a revolutionary aspect of the Acts: Parliament's contemporary valuing of Maori law. This is what creates the possibility for this new trans-jurisdiction era in this local place.

SOE legislation and case law, 2012–13

In 2012, a clash of trans-jurisdictional laws arose concerning the Crown's intention to privatize partially four hydroelectric state-owned enterprises. The first proposed sale was that of Mighty River Power Ltd, which had water permits issued pursuant to the RMA permitting it to operate hydro-generating power facilities along the Waikato River. When the government proposed this as policy, the New Zealand Maori Council successfully took a claim to the Waitangi Tribunal. The Tribunal agreed to hear the case under urgency and made interim recommendations in August 2012, which found that Maori still have residual proprietary rights in water and that the Crown would breach the principles of the Treaty of Waitangi if it went ahead with the intended share sale (Ruru, 2012). After a short consultation process with tribes, the Government proceeded to enact legislation to enable partial privatization. Parliament enacted amendment legislation to the State-Owned Enterprises Act

1986 to permit the Crown to sell up to 49% of its shares in Mighty River. The State-Owned Enterprises Amendment Act 2012, then scheduled to come into effect on 8 March 2013, would have removed Mighty River from being a state-owned enterprise in the State-Owned Enterprises Act 1986 and made it a mixed-ownership model company pursuant to the Public Finance Act 1989. The Public Finance (Mixed Ownership Model) Amendment Act 2012 inserted a new Part 5A into this 1989 Act. Part 5A includes s. 45Q which reads:

> (1) Nothing in this Part shall permit the Crown to act in a manner that is inconsistent with the principles of the Treaty of Waitangi (Te Tiriti o Waitangi); (2) For the avoidance of doubt, subsection (1) does not apply to persons other than the Crown.

Section 45Q(1) is identical to s. 9 of the State-Owned Enterprises Act 1986 (to view the leading case on s. 9, see *New Zealand Maori Council v. Attorney-General* (1987); Ruru, 2008).

In 2012, the New Zealand Maori Council applied to the High Court for judicial review of the Government's action to proceed with the partial share sales but late in 2012, the High Court held that the Crown's decision to proceed with the sales was not reviewable (see Ruru, 2012). A week later the Supreme Court granted the Maori Council and the Waikato River and Dams Claim Trust leave to appeal this judgement directly to the Supreme Court. In late February 2013, the Supreme Court issued its unanimous single judgement in *New Zealand Maori Council v. Attorney-General* (2013).

The overriding issue in the Supreme Court was whether the sale of up to 49% of the shares in Mighty River Power Ltd would be consistent with the principles of the Treaty of Waitangi. The New Zealand Maori Council argued 'no', the Crown argued 'yes'. The Supreme Court's judgement provided an overall win for the Crown: 'the partial privatisation of Mighty River Power will not impair to a material extent the Crown's ability to remedy any Treaty breach in respect of Maori interests in the river' (para. 8). However, there was also a substantive win for Maori. The Court stated that 'the proposed sale of shares (on which the claim of material prejudice is based) is reviewable for consistency with the Treaty principles (para. 7; see paras. 51, 89 and 90).

A second issue in the case was whether the Crown was in breach of s. 64 of the Waikato-Tainui Raupatu Claims (Waikato River) Settlement Act 2010. Section 64, in summary, records an acknowledgment that the Crown and Waikato-Tainui have different concepts and views regarding relationships with the Waikato River, and that the Crown, a Crown entity, state enterprise or mixed-ownership model company must engage with Waikato-Tainui in accordance with the Kiingitanga Accord principles if action is proposed to create or dispose of a property right or interest in the Waikato River. Thus the Maori appellants argued that the Crown must engage with Waikato-Tainui before pursuing any share sales. The Court disagreed. It observed:

> We are unable to accept that the partial privatization of Mighty River Power constitutes disposal of property or interests held by the Crown in the river. The rights and the associated land in the river bed held by Mighty River Power (which are undoubtedly 'property') are not being disposed of and will continue to be held by it (para. 82).

A third issue tested the adequacy of the consultation. The Maori appellants argued that the Crown's consultation following the recommendations of the Tribunal in August 2012 was 'too rushed' (para. 83). The Court disagreed, noting that a rejection of the Tribunal's recommendations was not evidence of 'empty or pre-determined' consultation' (para. 87).

In considering the main issue concerning whether the proposed sale of the Mighty River shares would breach the Treaty principles, the Court considered the existing law and the possible challenges to that law from a Maori perspective. It is this discussion of the different laws that makes this case interesting within the trans-jurisdictional context. For instance, the Court discussed a range of factors, including the evolution of the statutory regimes for controlling the use of water, and concluded that the current RMA 'substantially improved the recognition of Maori in relation to the management of waters' (para. 97). In addition, the Court discussed the initiatives that the Crown had implemented to engage with Maori about water management and water use. The Court was interested in the recorded statements made by the Crown, the Deputy Prime Minister and the Attorney-General, which accepted that Maori do 'have interests and rights in relation to particular waters' (para. 103) and the partial sale of shares 'will not have a "chilling effect" on the willingness of the Crown to provide appropriate rights recognition and redress' (para. 105). The Court then cited Treaty of Waitangi settlement legislation specific to water as proof that the Crown, in recent years, had been willing to negotiate new management and use regimes for water that better recognize Maori interests.

The Court then reflected: 'Such interests are however unascertained, including as to their nature and extent. How they may be given effect in modern conditions consistently with Treaty responsibilities and other government obligations is also far from clear' (para. 101). Moreover, the Court acknowledged that the Maori appellants did not rely on a common law Aboriginal title claim. The Court stated: 'So we propose to proceed on the basis that the proposed privatisation will not have any effect on such claims' (para. 131). The Court went on to emphasize that while the Crown can go ahead with the share sales, '[t]his is not to say that the water claims should be parked. The Waitangi Tribunal has emphasized the need for urgency in addressing proprietary claims. It appears from the policy initiatives and from the assurances given in the litigation that the message that there is need for action on these claims has been accepted [by the Crown]' (para. 148).

Ultimately, the Court dismissed the appeal, stating (para. 149):

> As is apparent, we are prepared to accept that privatization may limit the scope to provide some forms of redress which are currently at least

> theoretically possible. But in assessing whether this amounts to "material impairment", regard must be had to (a) the assurances given by the Crown, (b) the extent to which such options are substantially in prospect, (c) the capacity of the Crown to provide equivalent and meaningful redress, and (d) the proven willingness and ability of the Crown to provide such redress.

The Court did not award costs against the unsuccessful Maori appellants because they had succeeded on an 'important point of principle, namely that the Crown was bound to comply with the principles of the Treaty before deciding to sell the shares' (para. 151).

After the case, the Crown did proceed to partially sell shares in Mighty River Power Ltd. From a trans-jurisdictional governance and law perspective, this case is interesting for the interplay among different state laws – the RMA, the Waikato River Acts, and state-owned enterprise legislation – and the underlying recognition of Maori interests as protected in the principles of the Treaty of Waitangi.

Common law

Aotearoa New Zealand's state legal system has at its core inherited and adopted English common law. How does this common law apply in Aotearoa New Zealand? As the Court of Appeal held in *Ngati Apa v. Attorney-General* [2003] stated, the English common law applied in New Zealand from 1840 only so far as applicable to the circumstances of the local custom. There is much potential for cases to tease out the interplay among common law, Maori law and water as the Supreme Court acknowledged briefly in the Mighty River Power case discussed above. A significant case that has entered this terrain is the more recent Supreme Court decision in *Paki v. Attorney-General* [2014] (Laing *et al.*, 2014).

The *Paki v. Attorney General* case is legally complicated, and is discussed here briefly to highlight the continuing importance of common law as another aspect of trans-jurisdictional laws in a locality. The facts of this case concern a stretch of the Waikato River near Mangakino where, in the late 1800s, the Native Land Court issued Maori freehold titles to several riparian subdivisions from the Pouakani block. By the early 1900s the Crown had acquired most of these riparian blocks from the Maori owners. At issue is the operation of the common law *usque ad medium filum aquae* (mid-point) principle, which holds that a transfer of riparian land is presumed to transfer also the adjacent riverbed to the mid-point of the river. The descendants of the Maori owners argued in the contemporary courts that their ancestors were unaware of the consequence of transferring their riparian land to the Crown and did not realize that the Crown would also attain the riverbed to the mid-point. The descendants argued that the unexplained operation of this common law doctrine meant that the Crown had breached its fiduciary duties by failing to warn them of this. As explained by Elias CJ in her opening paragraphs in the 2014 Supreme Court judgement, the original owners would not have agreed to transfer their riparian lands if that meant a transfer of the riverbed as well, 'because the Waikato River was essential to their identity and was an important tribal property valued for its spiritual qualities as well as for sustenance provided by food resources obtained from it' (para. 3).

The descendants initially sought in the High Court a declaration that 'the Crown holds such riverbed of the Waikato River as constructive trustee' for the descendants of the original owners (para. 3). The Crown agreed that it owned the riverbed but free of any constructive or fiduciary duties. The case twisted and turned on preliminary issues such as river navigability in the lower courts and even an aspect of the case that went to the Supreme Court in 2012, but it is the final 2014 Supreme Court judgement that directly addressed the issue of fiduciary duties. While the Supreme Court held that the necessary foundation for the claim of breach of fiduciary duty in the Crown acquisition of the riparian lands had not been established, the comments made about possible Maori interests in water are significant for future cases.

For example, as Elias CJ stated, this case leaves open an important point about whether grants of riparian Maori freehold land 'extinguished any customary property in the river itself' (para. 59). Elias CJ is making the distinction here between general freehold land and Maori freehold land. Essentially she is querying whether the common law rule of *usque ad medium filum aquae* should apply when the owners of the riparian land are Maori. Should Maori owners of land in the late 19th century be expected to understand the intricacies of the inherited English common law? Elias CJ's statement in this regard is important for future cases and for understanding trans-jurisdiction within a local place, and is there quoted here at length here (para. 65):

> In the case of Maori land, the importation of this rule of the common law is not self-evidently dictated by similar considerations of 'common sense', even in circumstances where the riparian owners undoubtedly have the bed of the river or lake. Retention of an interest in a river or lake, as a valuable source of food even if on a seasonal basis (as appears to have been the case in respect of this stretch of the Waikato River) and as a focus of tribal identity, cannot be dismissed as 'wholly unprofitable' . . . The Pouakani people may have been sellers of part of their lands, but they retained other land and were not leaving the area. It may not have been the case that interests in the river according to Maori custom were confined to those whose occupation of the riparian lands entitled them to title to those lands in the Native Land Court.

Elias CJ also observed that an early New Zealand case also doubted '[w]hether Maori customary interests in features such as rivers and lakes is adequately rendered by concepts more familiar to English property law' (para. 70). Elias

CJ recognized that this doubt was critical to the creation of the Waikato River Treaty Claim settlements.

These comments illustrate that Maori custom is important and has a valid role to play in understanding other laws such as common law. Maori custom can influence the applicability of common law. Future cases will no doubt explore this in more detail.

Conclusion

Where the longest river in Aotearoa New Zealand flows, three laws intersect: Maori law, state legislation and arising case law, and common law. This chapter has provided a brief glimpse into these three different types of law and, in particular, how state law and common law are responding increasingly to Maori law. The Waikato River is a long river filled with water in high demand for competing and varied uses. The modern solutions for this river are testament to the possibilities of reconciliation and public participatory decision making. The contribution that this chapter has sought to make to this collection is that trans-jurisdictional law and governance are features of local places and waterways, and this is true even in the unitary island country of Aotearoa New Zealand.

Short glossary

hapu (subtribe/to be pregnant)
iwi (tribe/bone)
kaitiakitanga (guardianship)
mana (authority)
mana whakahaere (governance)
manaaki (hospitality)
mauri (spiritual life-principles/life force)
rahui (prohibition or conservation)
rangatiratanga (sovereignty/self-determination)
tangata whenua (the people of the land)
taniwha (water spirit/monster)
tapu (sacredness)
tikanga (correct procedure)
tupuna (ancestor)
utu (balance)
wai (water/memory/who)
waiata (song)
whakapapa (genealogy)
whakatauki (proverb)
whanau (extended family/to give birth)
whanaungatanga (family relationships)
whenua (land/afterbirth)

References

Benton, R., Frame A. and Meredith P. (2013) *Te Mātāpunenga. A Compendium of References to the Concepts and Institutions of Maori Customary Law*, Victoria University Press, Wellington, New Zealand

Calman, R. and Reed A. W. (2004) *Reed Book of Maori Mythology*, Reed Publishing, Auckland, New Zealand.

Co-Management Agreement for Waikato River Related Lands, Waikato Raupatu River Trust and Waikato Regional Council (2012), www.waikatoregion.govt.nz/PageFiles/15805/JMAs/2276497%20Tainui.pdf, accessed 8 March 2015.

Conservation Act 1987 (N.Z.).

Durie, E. T. (1996) 'Will the settlers settle? Cultural conciliation and law' *Otago Law Review*, 8, 4, 449–65.

Durie, E. T. (2000) 'Te Hono ki Hawai'iki: The bond with the Pacific', in M. Wilson and P. Hunt (eds) *Culture, Rights, and Cultural Rights. Perspectives from the South Pacific*, Huia Publishers, Wellington, New Zealand.

Environment Act 1986 (N.Z.).

Erueti, A. (2004) 'Maori customary law and land tenure: An analysis', in R. Boast, A. Erueti, D. McPhail and N. F. Smith, *Maori Land Law*, 2nd edn, LexisNexis, Wellington, New Zealand.

Fisheries Act 1996 (N.Z.).

Frame, A. (2010–11) 'A Few simple points about customary law', in R. Benton (ed.) *Tuhonohono: Custom and State, Yearbook of New Zealand Jurisprudence*, 13–14, 20–8.

Friedman, L. (1996) 'Borders: On the emerging sociology of transnational law', *Stanford Journal of International Law*, 32, 65–90.

Genesis Power Ltd v. Manawatu-Wanganui Regional Council [2006] NZRMA 536.

Genesis Power Limited v. Manawatu-Wanganui Regional Council (22 May 2007) High Court, Wellington, New Zealand, CIV-2004–485–1139.

Grinlinton, D. P. (2002) 'Contemporary environmental law in New Zealand', in K. Bosselmann and D. P. Grinlinton (eds) *Environmental Law for a Sustainable Society*, New Zealand Centre for Environmental Law, Auckland, New Zealand.

Her Majesty the Queen In right of New Zealand and Waikato-Tainui, Kiingitangi Accord (2008), http://nz01.terabyte.co.nz/ots/DocumentLibrary%5CWaikatoRiver KiingitangaAccord.pdf, accessed 8 March 2015.

Jackson, M. (1992) 'The colonization of Maori philosophy' in G. Oddie and R. W. Perrett (eds) *Justice, Ethics, and New Zealand Society*, Oxford University Press, Auckland, New Zealand.

Jones. C. (2012) 'High Court rejects Maori Council water rights case', *Maori Law Review*, December, 4–6.

Kawharu, M. (2000) 'Kaitiakitanga: A Maori anthropological perspective of the Maori socio-environmental ethic of resource management', *Journal of the Polynesian Society*, 109, 4, 349–70.

Laing, D., Lanning, G. and Winchester, J. (2014) 'Resource management – Paki (no 2) – The Supreme Court rules on riverbed ownership', Simpson Frierson, Wellington and Christchurch, New Zealand, http://simpsongrierson.com/resource-management-supreme-court-riverbed-ownership/, accessed 8 March 2015.

Lake Omapere (1929) 2 Bay of Islands MB 253.

Land Air Water Association v. Waikato Regional Council [2001] 7 NZED 26.

Local Government Act 2002 (N.Z).

Lord v. The Commissioners for the City of Sydney (1859) 14 ER 991 (PC).
Maori Fisheries Act 2004 (N.Z.).
Maori Language Act 1987 (N.Z).
Mead, H. M. (2003) *Tikanga Maori: Living by Maori Values*, Huia Publishers, Wellington, New Zealand.
Ministry for the Environment (2014) 'National Policy Statement for Freshwater Management 2014', Ministry for the Environment, Wellington, New Zealand.
Nelken, D. and Feest, J. (2001) *Adapting Legal Cultures*, Hart Publishing, Oxford, UK.
New Zealand Law Commission (2001) *Maori Custom and Values in New Zealand Law*, Study paper 9, New Zealand Law Commission, Wellington, New Zealand.
New Zealand Maori Council v. Attorney-General [1987] 1 NZLR 641.
New Zealand Maori Council v. Attorney-General [1994] 1 NZLR 513 (PC).
New Zealand Maori Council v. Attorney-General [2013] NZSC 6.
Ngati Apa v. Attorney-General [2003] 3 NZLR 643.
Ngati Rangi Trust v. Genesis Power Limited [2007] NZCA 378.
Ngati Rangi Trust v. Genesis (2009) 15 ELRNZ 164.
Ngati Rangi Trust v. Manawatu-Wanganui Regional Council (18 May 2004) Environment Court, Auckland, New Zealand, A67/2004.
Ngati Tuwharetoa, Raukawa, Te Arawa Iwi Waikato River Act 2010 (N.Z).
Nga Wai o Maniapoto (Waipa River) Act 2012 (N.Z).
Nolan, D. (ed.) (2014) *Environmental and Resource Management Law*, 5th edn, LexisNexis, Wellington, New Zealand.
Paki v. Attorney-General [2009] 1 NZLR 72.
Paki v. Attorney-General [2009] NZCA 584, [2011] 1 NZLR 125.
Paki v. Attorney-General [2012] NZCS 50, [2012] 3 NZLR 277.
Palmer, K. (2010) 'Reflections on the history and role of the Environment Court in New Zealand', *Environmental and Planning Law Journal*, 27, 1, 69–79.
Public Finance (Mixed Ownership Model) Amendment Act 2012 Treaty of Waitangi Act 1975 (N.Z).
Raukawa Settelement Trust and Taupo District Council (2013) 'Joint Management Agreement', http://taupodc.govt.nz/our-council/policies-plans-and-bylaws/joint-management-agreements/Documents/Raukawa-JMA.pdf, accessed 8 March 2015.
Raukawa Settlement Trust and Waikato Regional Council (2012) 'Joint Management Agreement', http://waikatoregion.govt.nz/PageFiles/15805/JMAs/Raukawa%20JMA%202201886.pdf, accessed 8 March 2015.
Te Arawa River Iwi Trust and Waikato Regional Council (2012) 'Joint Management Agreement', http://waikatoregion.govt.nz/PageFiles/15805/JMAs/2249272%20Te%20Arawa.pdf, accessed 8 March 2015.
Resource Management Act 1991 (N.Z).
Resource Management Amendment Act 2003 (N.Z).
Ruru, J. (ed.) (2008) *In Good Faith: Proceedings Marking the 20th Anniversary of the Lands Case*, New Zealand Law Foundation, Wellington, New Zealand.
Ruru, J. (2010) 'Undefined and unresolved: Exploring indigenous rights in New Zealand's freshwater legal regime', *Journal of Water Law*, 20, 5–6, 236–42.
Ruru, J. (2012) "Maori rights in water – the Waitangi Tribunal's interim report", *Maori Law Review*, September, 8–18.
Ruru, J. (2013) 'Indigenous restitution in settling water claims: The developing cultural and commercial redress opportunities in Aotearoa, New Zealand', *Pacific Rim Law & Policy Journal*, 22, 2, 311–52.

Scott, K. (2006) 'From the lakes to the oceans: Reforming water resource management regimes in New Zealand', *Journal of Water Law*, 17, 231–45.

State-Owned Enterprises Amendment Act 2012 (Mighty River Power Limited) Commencement Order 2013 (N.Z).

Stephens, M. and Boyce, M. (2013) *He Papakupu Reo Ture: A Dictionary of Maori Legal Terms*, LexisNexis, Wellington, New Zealand.

Te Aho, L. (2010) 'Indigenous challenges to enhance freshwater governance and management in New Zealand – the Waikato River settlement', *Journal of Water Law*, 20, 5, 285–92.

Te Aho, L. (2012) 'Ngā whakatunga waimāori: Freshwater settlements', in N. R. Wheen and J. Hayward (eds) *Treaty of Waitangi Settlements*, Bridget Williams Books, Wellington, New Zealand.

Te Arawa Lakes Settlement Act 2006 (N.Z).

Tohe, P. (1998) 'Maori jurisprudence: The neglect of Tapu', *Auckland University Law Review*, 8, 3, 884–92.

Waikato Raupatu River Trust and Waikato District Council (2010) 'Joint Management Agreement', http://waikatodistrict.govt.nz/CMSFiles/37/37ce0ceb-c0c5–4e0d-bbe4–62c14d10eb65.pdf, accessed 8 March 2015.

Waikato Raupatu River Trust and Waikato Regional Council (2012) 'Co-Management Agreement For Waikato River Related Lands', Hamilton, New Zealand, http://waikatoregion.govt.nz/PageFiles/15805/JMAs/2276497%20Tainui.pdf, accessed 8 March 2015.

Waikato-Tainui Raupatu Claims (Waikato River) Settlement Act 2010 (N.Z).

Waitangi Tribunal (2012) *The Interim Report on the National Freshwater and Geothermal Resources Claim*, Wai 2358, Wellington, New Zealand.

Williams, H. W. (1971) *A Dictionary of the Maori Language*, Government Printer, Wellington, New Zealand.

Part IV

Emerging challenges in trans-jurisdictional water law and governance

11 Diffuse source pollution and water quality law for the Great Barrier Reef

Why the reticence to regulate?

Marie Waschka and Alex Gardner

Introduction

Water pollution from diffuse (non-point) sources is the 'unfinished business of water quality regulation in Australia' (Nelson, 2011, p. 341). Why? Water quality is widely acknowledged as a critical water resource issue in Australia, but it has been a second-order priority on the national water reform agenda. Instead, priority has been given to establishing regulatory frameworks for the management of water quantity and its allocation under the National Water Initiative (NWI) (Council of Australian Governments (COAG), 2004), while 'the polycentric complexity of water quality management has stymied both State and Commonwealth government action' (Gardner, 2012, p. 269). To date, the regulation of water quality in Australia has been 'strewn over the legal and bureaucratic landscape or beyond its horizon' (Gardner, 2012, p. 269). Although the regulation of point source or 'end-of-pipe' pollution by state governments has largely been successful, water pollution from diffuse sources remains largely unregulated.

So, why the reticence by Australian governments to regulate diffuse water pollution? The nature of diffuse water pollution presents a number of challenges to the regulator. It is difficult to identify, measure and attribute (Campbell *et al.*, 2004; Gunningham and Sinclair, 2005a; Howarth, 2011). It is cumulative in nature and, while it may not be a problem at a property or regional level, it can accumulate at a landscape scale to create significant water quality impacts. This is where it presents a trans-jurisdictional water governance challenge; for example, where catchments extend across state boundaries or into marine areas, which in Australia are typically managed by the federal government. In these instances, the need to navigate the responsibilities and political agendas of multiple governments and agencies compounds the challenge of regulating diffuse water pollution.

This chapter explores these issues using the Great Barrier Reef (GBR) and its adjacent catchments as a case study. In the GBR, diffuse water pollution presents the largest threat to water quality and murky challenges for trans-jurisdictional water governance. Despite decades of cooperation between the

Queensland and Commonwealth governments investing in voluntary and regulatory mechanisms to address this problem, GBR water quality continues to decline (State of Queensland, 2013a, p. 1).

Our aim is to outline the basis of a regulatory framework that can address these trans-jurisdictional water governance challenges to reverse the water quality decline. We first define the problem of diffuse source pollution and provide an overview of 'regulation' in the Australian constitutional setting. Second, we review regulatory approaches to diffuse water pollution in a trans-jurisdictional setting, including the use of total maximum pollutant loads (TMPLs). Third, we consider the GBR case study and why there is a reticence to regulate diffuse source pollution. We argue that 'greater scientific understanding and measurement of "catchment to reef" processes' (State of Queensland, 2013a, p. 1) can provide the impetus for developing a 'TMPL' approach as the basis for an effective regulatory framework that supports regulatory and non-regulatory instruments to reduce diffuse source pollution of the GBR. The chapter concludes with a summary of the key arguments in support of cooperative state and Commonwealth legislation to mandate the adoption of the TMPL approach as the reticence to regulate recedes.

Defining the problem and 'regulation' in the Australian constitutional setting

Diffuse water pollution

Diffuse water pollution is defined as 'all sources of pollution that enter waters other than from identifiable entry points'; it can 'encompass contaminants that enter waters through surface water runoff or by percolation through soil, or wherever the point of entry cannot be precisely located' (Howarth, 2011, p. 129).

In comparison to point source pollution, there are three key characteristics of diffuse water pollution that make it difficult to regulate. First, diffuse water pollution is difficult to identify, measure and attribute, including at the single property scale (Campbell *et al.*, 2004; Gunningham and Sinclair, 2005a; Howarth, 2011). The movement of pollutants from single properties or areas into waterways (including groundwater) will depend on factors such as landform and land use, making comparable measurement and attribution of pollution between properties very difficult. In some catchments, there may also be a range of land-use activities concentrated in different parts of a catchment that contribute to diffuse source pollution in different ways that exacerbate problems of attribution. Second, the impacts of diffuse water pollution tend to be cumulative in nature, with significant impacts accumulating only on a landscape scale. Although the runoff of pollutants from one property may not significantly impact on the quality of water in a water course, the cumulative impacts of diffuse water pollution from many sources can be very significant. These two characteristics are exacerbated by a third – the episodic variability of the pollution load. The nature and rate of pollution emissions can be influenced by severe rainfall events that bring sudden high pollution loads and supplement regular sources of emissions with other sources.

In Australia, diffuse water pollution from agriculture poses a particular threat to water quality. It is estimated that 'in some jurisdictions, agriculture alone is now responsible for the majority of all river pollution' (Gunningham and Sinclair, 2005b, p. 221). Agriculture can be broadly defined to include cropping and pastoral grazing activities of variable intensity associated with raising livestock for food and fibre production. Agricultural and pastoral grazing activities, such as 'the application of fertilizers and pesticides, tillage practices, habitat alteration . . . and soil erosion', can all contribute to diffuse water pollution (Gunningham and Sinclair, 2005a, p. 52). These activities are widespread around Australia and impact on numerous riverine, groundwater, estuarine and coastal water systems. Diffuse water pollution 'enters waterways through runoff, infiltration, drainage, seepage, atmospheric deposition and leaching processes' (Australian Government, 2002, p. 15). However, the rate and nature of emissions or runoff of pollutants from agricultural and pastoral land can vary greatly based on climatic, geographic and geological factors, including the connectivity between surface and groundwaters (Campbell *et al.*, 2004). Understanding these hydrogeological interactions is important when considering which management responses are most appropriate in particular areas. These physical factors are further exacerbated by the complex economic and social circumstances of many agricultural landholders limiting their capacity to undertake management solutions. Addressing agricultural diffuse source pollution in Australia is a complex regulatory challenge of great political sensitivity.

Regulation

Traditionally, 'regulation' was used to describe governmental action supported by legislation to compel a change of behaviour permitted by inadequate common law principles. That view is giving way to an understanding that mandatory governmental controls may not be the best solution to a societal problem, so government may, instead, adopt policies aimed at inducing beneficial voluntary behaviour (Kennedy, 2010; Bates, 2013, paras. 2.43 and 3.5).

For contemporary regulatory theorists, 'regulation' encompasses all three of these spheres of societal influence (common law, legislation and policy inducing voluntary behaviour) and more. It can include non-legal mechanisms and non-state regulatory activity for the achievement of societal outcomes through behaviour modification (Black, 2002; Frieburg, 2010).

It is similar to 'governance', which may be defined as the totality of mechanisms and instruments available for influencing social change in certain directions (Fisher, 2006, p. 6). Governance may also involve broader arrangements that engage greater community and stakeholder involvement in decentralized decision making of government agencies (Holley and Sinclair, 2013, pp. 33–4).

Australian Governments have adopted narrower state-centred definitions of regulation in their regulatory impact assessment procedures. For example, the Australian Government Guide to Regulation 2014 says '[r]egulation is any rule endorsed by government where there is an expectation of compliance'. We adopt this governmental perspective on 'regulation'. This is not to deny the benefits of decentred (i.e. non-governmental) regulatory activities but, as discussed below, it has been shown before that non-legal voluntarism and education do not solve diffuse source pollution problems (Gunningham and Sinclair, 2004a, b; Holley and Sinclair, 2013). Yet, governments are reticent to adopt regulatory measures with an expectation of compliance to address diffuse source pollution from agriculture.

Such regulation may well be required because the common law generally does not adequately address the problem of cumulative diffuse source pollution. A brief analysis of common law riparian rights to water quality illustrates this. The classic statement of this property right is that:

> Every riparian proprietor [the owner of the bed and banks of a watercourse or wetland] is thus entitled to the water of his stream, in its natural flow, without sensible diminution or increase and without sensible alteration of its character or quality.
>
> (*John Young v. Bankier Distillery Ltd* [1893] AC 691 at 698, per Lord Macnaughton)

This definition of riparian rights has been adopted in Australia in *Jones v. Kingborough Corporation* (at 298–9; Gardner *et al.*, 2009, p, 163). A breach of these rights can be remedied by damages or injunction. While the common law rights in respect of water quantity have been largely overtaken by legislative reform, it is generally thought that the common law rights to water quality continue (Gardner *et al.*, 2009, para.10.13).

The three reasons identified above as making diffuse source pollution difficult to regulate also frustrate the application of this common law principle to diffuse source pollution: it is difficult to measure and attribute causation of the water quality decline. The cumulative impacts are really only felt at landscape scale and occur with the episodic variability of water flows. The common law also has rules of flood control and land drainage, which essentially permit the landholders to act reasonably to protect their own interests subject to the interaction with the rights to undiminished water quality (Gardner and Setter, 1998). Often, however, these problems of water flow and diffuse source pollution are seen as affecting public rather than the private rights protected by the common law of property and torts. Even if the private rights of many are affected and a class action in court is contemplated, it can be too difficult to take proceedings against many defendants with variable circumstances.

Tragically, the common law duties do not protect the public interest against private landowners degrading their own land or otherwise acting in ways that cumulatively have an adverse effect on the public interest in environmental sustainability (Gardner, 1998, p. 34). It is unlikely that the common law doctrine protecting water quality would be of much utility in addressing diffuse source pollution of the GBR. Common law judicial process has a limited capacity to address complex polycentric problems compared with executive government regulation. The executive has better resources and processes to tackle large and complex problems where the actions of many adopting similar practices cause cumulative adverse impacts on the interests of many others in disparate ways over a long period of time. So, why is there the reticence to regulate?

Constitutional setting

The constitutional setting for State, Territory and Commonwealth regulation of water resources, including of diffuse source pollution from agricultural and urban land use, has been discussed in a number of places and needs little repetition here (Wulf and Baird, 2006; Gardner *et al.*, 2009; Kildea and Williams, 2010; Gardner, 2012; Rothwell, 2012). The essential propositions are simple to summarize.

The primary constitutional authority to regulate terrestrial diffuse source pollution from agricultural and urban land use undoubtedly lies with the States and Territories. However, the Commonwealth has powers to supplement or override State and Territory laws by regulating activities that involve the use of or impacts upon water resources, including from diffuse source pollution. In particular, the Commonwealth Parliament may legislate to regulate directly such activities by foreign, trading and financial corporations; to regulate indirectly such activities where the products become the subject of interstate or overseas trade and commerce; and to legislate to implement the terms of international conventions aimed at the protection of biodiversity, wetlands of international importance and the offshore marine environment. The Commonwealth can also fund various activities by Commonwealth and State agencies and ordinary people within the scope of its legislative powers, which would capture most types of activities that could have a beneficial effect on diffuse source water pollution. There are limitations on the exercise of these powers, most notably constraints on State or Commonwealth legislation that disproportionately restricts freedom of interstate trade and commerce, and the s. 100 restriction on the Commonwealth's power to legislate with respect to trade and commerce that would abridge the right of a state or the residents therein to the reasonable use of the waters of the rivers for conservation or irrigation.

The recent experience of the Commonwealth's legislation for regulating impacts of land and water use activities on water resources shows that the Commonwealth could, if it chose, legislate to limit activities that adversely impact water quality, including land-use activities that cause diffuse source pollution. The Water Act 2007 (Cth) authorized the Murray-Darling Basin (MDB) Plan to set binding limits on the States' water resources quantity management based mainly on the Commonwealth's power to implement international treaty obligations (Rothwell, 2012). More recently, the Commonwealth has amended the Environment Protection and Biodiversity Conservation Act 1999 (Cth) (EPBC Act) to insert the 'water trigger' that requires Commonwealth environmental approval for coal seam gas development or large coal-mining development that has or is likely to have a significant impact on a water resource (EPBC Act ss. 24D, 24E and 528; Bice and Gilmour, 2013). The legislation clearly bases the trigger on the Commonwealth's legislative powers with respect to interstate and overseas trade and commerce, constitutional corporations and the territories' power.

The qualification to this picture is that certain elements of the Water Act regime have relied on a referral of State powers to the Commonwealth Parliament to support the full scope of that legislation; namely, in respect of 'referred subject matters' defined in s. 18B(9). They pertain mainly to the powers and functions conferred on Commonwealth agencies for the routine management of the physical assets and works that 'regulate' the river system, and the rights and liabilities in respect of those works, all of which were previously managed by the MDB States. This qualification indicates that the Commonwealth could not seek, in exercising its powers alone, to manage directly the physical natural resources by the operation of works and the administration of the State-based entitlement regimes relating to those natural resources. That does not deny the Commonwealth's authority to define limits on what the States authorize, but it may point to the limits on Commonwealth capacity to enforce those limits without State help.

Regulatory approaches in a trans-jurisdictional setting

Options for managing diffuse water pollution can be defined on the basis of legal enforceability as ranging from voluntary (e.g. education, training, information), at one end of the spectrum, to regulatory (legally binding) at the other, with self-regulation (e.g. industry codes) in between (Gunningham and Sinclair, 2005a). These approaches, and factors influencing their success, are as complex as the issue of diffuse water pollution itself, particularly in large interconnected systems such as the GBR, where there are trans-jurisdictional responsibilities for water governance. Specifically, the source of the diffuse pollution is land use within the territory of Queensland, governed mainly by State legislation, and the primary consequences are on the water quality of the GBR, which is mainly within the Commonwealth's jurisdiction. It is acknowledged that countries with federal systems, such as Australia, have the 'added complexity of finding the proper balance of national and state or provincial authority to control nonpoint source pollution most effectively, particularly when watersheds transcend state boundaries' (Graham *et al.*, 2011, p. 53).

There is evidence that, to be effective, policy and regulatory frameworks for diffuse water pollution must be 'purpose-built' and context specific, including a range of appropriate measures and instruments (voluntary, economic, regulatory and planning), and that no single approach will work in all situations (Chittock and Hughey, 2011, p. 543). The discussion of these approaches elsewhere by Gunningham and Sinclair (2005a, b), Chittock and Hughey (2011) and Suter *et al.* (2010) does not need repeating here. For our purposes, the key challenge is national trans-jurisdictional agreement on broad water quality objectives and targets to guide management and ensure that effective policy and regulatory tools are adequately implemented and enforced.

Australia does have a National Water Quality Management Strategy (NWQMS) (Australian Government, undated). It is an evolving set of guidelines, developed since 1992, that provides guidance on how water quality may be protected through decision making under State, Territory and Commonwealth laws that govern activities that may impact on water quality. However, the NWQMS neither sets national water quality targets nor mandates action to do so at state, regional or local levels.

For State action, two basic propositions may be distilled from the relevant NWQMS guidelines: the Fresh and Marine Water Quality Guidelines 2000 and the Framework for Marine and Estuarine Water Quality Protection 2002 (Australian Government, undated). First, it is for the States to develop a water quality planning framework and to use their own legal frameworks to make and implement those plans. There is no prescription as to what those legal frameworks should be or that the plan measures be implemented by binding regulatory tools. Second, in making the water quality plans, a number of factors should be considered, including the TMPLs that a water body can receive while still meeting the plan's objectives.

Establishing TMPLs for pollution regulation represents a shift in thinking about water quality targets, from simple measures of ambient concentrations of individual target substances in receiving waters at particular places and times to a measure that is a function of total flow and concentration of target substances from the whole catchment over a period of time (Nelson, 2011, p. 355; White, 2012, p. 16). As noted in the Framework for Marine and Estuarine Water Quality Protection Reference Document, 'in many instances, loads may give a better indication of the total amount of pollutants entering a water body than concentrations' (Australian Government, 2002, p. 13). TMPLs focus on the entire ecological health of a waterway and its capacity for assimilating pollutants from *all* sources (point, diffuse and natural), and loads also take into account pollutants that are delivered in 'pulses' or episodic rainfall and runoff events, which tend to be characteristic of diffuse water pollution (Australian Government, 2002).

TMPLs provide an indication of the reduction of pollutant loads required to meet the acceptable level for the water quality objectives. The next task is tricky: how to allocate the burden of achieving the pollutant load reductions. This raises sensitive issues of equity and relative costs of attaining the reductions.

What process should be adopted for allocating these burdens, taking into account established uses and legal rights? These allocative issues have confronted numerous other environmental regulatory challenges, and there are numerous examples to draw upon – not least, reductions in allocations under water access rights. In the case of water quality protection, the point sources have been relatively well tackled with regulatory measures but not in a way that is integrated with diffuse source pollution.

Although the TMPL approach has not been widely adopted in Australia, it is a key element of the United States' federal Clean Water Act 1972 (CWA) (referred to under the CWA as total maximum daily loads or TMDLs). Diffuse water pollution is the 'leading water quality problem' in the US and is considered the major obstacle in achieving water quality goals (Tobin, 2003, p. 808). The approach taken by the US towards managing diffuse water pollution is of relevance to Australia, because opportunities for addressing diffuse water pollution are influenced by similar federal water governance arrangements in both countries by which state governments have primary responsibility for natural resource management and land use regulation (Nelson, 2011, p.342). Unlike Australia, however, the US Federal Government has a major role in regulating water pollution through the federal CWA. The CWA provides a regulatory structure for managing both point source and diffuse source water pollution while preserving the roles of the states in water pollution control, particularly in relation to land use that contributes to diffuse source pollution (Nelson, 2011).

The establishment of TMDLs under the CWA involves establishing (for individual waterways) a 'maximum amount of a pollutant that can be "loaded" into the waters in question from all sources – point sources, nonpoint sources, and natural background sources' – before the health of that waterway is compromised (Nelson, 2011, p. 355). The TMDLs are developed by the states for degraded waterways and then submitted to the federal government's Environmental Protection Authority (EPA) for approval. They are a goal-setting mechanism intended to influence state planning processes and do not regulate diffuse pollution directly. However, TMDLs influence permit conditions for point source pollution, so 'including nonpoint pollutant loads in TMDLs means that point sources may be subject to more stringent requirements to compensate for the lack of regulation of nonpoint sources' (Nelson, 2011, p. 356). While the US federal government can require the states to develop TMDLs or even develop them on the behalf of states, they do not have the statutory power to enforce their implementation. The US Federal Government can only motivate states' participation by providing funding incentives (federal grant money) to compliant states (Nelson, 2011, p. 356). Although there are many flaws to the CWA approach, the CWA has provided the USA with a national, coordinated regulatory structure to address water pollution, including from diffuse sources. Australian jurisdictions could consider a similar approach to strengthen the regulatory frameworks for water quality.

The key difference between the US and the Australian approach is the lack of any Australian national legal mandate to set ambient water quality standards and to develop TMPLs. The National Environment Protection Council Act 1996 (Cth) and complementary State and Territory legislation authorizes the National Environment Protection Council to make 'national environment protection measures' for 'ambient marine, estuarine and freshwater quality' (National Environment Protection Council Act 1996 (Cth) s 14(1)(b)). The Council (so named) has not functioned under the statute for many years, and the most recent embodiment of a national ministerial council charged with the relevant statutory functions, the Standing Council on Environment and Water, was recently abolished (Council of Australian Governments 2013). No ambient water quality standards have been made under that legislation. Instead, Australia has relied on the non-statutory NWQMS, including the Fresh and Marine Water Quality Guidelines 2000 and the Framework for Marine and Estuarine Water Quality Protection 2002 mentioned above.

The NWQMS has recently been reviewed and recommended for continuation with significant reforms. A key issue identified in the 2011 final review report was that historical weakness in the links between water quality and water quantity in the key national water policies of the past 20 years has diminished the relevance of the NWQMS (Australian Government, 2011, pp. 17–20). In particular, the review found that there was a lack of a clear policy objective for the NWQMS; a need for clearer definition of roles for all parties, especially the Commonwealth Government; a more efficient process for review and approval of the guidelines; a greater capacity to address emerging issues; and a better process for reporting against water quality, policy and efficiency objectives (Australian Government, 2011, pp. 29–34). It is notable that the report highlighted that application of the NWQMS was seen as more effective in urban contexts, where there are more secure regulatory foundations for its implementation, and weaker in regional areas, particularly in respect of diffuse source pollution from farming and agricultural industries where there was scope for greater adoption of the guidelines (Australian Government, 2011, pp. 37–8). The report proposed a model of reform that would give the Commonwealth Government a greater role in a revised NWQMS (Australian Government, 2011, p. 46). In other words, there is still a national (Commonwealth and State) reticence to regulate diffuse water pollution.

It is time for Australia to reconsider the non-regulatory approach. The GBR case study explores the usefulness of the TMPL approach further, along with other regulatory tools for addressing diffuse water pollution.

A case study for regulation of diffuse water pollution: the GBR

Stretching 2,300 km along the Queensland coast, and covering an area of over 360,000 km^2 in total (Hutchings *et al.*, 2008, p. 1), the GBR is characterized by complex interdependencies between species and environments, including with its adjacent terrestrial catchments. The GBR is linked to, and receives runoff from, 38 major catchment areas that cover an area of land of approximately 424,000 km^2 in total (Great Barrier Reef Marine Park Authority (GBRMPA), 2009, p. 11). Under natural conditions, the runoff of freshwater and sediments from adjacent catchment areas plays an important role in supporting the productivity and growth of coral reefs (Furnas, 2003, p. 41). However, with vast modification of GBR catchments over the last 150 years, there has been at least a fourfold increase in the levels of sediment, nitrogen and phosphorus being discharged into the GBR lagoon (State of Queensland, 2007, pp. 6–7). Diffuse water pollution from agriculture in GBR catchments is the main cause of nutrients, fine sediments and pesticides impacting on GBR marine water quality, and poor water quality is the major cause of the declining health of GBR ecosystems (State of Queensland, 2013a, p. 1).

The Queensland and Commonwealth Governments have worked in partnership over decades to manage the GBR. This partnership is, legally, based on an agreement to 'establish joint consultative arrangements for the management and preservation' of the GBR under the 1979 Offshore Constitutional Settlement (Australian Government, 1980, p. 11), which was formalized in 1979 under the 'Emerald Agreement' and more recently by the 2009 Great Barrier Reef Intergovernmental Agreement (Commonwealth of Australia and State of Queensland, 2009). The Queensland Government has responsibility for managing the GBR catchments and their water resources, and its agencies deliver most of the day-to-day operational management of the GBR Marine Park (Osborne, 2011). The Commonwealth Government's management powers are enabled by the Great Barrier Reef Marine Park Act 1975 (Cth) (GBRMP Act). Among other provisions, the GBRMP Act provides for the establishment of this Marine Park and the Great Barrier Reef Marine Park Authority (GBRMP Authority), which is its key management agency. This Act also prohibits 'acts (whether in the Marine Park or elsewhere) that may pollute water in a manner harmful to animals and plants in the Marine Park' (GBRMP Act, s. 66 (2)(e)). The Commonwealth Government has used this provision to regulate the discharge of waste from aquaculture (Great Barrier Reef Marine Park (Aquaculture) Regulations 2000 (Cth)); however, it has not yet extended this regulation to diffuse water pollution from agriculture impacting on the GBR (Brodie, 2000). Its reluctance to do so is most likely due to perceptions of constitutional limitations facing the Commonwealth in relation to land-use management and regulation. The Commonwealth also has powers under the EPBC Act to 'provide for the protection of the environment, especially those aspects of the environment that are matters of national environmental significance' (EPBC Act s. 3(1)(a)–(g)). The GBR Marine Park was listed as a matter of national environmental significance in 2009, which means that any action that is likely to have a significant impact on the GBR Marine Park is subject to the approval processes under the EPBC Act (*Minister for the Environment and Heritage (Cth) v. Queensland Conservation Council – Nathan Dam Case* (2004) 139 FCR 24).

The primary instrument guiding the Commonwealth and Queensland Governments' programme of activities for improving the water quality of the GBR is the Reef Water Quality Protection Plan ('Reef Plan'), which was originally published in 2003 and updated in 2009 and 2013. The Reef Plan 'is a collaborative programme of coordinated projects and partnerships designed to improve the quality of water in the GBR' through improved land management in reef catchments (State of Queensland, 2013b, p. 10). The primary focus of the Reef Plan is diffuse source pollution from agriculture. The 2013 Reef Plan has a long-term goal: 'that by 2020 the quality of water entering the reef from broad scale land use has no detrimental impact on the health and resilience of the GBR' (State of Queensland, 2013b, p. 12). It states a number of water quality and catchment management targets. The 2013 Reef Plan adopted pollutant reduction load targets for end-of-catchment loads of nitrogen, sediment and pesticides based mainly on what was expected to be achievable with delivery of best management practices for land use. But, it conceded, there was not then sufficient knowledge to set maximum pollutant loads or the consequent levels of load reductions required to achieve reef-wide water quality guidelines for reef health. However, that knowledge was expected to become available within the 5-year life of the 2013 Reef Plan, which points to the possibility of moving towards the setting of 'ecologically based targets' that could use TMPLs for the GBR as the basis of future Reef Plans (State of Queensland, 2013b, pp. 18–19).

In late 2013, the Commonwealth Government announced its intention to build on the 2013 Reef Plan by developing a Reef 2050 Long-Term Sustainability Plan (2050 Reef Plan), the development of which is supported by numerous documents to which we will not refer. The 2050 Reef Plan is intended to be an 'overarching framework for protecting and managing the Great Barrier Reef from 2015 to 2050' (Australian Government, 2014, p. 16) and will become a schedule to The Great Barrier Reef Intergovernmental Agreement 2009. Whereas the 2013 Reef Plan focuses on addressing diffuse water pollution from agriculture, the draft 2050 Reef Plan also addresses water pollution from other sources, including port developments. The draft 2050 Reef Plan states that 'further catchment water quality improvements, both for point and non-point sources of pollution, will require innovative and cost-effective measures and mechanisms, for example market-based nutrient trading' (Australian Government, 2014, p. 21). It adopts the 2013 Reef Plan end of catchment pollutant load reduction targets for 2018, which arguably are a form of TMPL. Significantly, it also adds a new action: to develop reef-wide and locally relevant water quality targets for urban, industrial and port activities. Establishing TMPLs for the GBR catchments and the lagoon would be consistent with these targets, and with the framework approach to managing the GBR that the 2050 Reef Plan aims to achieve. A TMPL target for the GBR would also take the further step of trans-jurisdictional accounting for total pollution from all sources, considering the ecological health and pollution assimilation capacity of the whole GBR. Disappointingly, the 2050 Reef Plan expresses support for only *voluntary* trading programmes (along with other mechanisms) for reducing point and diffuse sources of pollution. Critically, TMPLs would provide a basis for coherent management of agricultural, urban, industrial and port activities and for using pollution trading programmes on a regulatory, not just a voluntary, basis.

In this regard, there is a State regulatory experience to draw upon. In 2009, the Queensland Government introduced the Great Barrier Reef Protection Amendment Act 2009 (Reef Protection Amendment Act) adopting, for the first time (McGrath, 2010, p. 216), regulatory measures to 'reduce the impact of agricultural activities on the quality of water entering the reef' (Reef Protection Amendment Act s. 6). The need for this regulatory tool emerged from the recognition that the range of voluntary activities undertaken under the 2003 Reef Plan had 'not been effective in solving the issue of declining water quality in the Reef' (State of Queensland, 2009, p. 5). This assessment is consistent with findings from other case studies that voluntary policy mechanisms alone are unlikely to be effective in addressing the issue of diffuse water pollution (Brull, 2006, p. 221; Nelson, 2011), and that reliance on voluntary measures alone is a 'manifestly unsuccessful' approach (Gunningham and Sinclair, 2005a, p. 80).

The Reef Protection Legislation requires (among other things) landholders who conduct certain land-use activities in identified priority catchments to complete 'Environmental Risk Management Plans' (ERMPs). Such plans must outline, first, actions for reducing runoff of sediment and nutrients at a farm level; and, second, measures for soil testing and reporting on nutrient inputs to ensure that optimal levels of nutrient application are not exceeded (Waschka and Gardner, 2012). In 2012, a new State government under Premier Newman suspended implementation of the Reef Protection Legislation in favour of an industry-led, best practice management approach combined with a suite of other measures outlined in the Reef Plan 2013. Why did the Newman government revert to a non-regulatory approach?

On one measure, the Reef Protection Legislation was successful. In 2012, a Reef Protection Package Impact Statement noted that there had been a high level of compliance with the Reef Protection Legislation, with submission of over 500 ERMPs. It found that the legislation was likely to have permanently reduced fertilizer nitrogen runoff from farms by at least 14% (State of Queensland, 2012, p. 2). Furthermore, the Reef Protection Legislation provides a regulatory tool to complement existing voluntary actions and incentives under the Reef Plan. It focuses on priority catchments and land uses with the largest impact on water quality, while providing the flexibility to expand these areas and activities at a later date (Waschka and Gardner, 2012). By regulating the input of nutrients, it provides a simple means of reporting on nutrient reductions and overcomes some of the challenges associated with attribution of runoff (Waschka and Gardner, 2012).

Perhaps the regulatory measures were successful. A GBR Report Card for 2012 and 2013, which reported on the impact of *all* Reef Plan activities from a 2009 baseline, documented a 'significant reduction' in annual average pollutant loads entering the reef, with reductions in average nitrogen, sediment and pesticide loads (State of Queensland, 2014, p. 1). It also reported 'major progress in adopting improved land management practices across the GBR catchment' from the 2009 baseline until June 2013. However, other than in the case of measuring reductions in nutrient inputs, it is difficult to attribute progress towards halting the decline of GBR water quality to any single action (regulatory or non-regulatory). Many actions are complementary and linked, and the momentum and capacity for undertaking action has built over many years.

Even so, why did the Newman Government suspend this relatively new regulatory approach in favour of an industry-led, voluntary, 'best management practice' approach? It was a political decision. The Reef Protection Legislation was very unpopular among landholders in GBR catchments and peak industry bodies who lobbied for its repeal during the 2012 State election. Industry bodies expressed concerns that the Reef Protection Legislation placed 'unnecessary regulatory burdens' on landholders, and that the regulations are 'excessive, complex and needlessly bureaucratic . . . overly complicated and difficult to understand by growers' (Canegrowers, 2012). The Newman Government supported the views of its constituency, promising to 'deliver, in partnership with the sugarcane and beef industries, a framework for a voluntary best management program to protect the Great Barrier Reef' (Queensland Government, 2012, p. 6). It did not enforce the Reef Protection Legislation but neither did it repeal it. Rather, the Newman Government described it as a 'reference point' that would remain until 'successful transition to BMP' (Queensland Government, 2013), though the criteria for that successful transition were not defined. The Government simply referred to the regulations as reflecting the 'minimum standard to protect the reef from pollutants', with the expectation that the BMPs will also 'cover aspects of farm management that focus on productivity and profitability' (Queensland Government, 2013).

It remains to be seen how effective this entirely voluntary scheme has been. There is some evidence that the 'threat' of a regulation (or other mandatory mechanism such as a tax) can induce adoption of voluntary programs and contribute to their success (Suter *et al.*, 2010). However, the lack of credible target setting as part of the BMP process, and the need for effective monitoring and reporting (Chittock and Hughey, 2011), raise questions about the potential effectiveness of this approach. It remains to be seen how the landholders' self-assessment of BMPs works in practice, what levels of adoption are achieved and how the impact of the BMPs is evaluated, monitored and reported by industry to government. Crucially, what difference will this approach make to improving GBR water quality? Both the 2012–2013 GBR Report Card and the *2013 Scientific Consensus Statement: Land use impacts on Great Barrier Reef water quality and ecosystem condition* found that water quality in the GBR was continuing to decline, that diffuse source pollution from agriculture continued to be the main source of excess nutrients and sediment, and that this situation had been exacerbated by recent extreme weather events (flooding, heavy rainfall and cyclones) in Queensland (State of Queensland, 2013a, p. 1). These findings are consistent with the Great Barrier Reef Outlook Report 2014, which includes poor water quality from diffuse water pollution as one of the main threats to the GBR (Great Barrier Reef Marine Park Authority, 2014, p. v).

The Reef Protection Legislation, 2013 Reef Plan activities and the BMP process all focus on reducing diffuse water pollution from agriculture, which has the largest impact on GBR water quality (State of Queensland, 2013a, p. 1). However, as recognized in the 2050 Reef Plan, other activities such as coastal urban development and port developments are also impacting on GBR water quality. The impact of port developments and associated dredging and spoil dumping activities on GBR water quality have recently received international attention, particularly the Abbot Point coal port terminal expansion programme. The significance of the Abbot Point proposal is that regulatory decision making about a proposed programme of dredging and dumping of dredge spoil has linked the pollutant loads from agriculture with those from the port expansion. The conditions imposed on offshore dumping of dredge spoil by the Commonwealth Government included the requirement for an 'offsets plan' that required (among other things):

- actions that will result in a net benefit for the GBR World Heritage Area;
- 150% of the total amount of fine sediments, potentially available for re-suspension into the marine environment from the dredging and disposal activities, must be offset by a reduction in the load of fine sediments entering the marine environment from the Burdekin and Don catchments.

(Australian Government, 2013, pp. 7–8)

There was no detail about how this offset plan would be achieved, and it is uncertain whether similar conditions would be imposed on any approval of onshore dumping, which has recently been proposed as an alternative to offshore dumping (Queensland Government, 2015). There are obvious questions about how such an offset would be achieved. How will such offset 'credits' be secured or acquired? Does this condition pass responsibility for mitigating impacts of the dredging activities onto farmers by improved farm management practices in the adjacent catchment area (as argued by Brodie, 2014)? How will the achievement of the offsets be measured and reported under State law? How will these conditions be enforced under Commonwealth law?

This is where TMPLs and implementing regulatory tools become important. A TMPL approach towards setting water quality targets and a regulatory framework to achieve them, including through voluntary mechanisms, could be an effective way of ensuring that inputs of pollution from *all* sources are considered in the management of GBR water quality. Establishing TMPLs for the GBR catchments and the GBR lagoon would better inform decision making in the case of developments like Abbot Point. It would provide a framework for appropriate trans-jurisdictional management responses by focusing on the entire ecological health of the GBR and its capacity to assimilate pollution. In effect, a TMPL approach should be seen as a precondition to adopting an 'offsets' approach more broadly across the GBR. It could enable the adoption of tools for achieving load limits – for example, a 'cap-and-trade' system on the load of pollutants (from all sources) that are discharged onto the GBR, or regions of it, to ensure that the total pollutant load is not exceeded. This approach has been adopted in the Lake Tahoe area of the US, where an innovative cap-and-trade system exists that allows municipalities to trade 'pollution credits' among themselves in order to meet TMDLs established under the CWA (Nelson, 2011, p. 358). It is difficult to see how a pollution-trading programme for the GBR could successfully be pursued without a regulatory framework to support it. This is the vision that the Commonwealth Government's draft 2050 Reef Plan should be pursuing.

Conclusion

Why is there this reticence to regulate diffuse source water pollution? Regulation, as we have defined it for this chapter, is the core element of water governance. It is difficult to tackle diffuse source water pollution, whether by traditional common law or by contemporary regulation: it is difficult to identify, measure and attribute because it is accumulative and episodic in nature, both at source and at the distant site of the impacts. The causes are often the exercise of private rights for everyday productive activities, and the impacts are typically on public interests with dispersed and marginal costs on private interests. Private and government responses are costly. These regulatory difficulties are compounded in large, interconnected systems where there are trans-jurisdictional water governance responsibilities and sensitivities, such as in the GBR where the sources are mainly State responsibility and the site of the impacts mainly Commonwealth.

The State and Commonwealth governments have cooperated in significant investments over decades to encourage voluntary measures to reduce diffuse water pollution from agriculture. In 2009, continuing water quality decline led the Queensland Parliament to legislate regulatory tools to support a range of land-use management activities, monitoring and reporting. In 2012, a new Queensland Government responded to pressure from landholders to suspend enforcement of the regulatory measures in favour of industry-led best practice management. By 2013, there were significant reductions in pollutant loads from a 2009 baseline, though it is difficult to attribute this to any particular regulatory or voluntary actions and water quality continues to decline. However, the 2013 Reef Plan adopted end-of-catchment pollutant load reduction targets. Although these are acknowledged not to be at ecologically sustainable levels, the science is improving and these targets are the initial attempts at TMPLs. The 2050 Reef Plan endorses these catchment targets and proposes that reef-wide locally relevant targets for urban, industrial and port activities are adopted by 2020. By 2014, the international and domestic political concerns were driving the

Commonwealth to link the regulation of pollutant loads from port expansion to further reductions in catchment loads, but with an uncertain regulatory framework.

It is time to adopt a TMPL approach for both the catchments and the GBR lagoon as the basis of a trans-jurisdictional regulatory framework. It would provide for appropriate management responses by focusing on the entire ecological health of the GBR and linking the management of coastal and port developments with catchment management. While the Commonwealth Government could legislate to limit activities that adversely affect GBR water quality, it is likely that a cooperative TMPL approach will work better. In the face of mounting concern for the ecological health of the GBR, the reticence to regulate should recede and the State and Commonwealth governments should enact a cooperative regulatory framework founded on a TMPL approach that supports a range of regulatory and non-regulatory instruments to reduce diffuse source pollution and achieve sustainable water quality outcomes. This approach for the GBR could also become a blueprint for a broader national regulatory framework for water quality.

References

Australian Government (undated) 'National Water Quality Management Strategy', Department of the Environment, Canberra, Australia, http://environment.gov.au/topics/water/water-quality/national-water-quality-management-strategy, accessed 26 January 2015.

Australian Government (1980) 'Offshore Constitutional Settlement: A milestone in cooperative federalism', Attorney-General's Department, Canberra, Australia.

Australian Government (2002) 'The framework for marine and estuarine water quality protection: a reference document', Department of the Environment, Canberra, Australia, http://environment.gov.au/resource/framework-marine-and-estuarine-water-quality-protection-reference-document, accessed 2 December 2013.

Australian Government (2011) 'Evaluation of the National Water Quality Management Strategy, Final Report', Department of Sustainability, Environment, Water Population and Communities, Canberra, Australia, http:// environment.gov.au/system/files/resources/963919a7-725d-4c5b-9219-96da308d5297/files/nwqms-final-report.pdf, accessed 26 January 2015.

Australian Government (2013) 'Approval Abbot Point Terminal 0, Terminal 2 and Terminal 3 Capital Dredging, Queensland (EPBC 2011/6213)', Department of the Environment, Canberra, Australia, http://environment.gov.au/epbc/notices/assessments/2011/6213/2011-6213-approval-decision.pdf, accessed 15 December 2013.

Australian Government (2014) 'Reef 2050 Long-Term Sustainability Plan: For Comment', Department of the Environment, Canberra, Australia, http://environment.gov.au/system/files/consultations/8b8f5023-3cfb-4310-bc51-1136aa5d875a/files/reef-2050-long-term-sustainaiblity-plan.pdf, accessed 20 January 2015.

Bates, G. (2013) *Environmental Law in Australia*, 8th edn, LexisNexis Butterworths, Chatswood, Australia.

Bice, M. and Gilmour, T. (2013) 'Controversial EPBC Act water resource trigger passed by Parliament, as recommended by Senate Committee', *Australian Environment Review*, 28, 6, 616–19.

Black, J. (2002), 'Critical Reflections on Regulation', *Australian Journal of Legal Philosophy*, 27, 1–46.

Brodie, J. (2000) 'Keeping the wolf from the door: Managing land-based threats to the Great Barrier Reef', in M. K. Moosa, S. Soemodihardjo, A. Soegiarto, K. Romimohtarto, A. Nontji, Soekarno and Suharsono (eds) *Proceedings of the 9th International Coral Reef Symposium, Bali Indonesia 23–27 October 2000, Volume 2*, Bali, Indonesia.

Brodie J. (2014) 'Dredging the Great Barrier Reef: Use and misuse of science', *Estuarine, Coastal and Shelf Science*, 142, 1–3.

Brull, S. (2006) 'An evaluation of non-point source pollution regulation in the Chesapeake Bay', *University of Baltimore Journal of Environmental Law*, 13, 2, 221–49.

Campbell, N., D'Arcy, B. J., Frost, A., Novotny, V. and Sanson, A. (eds) (2004) *Diffuse Pollution: An Introduction to the Problems and Solutions*, IWA Publishing, London.

Canegrowers (2012) 'Letter to the Chair of the Agriculture, Resources and Environment Committee, Canegrowers', Canegrowers, Brisbane, Australia, http://parliament.qld.gov.au/documents/tableoffice/commsubs/2012/qldindustries/017.pdf, accessed 5 October 2013.

Chittock, D. and Hughey, K. (2011) 'A review of international practice in the design of voluntary pollution prevention programs', *Journal of Cleaner Production*, 19, 5, 542–51.

Clean Water Act 33 USC (1972).

Commonwealth of Australia (2008) 'Regional Delivery Model for the Natural Heritage Trust and the National Action Plan for Salinity and Water Quality', Audit Report 21, Auditor-General, Canberra, Australia.

Commonwealth of Australia and State of Queensland (2009) 'The Great Barrier Reef Intergovernmental Agreement 2009', http://environment.gov.au/system/files/pages/7a85531d-9086-4c22-bdca-282491321e46/files/gbr-iga.pdf, accessed 26 January 2015.

Council of Australian Governments (COAG) (2004) 'Intergovernmental Agreement on a National Water Initiative 2004', COAG, Barton, Australia, http://nwc.gov.au/nwi, accessed 26 January 2015.

Council of Australian Governments (2013) 'Meeting Outcomes, 13 December 2013', COAG, Barton, Australia, http://coag.gov.au/meeting_outcomes, accessed 1 February 2015.

Environment Protection and Biodiversity Conservation Act 1999 (Cth).

Fisher, D. (2006) 'Water resources governance and the law', *Australasian Journal of Natural Resources Law and Policy*, 11, 1, 1–41.

Frieburg, A. (2010) *The Tools of Regulation*, Federation Press, Annandale, Australia.

Furnas, M. (2003) 'Catchments and Corals: Terrestrial Runoff to the Great Barrier Reef', Australian Institute of Marine Science and CRC Reef Research Centre, Cairns and Townsville, Australia.

Gardner, A. (1998) 'The duty of care for sustainable land management', *The Australasian Journal of Natural Resources Law & Policy*, 5, 1, 29–63.

Gardner, A. (2012) 'Water reform and the federal system', in P. Kildea, G. Williams and A. Lynch (eds) *Tomorrow's Federation: Reforming Australian Government*, Federation Press, Annandale, Australia.

Gardner, A. and Setter, T. (1998), 'Legal powers and responsibilities for managing diffuse pollution from agriculture: regulating the flow of water – "the common enemy"', *The Australasian Journal of Natural Resources Law and Policy*, 5, 2, 187–224.

Gardner, A. Bartlett, R. and Gray, J. (2009) *Water Resources Law*, LexisNexis Butterworths, Chatswood, Australia.

Graham, S., Schempp, A. and Troell, J. (2011) 'Regulating nonpoint source water pollution in a federal government: four case studies', *Water Resources Development*, 27, 1, 53–69.

Great Barrier Reef Marine Park Act 1975 (Cth).

Great Barrier Reef Marine Park (Aquaculture) Regulations 2000 (Cth).

Great Barrier Reef Marine Park Authority (GBRMPA) (2009) 'Great Barrier Reef Outlook Report 2009: In Brief', Great Barrier Reef Marine Park Authority, Townsville, Australia.

Great Barrier Reef Marine Park Authority (2014) Great Barrier Reef Outlook Report 2014, GBRMPA, Townsville, Australia.

Great Barrier Reef Protection Amendment Act 2009 (Qld).

Gunningham, N. and Sinclair. D. (2004a) 'Non-point pollution, voluntarism and policy failure: lessons for the Swan-Canning', *Environmental and Planning Law Journal*, 21, 2, 93–104.

Gunningham, N. and Sinclair, D. (2004b) 'Curbing non-point pollution: lessons for the Swan-Canning', *Environmental and Planning Law Journal*, 21, 3, 181–99.

Gunningham, N. and Sinclair, D. (2005a) 'Policy instrument choice and diffuse source pollution', *Journal of Environmental Law*, 7, 1, 51–81.

Gunningham, N. and Sinclair, D. (2005b) 'Regulating intensive agricultural pollution', *Australasian Journal of Environmental Management*, 12, 4, 221–31.

Holley, C. and Sinclair, D. (2013) 'Deliberative participation, environmental law and collaborative governance: Insights from surface and groundwater studies', *Environmental and Planning Law Journal*, 30, 1, 32–55.

Howarth, W. (2011) 'Diffuse water pollution and diffuse environmental laws', *Journal of Environmental Law*, 23, 1, 129–41.

Hutchings, P., Kingsford, M. and Hoegh-Guldberg, O. (2008) *The Great Barrier Reef: Biology, Environment and Management*, CSIRO Publishing, Melbourne.

Jones v. Kingborough Corporation (1950) 82 CLR 282.

Kennedy, A. (2010) 'Using community-based social marketing techniques to enhance environmental regulation', *Sustainability*, 2, 4, 1138–60.

Kildea, P. and Williams, G. (2010) 'The Constitution and the management of water in Australia's rivers', *Sydney Law Review*, 32, 4, 595–616.

McGrath C. (2010) Does Environmental Law Work? How to evaluate the effectiveness of an environmental legal system, Lambert Academic Publishing, Saarbruckenen, Germany.

Minister for the Environment and Heritage (Cth) v. Queensland Conservation Council – Nathan Dam Case (2004) 139 FCR 24.

National Environment Protection Council Act 1996 (Cth).

Nelson, R. (2011) 'Regulating nonpoint source pollution in the US: a regulatory theory approach to lessons and research paths for Australia', *University of Western Australia Law Review*, 35, 2, 340–84.

Osborne, M. (2011) 'Commonwealth and State Marine Environmental Management', in R. Baird and D. R. Rothwell (eds) *Australian Coastal and Marine Law*, Federation Press, Annandale, Australia.

Queensland Conservation Council v. Minister for the Environment and Heritage [2003] FCA 1463.

Queensland Government (2012) 'Six Month Action Plan July-December 2012', Queensland Government, Rockhampton, Australia, http://thepremier.qld.gov.au/plans-and-progress/plans/assets/6-month-action-plam.pdf, accessed 3 October 2013.

Queensland Government (2013) 'Farming in reef catchments – reef regulations', Queensland Government, Rockhampton, Australia, http://qld.gov.au/environment/agricultural/sustainable-farming/reef-legislation, accessed 5 October 2013.

Queensland Government (2015) 'Abbot Point Port and Wetlands Strategy', Department of State Development, Infrastructure and Planning, Rockhampton, Australia, http://dsdip.qld.gov.au/infrastructure-planning/abbot-point-port-and-wetlands-strategy.html, accessed 30 January 2015.

Rothwell, D. (2012) 'International law and the Murray-Darling Basin Plan', *Environmental and Planning Law Journal*, 29, 4, 268–80.

State of Queensland (2007) 'Water Quality Report: Great Barrier Reef Catchments and Inshore Ecosystems', Environmental Protection Agency, Brisbane, Australia.

State of Queensland (2009) 'Reef Water Quality Protection Plan 2009 for the Great Barrier Reef World Heritage Area and Adjacent Catchments', Department of Premier and Cabinet Reef Water Quality Protection Plan Secretariat, Brisbane, Australia.

State of Queensland (2013a) '2013 Scientific Consensus Statement: Land use impacts on Great Barrier Reef Water Quality and ecosystem condition', State of Queensland Reef Water Quality Protection Plan Secretariat, Brisbane, Australia.

State of Queensland (2013b) 'Reef Water Quality Protection Plan 2013', Reef Water Quality Protection Plan Secretariat, Brisbane, Australia.

State of Queensland (2014) 'Great Barrier Reef Report Card 2012 and 2013: Reef Water Quality Protection Plan', Reef Water Quality Protection Plan Secretariat, Brisbane, Australia.

Suter, J., Segerson, K., Vossler, C. A., and Poe, G. (2010) 'Voluntary-Threat approaches to reduce ambient water pollution', *American Journal of Agricultural Economics*, 92, 4, 1195–213.

Tobin, E. (2003) 'Pronsolino v. Nastri: Are TMDLs for nonpoint sources the key to controlling the "unregulated" half of water pollution?' *Environmental Law*, 33, 3, 807–40.

Waschka, M. and Gardner, A. (2012) 'Using regulation to tackle the challenge of diffuse water pollution and its impact on the Great Barrier Reef', *The Australasian Journal of Natural Resources Law and Policy*, 15, 2, 109–47.

Water Act 2007 (Cth).

White, K. S. (2012) 'Hardy Inlet water quality improvement plan: Stage one – the Scott River Catchment', Department of Water, Western Australia.

Wulf, P. and Baird, R. (2006) 'Regulatory arrangements for the management of diffuse coastal marine pollution in Australia', Technical report 91, Cooperative Research Centre for Coastal Zone, Estuary & Waterway Management, Canberra, Australia.

12 Trans-jurisdictional water law and governance in the context of unconventional gas mining

The Australian experience

Janice Gray

Introduction

Governing water effectively is a challenging task anywhere. This is particularly so in Australia which, on the basis of its 'hydrological balance sheet', is commonly regarded as the driest inhabited land mass in the world (Gardner *et al.*, 2009, p. 3).[1] Accordingly, there is often a shortage of water in places where it is needed and disagreement about how limited water resources should be allocated and protected. The impacts of climate change have also exacerbated the difficulties of predicting weather patterns, making effective water management an even greater challenge (Intergovernmental Panel on Climate Change, 2014).

Additionally, the very nature of water itself makes it difficult to govern. It is a fugacious resource that is not respectful of boundaries. Indeed it pays no heed to artificial constructs such as national, state or local divisions, yet the sustained record of interstate and Commonwealth versus state water tensions (evident in the 1890s constitutional debates, the proposed Franklin Dam project in Tasmania,[2] the ongoing governance of the Murray-Darling Basin and disputes about the native title right to use water to fish[3]) speak to the significance of those jurisdictional boundaries which water itself ignores (see also Ghosh *et al.*, 2009). Arguably such tensions are likely to be more pronounced in a federated system such as Australia, as opposed to a unitary system. For a start there are simply more boundaries and divisions with which to deal, creating greater opportunities for the fragmentation of water resources management (Fisher, 2000, p. 61).

However, water does not just cross legally defined jurisdictional boundaries. It also crosses political and social boundaries. The difficulties of managing water are compounded when its governance interacts with the governance of other fields, sectors, disciplines, realms or bounded areas such as those relating to energy, land, agriculture or health. In such circumstances it becomes important to consider the effects of water law and governance on the law and governance of those related fields and vice versa (Connell, 2014, ch. 9).

One approach to overcoming fragmented and siloed governance (a form of governance which tends to ignore the effects of decisions in and on other fields) has been to employ what is commonly termed a trans-jurisdictional approach. While aspects of that approach are often positive, this chapter will argue that trans-jurisdictional approaches to water governance are not, by and of themselves, always positive. In practical terms, effective trans-jurisdictional water law and governance are often difficult to achieve and, where the particular governance tool employed is flawed or problematic, it is unlikely that simply applying it across jurisdictions will ameliorate the problems. Governance will remain unsatisfactory but will be so in more than one jurisdiction or sector. In fact, poor governance mechanisms applied across jurisdictions may end up exacerbating problems.

This chapter considers examples of where water and unconventional gas are presently managed, or could be managed, by way of a trans-jurisdictional approach, analysing that approach's effectiveness where relevant. The particular unconventional gas on which the chapter focuses is coal seam gas (CSG) or, as it is known in the United States and parts of the United Kingdom, coal bed methane. Accordingly three preliminary questions arise:

- What is meant by the term 'trans-jurisdictional law and governance'?
- Why focus on Australia?
- Why link a discussion of water law and governance to coal seam gas law and governance?

Following these preliminary questions the chapter will proceed by:

- mapping the legal and governance framework for water in one key Australian jurisdiction, New South Wales (NSW), and highlighting some sites where a trans-jurisdictional approach has been taken or attempted;
- briefly outlining the processes of unconventional gas mining, specifically CSG, and setting out the legal and governance framework in which they operate; and
- identifying some key intersections between the water and CSG sectors, isolating the governance mechanisms used at those intersections and analysing their effectiveness.

The meaning of trans-jurisdictional law and governance

The term trans jurisdictional water law and governance contains six elements that require consideration.

The 'trans-' element is taken to mean 'across' and is, therefore, not to be confused with the prefixes 'multi-', 'inter-' or 'intra-', which are commonly used in association with the word jurisdiction but which convey slightly different meanings (Klein, 1998, 2000; Dykes *et al.*, 2009, p. 103). Implicit in the prefix 'trans-' is the idea that bounded areas, jurisdictions or legal disciplines may be transcended or alternatively bored through from one to the other.

The term 'jurisdictional' is interpreted broadly in this chapter and extends beyond a meaning confined to the area in which a court has authority to decide a case or matter. For the purposes of this chapter the term also includes other bounded areas, disciplines, scales and spaces, such as institutional spaces in which decisions are made. Meanwhile, the term 'trans-jurisdictional' is taken to be more expansive than the term 'trans-boundary', which has historically been used to define water bodies that cross nation state borders and which usually involve elements of trans-jurisdictionality in their governance. The term 'water' is limited to terrestrial water both on and below the earth's surface.

'Law' is taken to include statute, the common law and customary law as recognized through Native Title, while 'governance' is a broader term. Governance is used to cover the wide spectrum of methods and tools employed to organize and control behaviours in society with a view to directing outcomes (Bevir and Rhodes in Braithwaite, 2008, p. 1; Gunningham, 2009, p. 145; Gray *et al.*, ch. 1 in this volume). Technically, law is, therefore, a subset of governance; it represents one of the tools in the governance toolkit.

It is also worth noting that in recent years, the definitions above have operated in a space where the very idea of established, distinct jurisdictional or sectoral boundaries has itself been revisited. Knowledge has been repackaged and criss-crossing between traditional boundaries, disciplines and limits has become more prevalent. Hence, new jurisdictional spaces like water law, which blend established legal categories or disciplines such as tort, property, administrative law and contract, have emerged. Furthermore, in some cases, questions of public law have become questions of private law and vice versa. With these changes, some of the divides have become less sharp. Nevertheless discipline, sectoral and jurisdictional spaces still exist albeit that the boundaries delineating them have not only shifted but also become more blurred.

This chapter proceeds on the basis that application of a trans-jurisdictional approach aims to govern *across* jurisdictions, providing an integrative response which is designed to lead to better, holistic management. However, with the transcendence of boundaries and the associated discipline or jurisdictional rules, norms and measures, there may also be challenges. Without a focus on the limits of jurisdictional and discipline boundaries, the benefits offered by categorization, systematization, order and predictability about consequences and outcomes may be diminished, making remedies for infringing behaviours more difficult to enforce.

Furthermore, application of a trans-jurisdictional approach to law and governance will not achieve desired ends if the fundamental mechanisms and tools on which it relies are flawed or inappropriate. If they do not work well, simply applying them across jurisdictions will not cause them to improve. It will, as noted above, merely spread bad governance. In other cases, however, transcending or cutting across and through jurisdictional limits may allow a subject to be governed without the disadvantages of fragmentation and silo-oriented governance, which are commonly attributed to be the 'at the root of the mismanagement of water resources' (Louka, 2008, p. 1).

Why Australia as a case study?

Given Australia's remote island status and the fact that its antipodean terrain is well removed from international geographic boundaries, save those delineated by the sea, one might ask why Australia should form the subject of a useful case study about water, unconventional gas and trans-jurisdictional law and governance. The answer to that question lies in the fact that there are other significant divisions, borders, boundaries and limits which relate to terrestrial water, particularly those formed by states, regions, municipalities, agencies and institutions, as well as those marked by political, cultural, social and/or economic boundaries which are pertinent. In fact, interactions across and between those other boundaries commonly introduce multi-scalar approaches to governance which arguably help build environmental resilience, particularly through collaboration and community building (Goldstein, 2012; Beilin *et al.*, 2013).

Examination of the specific Australian experience is also helpful because Australian water law has developed a reputation for responding creatively to a range of challenges, particularly those emerging from its status as the driest inhabited continent in the world (Smith, 1998, p. xvii). Over the last 15 years or so, Australian water law and governance has been widely recognized as being at the cutting edge. The majority of Australian state and territory jurisdictions, for example, moved to separating rights in water from rights in land; introduced comprehensive water planning regimes; and incorporated trading mechanisms long before many other international jurisdictions.[4] How Australia's water governance, with its long history of innovation, interacts with the governance of CSG mining (and vice versa) is, therefore, worthy of consideration.

Furthermore, the CSG industry in Australia has also grown from about 2010 onwards (McHugh, 2013; NSW Industry, undated).[5] Its good governance may also indirectly govern water in a positive manner. Hence mechanisms that help protect aquifers and surface water from negative impacts, including impacts from hydraulic fracturing, commonly called fracking (discussed below) are important,[6] along with mechanisms addressing issues such as salinity, pollution, altered flow patterns, formation water storage and the degradation of neighbouring land from CSG activities. Accordingly policies, codes, reports and plans that attempt to achieve these ends, such as the Aquifer Interference Policy (AIP) (NSW), the NSW Code of Practice for Coal Seam Gas Fracture Stimulation Activities (Fracking Code), the NSW Chief Scientist and Engineer's Report into CSG 2014 (NSWCSE) and the NSW Gas Plan, together with relevant water and CSG mining legislation, are relevant to water governance and may have comparative value especially for governments, water managers, policy makers, environmentalists and citizens, both domestically and internationally.

Water law and governance in Australia

Constitutional issues

Due to constitutional constraints, terrestrial water in Australia has largely been governed by the states and territories rather than the Commonwealth (Commonwealth of Australia Constitution Act; Gardner *et al.*, 2009, chs 5–7), although there has been a shift in recent years towards significant federal involvement in water management, law and policy so as to provide a more nationally coordinated response to pressing water quality and quantity issues (See Waschka and Gardner, ch. 11 in this volume).

One example of the shift towards greater Commonwealth involvement may be seen in the introduction of the Water Act 2007 (WA), an Act largely dependent on a fresh assertion of Commonwealth legislative powers (Rothwell, 2012) and a referral of states' powers to the Commonwealth (Kildea and Williams, 2010, p. 596). However, the constitutional challenge to the WA (*Lee v. Commonwealth of Australia*) serves to highlight some of the difficulties associated with governing across state and federal jurisdictional boundaries in the Australian water sector and is emblematic of a longer, fraught water history of Commonwealth–state relations. Lee's case was a response to the development of the Murray-Darling Basin Plan, a plan which was required by the Water Act 2007. Fearing that the Plan would result in hardship for farmers and irrigators who had to contend with reduced water diversions, two horticultural farmers led a challenge to the validity of the Commonwealth legislation responsible for the creation of the Plan.

The applicants argued that the constitutional provisions: (1) establishing that interstate trade within the Commonwealth is to be free (s 92); (2) preventing the Commonwealth from making any law 'in trade, commerce or revenue' that gives preference to one state over another (s99); and (3) limiting the Commonwealth's ability to abridge the right of a state or its residents to use river waters for conservation or irrigation, had been breached. Clearly all these provisions hinged on borders and divisions between the states and the Commonwealth. Ultimately the High Court found the legislation was not unconstitutional, deciding the matter on the basis of the external affairs power (s. 51(xxix)) and so introducing an extra-national element to the issue. The case serves to highlight how national, state and external (i.e. beyond national) borders and limits have been part of a long and often fraught Australian water history.

The shift towards a nationally oriented water policy, therefore, appears a particularly ambitious attempt to reset the course of water management (Gardner *et al.*, 2009, p. 561). The stages of that fraught history have been well conceptualized elsewhere (Wallis and Ison, 2011, p. 85).

In 2014, another attempt to assert federal jurisdiction in the water sector emerged with the inclusion of the 'water trigger' in the Environment Protection and Biodiversity Act 1999 (EPBC). The trigger requires that CSG proposals (or large coal-mining developments) likely to impact significantly on water resources be referred to the Commonwealth for environmental approval.

Beyond constitutional issues

Turning to domestic water law and governance outside the constitutional arena, two main categories emerge: (1) water resources and (2) water quality.

Water resources

Although water resources governance differs across Australian jurisdictions, all generally rely on the interdependence of: (1) a water access entitlement and trading regime and (2) a water planning regime, which provides an avenue for 'settling the trade-offs between competing outcomes for water systems' (Council of Australian Governments, 2004, para. 36; Gardner *et al.*, 2009, ch. 12).

In New South Wales (NSW), the key focus of this chapter, access to water (excluding that for domestic or stock purposes) must (as in other states) be licensed. Each water access licence (WAL) holder receives a perpetual share of a variable consumptive pool. The actual volume in the pool is determined by the relevant Minister each water year (Water Management Act 2000 s59 (WMA)). Both (ongoing) entitlements and (annual) allocations are tradeable (WMA s71), so as to provide mechanisms for moving water to places where it is needed. Because the boundaries or ambit of the access rights are slightly differently defined and subject to different rules in each jurisdiction, trade across state and territory boundaries is often complex (Gray, 2013). This is particularly so in relation to trade in the Murray-Darling Basin (MDB), where many current governance arrangements are set out in Commonwealth legislation (WA) but operate simultaneously with state legislation.

Wherever interstate trade occurs, a form of fragmentation is encountered and trans-jurisdictionality becomes an issue. Unsurprisingly, the development of effective governance regimes for trade in the MDB has been the subject of comprehensive reports, advices, plans and reforming legislation (Gray, 2012), which seek to help harmonize water law and governance across a wide array of jurisdictions, institutions, sectors, cultural spaces and hydrological areas.

'Unbundling', whereby entitlements to water were de-coupled from entitlements in land, has helped facilitate water trading and markets and was a key tenet of the 1994 COAG Water Directive Framework (National Competition Council, 1993; Gardner *et al.*, 2009; Stoeckel *et al.*, 2012). The reforms based on the framework were ultimately better harmonized after agreement of the Intergovernmental Agreement on a National Water Initiative (NWI) 2004, a Commonwealth blueprint for water reform.[7] The Commonwealth financial incentive payments made to states and territories, if they unbundled water rights and introduced new legislation to help facilitate trade for example, formed part of this reform process. The incentives represent a non-legal aspect of 'trans-' (i.e. across, state and federal) governance, which helped steer behaviour towards desired social and environmental outcomes.

Water quality

Water quality regulation and governance has been assisted by all levels of government (Stoeckel *et al.*, 2012, p. 532). The National Water Quality Management Strategy (NWQMS) applies to both potable and non-potable water quality, and national policies, guidelines and principles such as those on recycled water, drinking water and water quality monitoring and reporting come under this strategy (Australian Government, 1994).[8] Their implementation is, however, left up to individual state or territory jurisdictions, revealing yet another example of trans-jurisdictional water governance. The relevant NSW legislation in this field is the Public Health Act 1991 (Pt. 2B).

Non-potable water quality is governed by the Protection of the Environment Operations Act 1999 (NSW) (POEA), which relies on four key tools: state environmental protection policies (SEPPs); licensing controls over potentially environmentally harmful activities; notices to stop polluting or harming the environment with or without a requirement to remedy; and pollution offences (Stoeckel *et al.*, 2012, p. 557).

Unconventional gas mining

Technical aspects

In order to appreciate better the opportunities for, and examples of, trans-jurisdictional governance in the water and CSG sectors, it is useful to: (1) explain the basic, technical aspects of CSG mining and (2) briefly outline the CSG law and governance framework in NSW. The latter is particularly helpful because trans-jurisdictional governance is capable of operating in two directions: from water to CSG and from CSG to water. Hence water may be partly governed by the impacts of legislation, common law and policy in the CSG field.

CSG (a type of petroleum) is trapped underground in complex geological structures by the pressure of (usually very brackish) water. The gas bonds to the surface of coal, and only when the seam or 'formation' water pressure keeping the gas in place is released by drilling down a well (and creating an avenue through which the water can escape towards the surface) is the CSG also released (Vidas and Hugman, 2008, p. 9; Australian Government, 2013). This process may also release naturally occurring contaminants which had been safely stored underground, but which on release may pollute water (and land) and pose potential health risks (Gray, 2014; St John, undated). As unconventional methods are required to release CSG, it is known as an unconventional gas.

Fracking helps release unconventional gases. It involves water, sand and/or other chemical additives being pumped into a well at high pressure, causing the coal cleats to fracture, so assisting the depressurization process and, in turn, allowing the unconventional gas to flow back up through the well (Kyaw *et al.*, 2012).

Large volumes of water are commonly taken out of wells during CSG mining. These include the fracking water (originally pumped in) and the formation water. The average volume of formation water taken out of wells in the state of Queensland is 10,000 litres per day (St John, undated). Meanwhile the National Water Commission estimated that 7,500 gigalitres of water from groundwater systems could be extracted by the CSG industry over the next 25 years, or approximately 300 gigalitres per year (NSW Legislative Council, 2012, p. 52). By comparison, the amount of water extracted annually from the Great Artesian Basin is 540 gigalitres (NSW Legislative Council, 2012, p. 43, fn. 299).

Water-related problems may arise at a number of intersections between water, land and CSG mining. Problems include: CSG bore to aquifer contamination; cross-aquifer contamination; land contamination from CSG-produced water; the disposal and/or reuse of produced water, including by-products such as salt; aquifer depressurization; and the competition between CSG and non-CSG interests (such as agricultural/farming and drinking water) in relation to the allocation/sharing of water resources (Poisel, 2012). Of particular concern is the cumulative impact of such high levels of water extraction and use on water resources and water quality, ecosystems, human health and food security.

In summary, unconventional gas mining both uses *and* produces large volumes of water, giving rise to a range of opportunities where trans-jurisdictional approaches may play a role.

Governance framework for unconventional gas

In NSW, four main Acts form the legal framework for CSG governance. These are: the Petroleum (Onshore) Act 1991 (P(O)A); the Environmental Planning and Assessment Act 1979 (EPAA); the POEA and; the WMA. The (P(O)A) governs CSG prospecting, exploration, assessment and production and provides for petroleum titles including petroleum exploration licences (PELs) and petroleum production leases (PPLs) (P(O)A ss29, 41). As s60I (which requires that a person involved in CSG activities holds a WAL) demonstrates, the P(O)A operates across jurisdictions, in this case, across the water sector.

Under the P(O)A, all petroleum is the property of the Crown (s. 6(1)) and all Crown grants, leases, licences or other tenures over land held by the Crown are treated as having reserved to the Crown all petroleum existing in a natural state (s. 6(2)). Petroleum titles are deemed personal rather than real property and may pass inter-vivos, by way of testamentary disposition or on intestacy (s. 26).

PELs are normally conditioned by the creation of environmental management obligations, and holders must adhere to environmental standards and progressively monitor their activities. These standards may affect water quality.[9]

The NSWCSE Report has impacted on CSG governance and indirectly also on water governance. The Report ultimately supported the CSG industry but called for a better understanding of how coal, CSG and *water*, and the regimes governing them, interact. It also called for appropriate solutions 'to manage storage, transport, re-use or disposal of *produced water* and salts' and noted that 'in *drinking water catchments risk management needs to be of a high order*' (italics added) (NSW Government, 2014, pp. 10 rec. 6). These recommendations represent a clear attempt to govern water and CSG across boundaries. They recognize the high levels of inter-connectedness between the fields and the impacts of activities in one area on another.

The recommendations in the NSWCSE Report were embraced by the NSW Gas Plan (NSW Industry, 2014), a plan which, in part, responds to concerns about the high proportion of land covered by PELs and the consequent opportunities to engage in CSG activities. The negative impacts of CSG mining on water, particularly aquifers, were cause for concern (Owens, 2012; Poisel, 2012) and at the time the NSWCSE Report was announced, PELs had been issued across 45% of the state (Camenzuli and Lee, 2015).[10] The Plan is, however, a policy instrument and is, therefore, dependent on legislative reform (including legislation to expunge 16 formerly current PEL applications) for much of its implementation (P(O)A sch. 2; Petroleum (Onshore) Amendment (NSW Gas Plan) Act 2014; State Environmental Planning Policy Amendment (Gas Exploration and Mining) Act 2014).

The Plan operates by strategies that include a licence buy-back scheme, a Strategic Release Framework limiting CSG development to approximately 15% of the state, and the prohibition of PELs in National Parks. At the time of writing, the NSW government had bought back 15 PELs including all PELs in the Illawarra drinking water catchment (Arnold, 2015).

The Plan falls short of supporting a right of landholders to refuse CSG titleholders access to land, the denial of access being a form of resistance employed by many landholders keen to protect water resources (Gray, 2014; Lock the Gate Alliance, undated (a)). However, a Land Access Code is being developed (Hepburn, 2015, 213).

Strategic regional land use policy

Despite its name, the Strategic Regional Land Use Policy (SRLUP) is a governance tool that applies to water as well as land (Owens, 2012). It is designed to provide greater protection for valuable agricultural land and water from the impacts of CSG and other mining activities, and is part of a broader planning framework being implemented through initiatives including the AIP, the establishment of a Land and Water Commissioner, a Fracking Code and a Well Integrity Code (Camenzuli and Lee, 2015).

Intersections between water and CSG mining: How trans-jurisdictional and how effective is water governance at its intersections with CSG?

Points of interaction (here, between water and CSG) are commonly the focus of trans-boundary or trans-jurisdictional conflict (Coase, 1937). Hence consideration of interactions and intersections between water and CSG and their governance serves to highlight areas where nuanced and inclusive governance that is mindful of the effect of decisions in one area on another area is important.

This section, therefore, turns to examples of governance which touch both the water and CSG sectors and it considers the methods, tools and arrangements employed. It reflects on whether a fragmented approach, reinforcing (arbitrary) divisions imposed by subject matter-oriented legal and other categories and characterizations, has been employed or whether the governance has stretched across jurisdictions, disciplines and/or institutions to recognize, incorporate and respect other fields, as part of the quest to reach desired outcomes through guided conduct.

The section focuses on issues such as CSG ownership and land access; competition, allocation and water trading in the CSG sector; the water trigger; and the AIP. It considers their relationship to, and impacts on, water governance bearing in mind (as previously noted) that water may be governed from outside the water domain as well as from within. Hence CSG governance may equally reach across boundaries and limits to govern water as much as water governance may reach into the CSG domain.

In this section we observe that even where a trans-jurisdictional approach has been employed, desired outcomes will not necessarily result. The success of the governance approach taken will, in part, depend on the integrity of the approach itself – not just that it operates across jurisdictions. However, where fundamentally effective tools are used, their application across jurisdictions and other boundaries may alleviate the fragmented and siloed governance which commonly prevents an appreciation of the impacts of decisions in one field on another. The first area to which we turn deals with how water may be governed indirectly by way of CSG ownership, and access to land for CSG mining purposes.

Ownership of petroleum, land access and water protection

In NSW, unlike in many parts of the USA, the Crown rather than the landholder owns all CSG resources (P(O)A s. 6(1)). Hence it is not legally possible for a landholder to decide individually whether CSG resources will be explored or exploited. That decision is instead made by the relevant government Minister, acting in accordance with statute, and issuing licences for CSG activities where appropriate (P(O)A ss. 8–9). The NSW position, therefore, prevents landholders from making private decisions about protecting water resources by way of preventing the exploitation of CSG under their land. Landholders (without CSG titles) simply do not have any private property rights in CSG which they can deny others. Their position represents the flip side of the *nemo dat non quod habet* principle.[11] Nor are landholders legally able to prevent the entry of petroleum title-holders onto their land. Statute requires that petroleum title-holders and landholders enter into a land access agreement. If they cannot reach agreement, ultimately a determination may be imposed by an arbitrator (P(O)A pt. 4A). Hence landholders are not in the position to respond to concerns about the potential negative effects of CSG activities on water, by preventing CSG mining through an assertion of private land property rights and a consequent denial of access to their land, *or* by an assertion of private, property rights in CSG itself. Yet, if negative impacts on water result, those impacts will be felt by landholders whose agricultural, farming and drinking water interests may be affected.

The extent of restrictions on landholder rights came as a surprise to many and led to much misunderstanding (Boulle *et al.*, 2014) but private, land property rights in NSW have always been far from absolute (Alexander and Penalver, 2012; Gray *et al.*, 2012). Those rights are, for example, subject to competing rights (and reservations) including those of the Crown (such as compulsory acquisition) and rights (usually statutorily created) in favour of others. The capacity of CSG title-holders to trump landholders' rights and gain entry onto land that is the subject of a PEL or PPL, for example, should, in some ways, therefore, not come as a surprise. However, because land, water and gas are so inter-connected, it is troubling that their governance (in the land access context) has been treated in a largely fragmented manner with little regard as to the impacts of CSG title-holder dominance on water (and land) management.

In practice, many CSG title-holders have simply asserted their legal rights with little negotiation and arguably insufficient concern for the known and unknown impacts of their activities on natural resources, including water (Higginson, 2014). The insular and independent exercise of such rights that has marginalized or ignored effects on water quality and quantity has, in turn, sounded in burgeoning social resistance and protest (Gray, 2014; Hannam, 2015; Lock the Gate Alliance, undated(b)).

Landholders (without the legal right to do so) have responded by simply denying access to CSG proponents as part of what they see as their developing role as protectors of water (and land) resources and water quality – a role which they argue has been usurped by the state (Lock the Gate Alliance, undated(b); Stop CSG Sydney, undated; NSW Government, 2014, p. 7). Their actions arguably represent a form of trans- (across land, water and mining) governance, although it is a form of governance not directed or guided by government but rather guided and shaped by grassroots-level protest and specific interest groups. Whether it is an example of effective trans-jurisdictional governance is a moot point. Protesters certainly have been at the vanguard of calls for legislative and policy review concerning water in the CSG sector, and they have been instrumental in slowing the progress of CSG mining (Turnbull, 2015). That a report into CSG by the NSWCSE was commissioned in the first place and a Gas Plan developed is also, in large part, attributable to the CSG protest movement. However, this form of governance, which clearly attempts to emphasize the impacts of the inter-relatedness of water, mining and land, is somewhat cumbersome, time consuming, expensive and inefficient. It requires the regular mobilization of forces contrary to the position taken in legislation, policy and government, on each issue of dispute. The shift to a direct, grassroots-level form of governance also goes to the heart of the modern democratic process where elected representatives are responsible for setting direction, albeit in collaboration with communities and the market, for example. In the CSG context, protest governance is based on the view that the state (through elected representatives) cannot be relied upon to protect the water (and other resources) of its citizens. Accordingly, protection, it is argued, is left up to protesting private landholders and their supporters (Lock the Gate Alliance, undated(b)).

While public protest is an important part of freedom of expression in democratic societies, whether it should be relied on as a primary governance tool is perhaps more problematic. A more stable mechanism may be a nuanced legislative, regulatory and/or policy response which takes heed of community (particularly water advocates') concerns and relies on evidence-based knowledge.

The NSW Gas Plan, with its promise to 'keep water pure' (NSW Industry, 2014), represents an attempt by government to develop such a mechanism. The Plan, outlined above, reaches from the CSG sector across to the water sector and seeks to ensure that the assessment of all planned gas-producing activities and their impacts on water resources are rigorous and clear. It also sets stringent controls for CSG producers and reinforces the WAL requirement; expands the relevant government department's surface water quality monitoring network; and develops a website that brings together environmental data held by regulators, thus making it easier for the community to access information including water data (NSW Industry, 2014). It is envisaged that water governance will be shaped by the Plan.

The execution of the 2014 Agreed Principles of Land Access instrument entered into by CSG mining companies (AGL and Santos) and landholder representatives (NSW Farmers, Cotton Australia and the NSW Irrigators' Council) may also represent an attempt to help govern water (at least indirectly) in a more trans-jurisdictional manner. It states that if landholders clearly express the view that CSG mining operations on their land would be unwelcome, the companies will not enter the land (NSW Industry, undated). Hence water resources *could* be indirectly protected from the potential and/or purported ill effects of CSG mining through the denial of access to land. Yet, this form of private and limited arrangement that bridges land, water and CSG governance may ultimately prove merely symbolic if CSG title-holders turn instead to purchasing the relevant land below which the CSG resources exist, rather than relying on access agreements under the P(O)A (McCarthy, 2011). CSG titleholders would be able to circumvent landholder opposition based on water concerns. Furthermore, as only two of the many CSG companies have entered into the above agreement, its impact is limited. The benefits of this tool in terms of trans-jurisdictional water governance therefore remain somewhat ambiguous.

Licensing, trading and water sharing in the CSG context

Competition for access to water remains a controversial issue, because water is a finite resource and CSG mining has placed increased demands on available water resources. As noted above, water taken for consumptive use and/or water taken incidentally as part of a CSG activity must, in NSW, if amounting to more than three megalitres per year,[12] be taken under licence, irrespective of whether the water is later returned (WMA s. 60I; EPAA s. 89J(1)(g); Water Management (General) Regulation 2011 cl. 7). The interlocking operation of the instruments leading to this outcome represents another example of trans-jurisdictional governance, but is it effective governance?

The requirement to hold a WAL is positive in that it brings water used in CSG activities within the overall water-sharing arrangements. However, the facility for the Minister to create additional WALs (which could be used for CSG purposes) through a declaration of a controlled application order over unassigned water needs to be monitored carefully (P(O)A s. 65). It is important to ensure that the Minister does not create additional WALs for CSG proponents if unassigned water is, from an environmental perspective, better left unassigned.

Where water has already been fully allocated and no additional WALs are available, a WAL with sufficient units must be purchased on the market to fulfil the requirements of s. 60I. Trading (and the market) then become integrated tools of water governance in the CSG sphere. They are regulated via statute and in conjunction with water-sharing plans, trading rules and the AIP (WMA ch. 3 pt. 2; ch. 2). Accordingly a WAL-holder will need to comply with a range of requirements designed to help ensure that water taken in the course of CSG activities is both protected and monitored. Examples of such requirements include: identifying the water source from which the water will be taken; predicting or recording details of water take and possible future variations of that take; and identifying how water take may impact on existing users (NSW Department of Primary Industries, 2013a).

The above demonstrates that NSW water licensing and trading provisions operate across water and CSG boundaries. By enveloping aspects of water governance into CSG governance and vice versa and by linking water sharing, trading and CSG mining, a more integrated approach to resource management is potentially achievable. However, water licensing and trading are based on the unbundling concept which, by its very nature, constitutes a form of fragmentation permitting the component parts of an entity to be separated and dealt with discretely. In being broken down into separate elements, it is possible to lose sight of what property lawyers commonly call 'thingness' (in this case, water itself) and the sense of place associated with all water. If water is physically moved into new downstream areas for CSG purposes (as the result of water access trading), but this physical movement is downplayed in favour of a focus on the legal tools (and their legitimacy) which enabled the movement, then it is likely that water's 'thingness' may be overlooked. Put another way, focusing on abstract and disaggregated rights rather than the reality of the physical movement of water may cause real environment impacts to be sidelined. The result is arguably weakened governance that fails to recognize the interconnectedness of nature and culture and, instead, construes water primarily as a commodity separate from the world around it (see Graham, 2011).

Hence unbundling constitutes another example of where trans-jurisdictional governance is not of, and by itself, positive. If the underlying governance tool is flawed, then merely causing it to operate across jurisdictions will not make it positive. To use a medical analogy, it may just spread the illness.

Water trigger

The water trigger in the EPBC is potentially another trans-jurisdictional tool. It was originally designed to operate across boundaries between; the Commonwealth and state; water and CSG law; and associated institutions. It was envisaged that it would provide more robust protection for water, because CSG proposals likely to impact significantly on a water resource were to be referred to the Commonwealth Minister for assessment, as well as having to pass the state approvals process. It was also anticipated that the trigger would help states avoid a conflict of interests because it took part of the approvals decision-making process out of their hands so that they were not approving the projects from which they would also reap royalties. However, the benefits of the trigger have been questioned in the face of changes to the review process introduced by way of an Assessment Bi-lateral Agreement (effective from 2015) between the Commonwealth and NSW. The Agreement means that the Commonwealth is able to accredit state planning systems under national environmental law, and a single environmental assessment and approvals process can be introduced. In particular, it means that if the accreditation process is in compliance with the Agreement, the Commonwealth Minister is able to rely on the environmental impact assessment for the purpose of deciding whether or not to approve a controlled action (EPBC s. 82).

On one hand, the reforms may harmonize governance arrangements between the states and the Commonwealth, exemplifying a trans-jurisdictional approach, but whether they would result in effective and rigorous water governance is contestable. While the efficiencies of a 'One Stop Shop' for environmental approvals have been expounded, the additional protection offered by Commonwealth assessment may be lost (Waddell *et al.*, 2014, p. 10). State legislation commonly does not provide adequate detail setting out how Matters of

National Environmental Significance (MNES) (of which water in the CSG context is one) are to be approached. Furthermore, the Commonwealth's role in relation to MNES usually arises by way of international environmental law obligations, obligations with which the states do not usually concern themselves. Accordingly, 'there is limited scope for the functions at each level of government to overlap even where they assess the same project' (Waddell *et al.*, 2014, p. 10). Additionally, the state and territory development assessment processes are primarily overseen by Planning Ministers while State Environment Ministers play more of a secondary role, commonly only giving advice or granting approvals within the narrow confines of state environmental laws. By comparison, under federal law, the Environment Minister is the ultimate decision maker, revealing that the mandates of the decision makers in the different jurisdictions are markedly different. These aspects of the water trigger indicate that, although employing a trans-jurisdictional approach, there may be limits to its effectiveness.

SRLUP

Although SRLUP is a state-based policy, it operates through regionalism causing it to cross vertical, jurisdictional divisions and boundaries (NSW Government, 2013). Some of its measures address water quality and quantity directly, while others do so indirectly by generally restricting the circumstances where CSG mining (which potentially harms or diminishes water resources) may take place.

Given that water degradation caused by (1) the release of previously well-secured, underground, naturally occurring toxic chemicals and (2) the management of by-products such as salt are considered serious concerns, SRLUP codes such as those on well integrity and hydraulic fracturing, for example, are relevant. However, because of space constraints discussion is limited to one SRLUP measure, the AIP.

AIP

The AIP operates across the water, CSG and land sectors. In conjunction with the WMA, it defines a regime for protecting and managing interferences with aquifers and is mooted to provide a balance between competing interests such as the 'water needs of towns, farmers, industry and the environment' (NSW Department of Primary Industries, 2012). It provides for the assessment and management of risk, water accounting and consideration of the impacts of CSG activities on aquifers. In particular, it spells out how potential impacts on aquifers will be assessed by the relevant agency and how advice from the NSW Office of Water (NSWOW) on this assessment will be provided to other bodies such as the Gateway Panel or the Planning Assessment Commission (NSW Department of Primary Industries, 2012).[13] The NSWOW's advice must be based on the AIP, thus integrating the provisions of the WMA and

EPAA and providing yet another example of trans-jurisdictional governance (Townsend, 2012). However, the AIP does not address the use of fracking chemicals that might affect water quality. At present these are banned under the Fracking Code, and CSG activities that pollute water (including aquifers) are dealt with under the POEA, where a pollution licence is required.

The AIP requires a proponent consider: the risks of the proposed project (using minimal impact assessment criteria but limited to groundwater sources, connected water sources and their dependent ecosystems, culturally significant sites and water users); where the main uncertainties of prediction lie; and the likely impacts both during and after the life of the project if the predictions are incorrect. The proponent is then required to develop strategies to mitigate risk, plan for situations where greater than predicated impacts occur and plan monitoring procedures.

Government literature suggests that there is a range of ways in which mitigation may be achieved, including by the purchase of further WALs or the use of off-sets (NSW Department of Primary Industries, 2013(b)). One of the problems that most mitigation strategies face is that the environmental impacts of aquifer interference may not emerge until well into the future, while risk mitigation plans may be based on impacts emerging over shorter time-frames. Where the time cut-offs should be set is at best tricky and at worst, unknown. The problem potentially introduces aspects of Collingridge's dilemma, whereby the negative impacts of the technologies and activities which receive approval cannot easily be predicted until the approval is given and the CSG technology is developed and widely used, in turn making it difficult to alter because it has become so entrenched (Collingridge,1980).

Furthermore, interpreting predicted impacts to ascertain whether they only slightly exceed Level I minimal impact considerations becomes an important task. If they exceed by no more than 'the accuracy of an otherwise robust model' (that is, within statistical tolerance), then they may be permitted (Townsend, 2012). Hence there is flexibility in this space and it is important that such flexibility is not abused. It is, therefore, important to ensure that appropriate accountability mechanisms are in place.

Additionally, if allowable risk mitigation involves the use of off-sets, further hydrological damage may, in some cases, occur because quite simply no two ecosystems are the same and improving one ecosystem will not necessarily redress damage caused to another. This will need to be guarded against. The kinds of possible impacts on water to which approved CSG activities involving fracking may lead, and which off-sets may not overcome, include: effects on ground and surface water volume and availability caused by high rates of water withdrawal; surface chemical spills at well sites that pollute water resources; seismic activity caused by formation water re-injection leading to altered water conditions; surface spills of flowback or formation water on drinking water sources; and pollution and water quality degradation – particularly of drinking water – from the inadequate treatment of fracking wastewater (see Lyster, 2013;

United States Environmental Protection Agency, 2015). In such cases it will be necessary to develop and/or implement alternative mechanisms through the AIP and/or other policies so as to redress these problems.

Exemptions for CSG activities at the exploration stage may also represent a weakness in the AIP as a tool of trans-jurisdictional governance. Although environmental impacts at this stage are still assessed under the EPAA, Pt 5 and a Review of Environmental Factors may be required, the process is less robust than that required by the AIP. Arguably exemptions should be linked more closely to environmental impacts and risk so as to protect water resources more effectively.

Returning to the issue of water sharing, on one reading, by operating in conjunction with the WMA's licensing provisions and causing CSG miners to account for water taken from groundwater and surface water sources, the AIP helps ensure that extraction limits set by water plans are met. Those plans in turn, assist water sharing among the environment, domestic and stock users, irrigators, town suppliers and industries, including CSG mining. In theory, therefore, the licensing requirements, as supported by the AIP, work well as a resource distribution mechanism but, on another reading, those same requirements could serve to marginalize or postpone the water needs of important competing (non-CSG) users if those non-CSG users operate in less profitable (but still highly significant) sectors, such as irrigated agriculture. To explain, if CSG companies are, for example, financially strong, they may be able to engage in profligate water use through the acquisition of more WALs (and allocations under them) while other less well-off competing water users may only be able to purchase fewer WALs and so be entitled to access smaller volumes of water.

Furthermore, if CSG companies are in a position to purchase the necessary WALs (and/or allocations under them) but agriculturalists are not, questions of food security and the public good are likely to arise. It may be that water governance tools end up supporting CSG mining but not food production, which is also highly water dependent. Such hypothetical scenarios also raise once more the normative question of whether a licensing and trading scheme is an effective governance tool to share water resources across jurisdictions. If it is not, merely adjusting the settings will not overcome water-sharing problems. If the governance mechanism is itself seriously flawed, then it is best replaced (Gray, 2012). In that regard it is worth noting that in other sectors a backlash against the market as a governance mechanism is emerging, and privatization is being represented as a form of fragmentation – the very thing that trans-jurisdictional governance seeks to avoid.[14]

Therefore, although the AIP represents an ambitious attempt to govern water across jurisdictions, connecting as it does with the WMA, the EPAA, POEA and the CSG sector itself, and demonstrating considerable success in that regard, it also reveals several weaknesses. If greater success is to be achieved in this area, those weaknesses need to be addressed.

Drinking water

Drinking water represents another site where trans-jurisdictional governance is potentially relevant. The management of drinking water catchments and resources has come under close scrutiny, resulting in the imposition of a NSW government moratorium on CSG activities in such catchments (James and Daniel, 2013; Lock the Gate Alliance, undated(b0). The very need to impose a moratorium on CSG mining in drinking water catchments, a moratorium which followed strident community concern and protest, suggests that governance across water and CSG in those catchments was not adequate to confidently protect water (and other natural) resources (Australian Drinking Water Guidelines, 2011; NSW Ministry of Health, 2013).

By May, 2015, PELs either covered or encroached on all five catchments (Warragamba, Woronora, Upper Nepean, Blue Mountains and Shoalhaven) supplying drinking water to 4.5 million households and other users, in the greater Sydney area, or 60% of the state's population (Hasham, 2013).

Development consent to drill and run CSG wells had also been granted in NSW drinking water catchments (see, e.g. Apex Burragarong DDH1). Because the catchments cover only 2% of the state's land and as the condition of that land and the water resources in it is considered of critical importance, questions were asked as to how law and governance had let this happen. How was it that CSG titles had been granted in the first place in these water-important areas?

Indeed, one respondent to the Legislative Council Inquiry noted, that '[w]e have Special Areas, which are gated and have padlocks and you get fined $11,000 or more if we [sic] set foot in them, but they [the drinking water catchments] are apparently being approved for coal seam gas extraction' (NSW Legislative Council, 2012, p. 51).

Such a submission serves to highlight the perceived inconsistency between restricting public access to drinking water catchments, on one hand and permitting potentially harmful CSG activities to take place on them, on the other.

One response has been the progressive government buy-back of PELs discussed above (ABC News, 2015). By July, 2015, 15 PELs including an extensive PEL held by AGL for exploration in the Illawarra drinking water catchment had been purchased. The buy-backs reduced the percentage of PELs across NSW to nine. On the surface, this redirection represents a positive approach to governing water by way of a policy shift and tighter controls on PELs. However, the legislation which permits the granting of PELs in the first place remains intact, meaning that further PELs could be granted and water resources could again be threatened unless the SRLUP and the strategic review of land, for example, operate in a robust, predictable and reliable manner. The relevant Minister's words that the state 'was not closing the door on CSG development' (ABC News, 2015), together with the defeat of a bill introducing a state-wide ban on all CSG activity have, however, been cause for concern in some circles, with one politician stating that '[u]ntil there is a scientific consensus that CSG mining will not damage aquifers and groundwater systems, we need to hit the pause button' (*Business Spectator*, 2015).

Conclusion

The intersections between water, CSG mining and land provide many examples of where effective trans-jurisdictional water governance is required and where it is necessary to address the challenges imposed by federal, state, regional, local and other boundaries, while remaining cognisant of the fact that although only federal and state jurisdictions have legislative power, regional, local and community-based knowledge and concerns need to be incorporated into governance structures. Failing to do so makes adaptive, inclusive and respectful trans-jurisdictional water governance difficult to achieve and, as this chapter demonstrates, may lead to public protest and resistance. At times that resistance may go on to become a form of grassroots-level governance or the driver of government-directed governance.

One of the key challenges with which water governance struggles is how best to design governance measures that respond effectively to water quality and quantity issues, bearing in mind: (1) the need to balance environmental, social and economic considerations; and (2) the restrictions imposed by jurisdictional boundaries be they vertical, such as those between federal, state regional or local divisions, or horizontal, such as those existing between different government agencies, institutions, communities, subject matters or disciplines? One answer is to devise governance that is trans-jurisdictional– that crosses defined areas, disciplines and/or other boundaries. This chapter took that idea and explored examples of where trans-jurisdictional approaches in relation to water law and governance in the CSG mining context have, or could have, been taken. Where they have been taken, it assessed their effectiveness. Hence the chapter considered how law and governance applied at selected intersections between water, CSG mining and land. Accordingly it reviewed and analysed specific policy and legislative applications to reflect on the effectiveness of mechanisms used. These included how the ownership of petroleum and access to land impact on water governance; the potential effect of water licences and trading in the CSG context; the role of the water trigger; the implications of the AIP and the moratorium on CSG mining in drinking water catchments.

In order to engage with the ideas and questions that these examples raised, the chapter began by outlining the law and governance frameworks for both water and CSG mining because, although water governance is able to reach across into the CSG sector, CSG governance is equally able to reach back across into the water sector. When CSG governance does this, it can directly and indirectly govern water resources and quality.

In its consideration of trans-jurisdictional water governance the chapter posed a fundamental question: if we institute trans-jurisdictional water governance will water management necessarily improve? Its conclusion was negative. Merely designing a governance system that reaches across water and CSG boundaries will not necessarily promote better water outcomes. Hence if, for example, the system of water licences and water trading is found wanting, the fact that it spans jurisdictions will not cure fundamental defects based on the effects of unbundling and the creation of disaggregated rights. Nor will designing governance systems whereby federal law simply incorporates or embraces state-based mechanisms (as is possible under Commonwealth–state bilateral agreements) be a solution to water governance if checks and balances, standards or rigour are reduced in the process. The basic trans-jurisdictional mechanism needs to be sound.

Governance needs to be adaptive and collaborative, taking into account actors and impacts in other jurisdictions, but it needs to be more than that. It also needs to operate effectively *within* known divisions and boundaries which jurisdictions, institutions and disciplines define because those boundaries are not likely to disappear any time soon. They may become fuzzier at the edges, as the NWI and the Water Act 2007 (which permit the Commonwealth a governance role in predominantly state matters) demonstrate, but State and federal jurisdictions, for example, will continue to exist. Hence good law and governance needs to be trans-jurisdictional at times, while at others it needs to demonstrate the capacity to operate within boundaries not just across them. Good law and water governance needs to select and rely on the benefits that divisions, borders, bounded subjects, disciplines and jurisdictions offer but, where these inhibit effective outcomes, it needs to embrace the concept of reaching across those very same boundaries. This may involve relying on a greater range of regulatory approaches than is currently employed, including for example, economic instruments, social contracts, community participation platforms, and perhaps some command and control regulation, and law, for example (see Godden, Ch. 2 in this volume).

Governance, in all the scenarios of water and CSG mining interaction outlined above, needs to respond to protect water quality and quantity through realistic risk assessment, and the design and implementation of effective regimes of management which engage a range of stakeholders, are flexible and have the capacity to integrate entities on both vertical and horizontal axes. Such governance regimes will, therefore, necessarily need to be able to move between both the water and CSG jurisdictions as well as integrate the operation of federal, state and local governance as a starting point. Beyond that they will need to respect and permit collaboration between the institutions and agencies which help operationalize law and policy. However, crucially, they will need to rely on mechanisms and tools which are fundamentally appropriate and effective. If those tools prove to be essentially flawed from the outset because they are, for example, mono-dimensional (Wallis and Ison, 2011, p. 1) or are grounded in a rationale which is increasingly, globally demonstrating weaknesses (Klein, 2014), then they will not improve governance even if they are applied across jurisdictions. They will, as noted above, simply taint another jurisdiction.

Effective trans-jurisdictional water law and governance is a tall order. Indeed it may be too tall, but if appropriate governance tools were employed *and* the holistic and inter-connected nature of the environment itself were echoed in the governance that sought to protect water and related fields, the chances of success would be much enhanced.

Notes

1 The balance sheet measures inputs of precipitation against river run-off, recharge to groundwater and evapotranspiration.
2 *Commonwealth v. Tasmania* [1983] HCA 158 known as The Tasmanian Dam Case (1 July 1983).
3 Karpany v Dietman [2013] HCA 47.
4 See legislative reforms based on Council of Australian Government's Framework (COAG) 1994. Framework discussed in Gardner *et al.* (2009, chs 5, 25).
5 For NSW projects see NSW Department of Industry, Resources & Energy(undated) Coal Seam Gas Locations Map.
6 Shale gas (which is another unconventional gas found at deeper depths than CSG) and CSG may both be fracked. Australia presently mines more CSG, but its SG deposits are ranked 6th in the world. See United States Energy Information Administration (2013).
7 The key legislation that was introduced or amended to accommodate the reforms was Water Management Act 2000 (NSW); Water Act 2000 (Qld); Natural Resources Management Act 2004 (SA); Water Management Act 1999 (Tas); Water Act 1989 (Vic); Water Resources Act 2007 (ACT); Rights in Water and Irrigation Act 1914 (WA); Water Act 1992 (NT).
8 The Australian Government *National Water Quality Management Strategy* website provides a portal to the specific policies and guidelines. See *Australian Drinking Water Guidelines* 2011 available on this site.
9 Note the EPA issued Santos a $1,500.00 fine for a 'pollution incident' in 2014.
10 *6761 Compiled A.doc* CSG titles as at 1 September, 2015 are available at NSW Industry (2015) 'Petroleum titles'.
11 One cannot give what one does not have.
12 See Water Management (General) Regulation cl 7.
13 The Gateway Panel is an independent body which assesses the agricultural and water impacts of CSG proposals located on Strategic Agricultural Land. The Commission is another independent body that determines development applications delegated to it by the relevant Minister.
14 See the increasing support for the renationalization of the rail network in the UK and the developing trend away from infrastructure privatization (BBC News, 2015).

References

ABC News (2015) 'Coal seam gas licences cancelled after AGL reaches buy-back deal with NSW Government', *ABC News*, 6 July, http://abc.net.au/news/2015-07-06/agl-csg-licences-bought-back-hunter-sydney-illawarra/6597714, accessed 2 August 2015.

Alexander, G. and Penalver, E. (2012) *An Introduction to Property Theory*, Cambridge University Press, New York.

Arnold, A. (2015) 'Illawarra now CSG free: PEL 2 cancelled', *Illawarra Mercury*, 6 July, http://illawarramercury.com.au/story/3193071/illawarra-now-csg-free-pel-2-cancelled/, accessed 3 September 2015.

Australian Drinking Water Guidelines (2011) Department of the Environment, Parkes, Australia, http://environment.gov.au/topics/water/water-quality/national-water-quality-management-strategy, accessed 7 December 2013.

Australian Government (1994) National Water Quality Management Strategy, Department of the Environment, Parkes, Australia, www.environment.gov.au/topics/water/water-quality/national-water-quality-management-strategy, accessed 5 December 2013.

BBC News (2015) 'Andy Burnham pledges to renationalise railway network', *BBC News*, 4 August, http://bbc.com/news/uk-politics-33780754, accessed 5 May 2015.

Beilin, R., Reichelt, N. T., King, B. J., Long, A. and Cam, S. (2013) 'Transition Landscapes and Social Networks: Examining On-Ground Community Resilience and its Implications for Policy Settings in Multiscalar Systems', *Ecology and Society*, 18, 2, 30, http://dx.doi.org/10.5751/ES-05360-180230, available via link at www.ecologyandsociety.org/issues/view.php?sf=54, accessed 4 December 2013.

Boulle, L., Hunter, T., Weir, M. and Curnow, K. (2014) 'Negotiating conduct and compensation agreements for coal seam gas operations: Developing the Queensland regulatory framework', *The Australasian Journal of Natural Resources Law and Policy*, 17, 1, 43–74.

Braithwaite, J. (2008) *Regulatory Capitalism*, Edward Elgar, Cheltenham, UK.

Business Spectator (2015) 'CSG Moratorium Bill shot down in NSW', *Business Spectator*, 13 August, http://businessspectator.com.au/news/2015/8/13/resources-and-energy/csg-moratorium-bill-shot-down-nsw, accessed 2 September 2015.

Camenzuli, L. and Lee, L. (2015) 'Pause, set and recommence: Release of the NSW Gas Plan and where to for the onshore gas industry in NSW', http://corrs.com.au/publications/corrs-in-brief/pause-reset-and-recommence-release-of-the-nsw-gas-plan-and-where-to-for-the-onshore-gas-industry-in-new-south-wales, accessed 24 May 2015.

Coase, R. H. (1937) 'The Nature of the Firm', *Economica*, 4, 16, 386–405.

Collingridge, D. (1980) *The Social Control of Technology*, St Martin's Press, London.

Commonwealth of Australia Constitutional Act 1901 (Cth).

Connell, D. (2014) 'Trans-boundary water governance', in R. Q. Grafton, P. Wyrwoll, C. White and D. Allendes (eds) *Global Water- Issues and Insight*, ANU EPress, Acton, Australia, http://press.anu.edu.au/titles/global-water/, accessed 21 July 2015.

Council of Australian Governments (COAG) (2004) 'Intergovernmental Agreement on a National Water Initiative 2004', COAG, Barton, Australia, http://nwc.gov.au/nwi, accessed 5 December 2013.

Dykes, T., Rogers, P. and Smyth, M. (2009) 'Towards a new disciplinary framework for creative design', *CoDesign*, 5, 2, 99–116.

Environmental Planning and Assessment Act 1979 (NSW).

Environmental Protection and Biodiversity Act 1999 (Cth).

Fisher, D. (2000) *Water Law*, LBC Information Services, North Ryde, Australia.

Gardner, A., Bartlett, R. and Gray, J. (2009) *Water Resources Law*, Lexis Nexis, Sydney, Australia.

Ghosh, D., Goodall, H. and Hemelryk, D. (eds) (2009) *Sovereignty and Borders in Asia and Oceania*, Routledge Studies in Physical Geography and Environment Series, Routledge, New York.

Goldstein, B. E. (2012) *Collaborative Resilience: Moving Through Crisis to Opportunity*, The MIT Press, Cambridge, MA.

Graham, N. (2011) *Lawscape*, Routledge Glasshouse, Abingdon, UK.

Gray, J. (2012) 'The legal framework for water trading in the Murray-Darling Basin: An overwhelming success?', *Environmental, Planning and Law Journal*, 29, 4, 328–48.

Gray, J. (2013) 'Dollars and dreams: Legal aspirations and report cards in the Murray-Darling Basin of Australia', in L. Westra, P. Taylor and A. Michelot (eds) *Confronting Ecological and Economic Collapse*, Routledge Earthscan, Abingdon, Oxon, UK.

Gray, J. (2014) 'Frack Off! Law, policy, social resistance, coal seam gas mining and the Earth Charter', in L. Westra and M. Vilela (eds) *The Earth Charter, Ecological Integrity and Social Movements*, Routledge Earthscan, Oxon, UK.

Gray, J., Edgeworth, B., Foster, N. and Dorsett, S. (2012) *Property Law in New South Wales*, 3rd edn, LexisNexis, Chatswood, Australia.

Gunningham, N. (2009) 'The new collaborative environmental governance: The localisation of regulation', *Journal of Law and Society*, 36, 1, 145–66.

Hannam, P. (2015) 'AGL in dam storage row as anti-coal seam gas projects hit 100 weeks', *Sydney Morning Herald*, 2 September, http://smh.com.au/environment/water-issues/agl-in-dam-storage-row-as-anticoal-seam-gas-protests-hit-100-weeks-20150902-gjdcke, accessed 7 September 2015.

Hasham, N. (2013) 'Labor proposes ban on coal seam gas mining near water catchments', *Sydney Morning Herald*, 17 October, www.smh.com.au/nsw/labor-proposes-ban-on-coal-seam-gas-mining-near-water-catchments-20131017–2vnve.html, accessed 7 December 2013.

Hepburn, S. (2015) *Mining and Energy Law*, Cambridge University Press, Port Melbourne, Australia.

Higginson, S. (2014) 'CSG licence cancellations show need for regulatory overhaul', Environmental Defender's Office, Sydney, Australia, http://edonsw.org.au/csg_licence_cancellations_show_need_for_overhaul, accessed 16 October 2014.

Intergovernmental Panel on Climate Change (2014) 'Climate changes 2014 – Mitigation of climate change', Intergovernmental Panel on Climate Change, Geneva, Switzerland, http://ipcc.ch/report/ar5/wg3/, accessed 24 March 2015.

James, M. and Daniel, S. (2013) 'Moratorium on coal seam gas extraction in Sydney's drinking water zone', *ABC News*, 12 November, www.abc.net.au/news/2013–11–12/moratorium-placed-on-coal-seam-gas-extraction-in-sydney27s-dri/5087252, accessed 18 January 2015.

Kildea, P. and Williams, G. (2010) 'The Constitution and the management of water in Australia's rivers', *Sydney Law Review*, 32, 4, 268–80.

Klein, J. T. (1998) 'Notes toward a social epistemology of transdisciplinarity', *Interactive Bulletin of the International Centre for Research and Studies Transdisciplinary*, no 12.

Klein, J. T. (2000) 'Integration, evaluation, and disciplinarity', in M. A. Sommerville and D. J. Rapport (eds) *Transdisciplinarity: Recreating integrated knowledge*, EOLSS Publishers Co Ltd, London.

Klein, N. (2014) *This Changes Everything. Capitalism v. The Climate*, Simon and Schuster, Sydney, Australia.

Kyaw, A., Azafar, B. S. B. N. and Tunio, S. Q. (2012) 'Fracturing Fluid (Guar Polymer Gel) Degradation Study by using Oxidative and Enzyme Breaker', *Research Journal of Applied Sciences, Engineering and Technology*, 4, 2, 1667–71.

Lee v Commonwealth of Australia [2014] FCA 432.

Lock the Gate Alliance (undated (a)) 'Lock Your Gate', Lock Your Gate Alliance, South Lismore, Australia, http://new-lockthegate.nationbuilder.com/lock_your_gate, accessed 7 September 2015.

Lock the Gate Alliance (undated (b)) 'Lock Your Gate Alliance', Lock Your Gate Alliance, South Lismore, Australia, www.lockthegate.org.au, accessed 31 January 2015.

Louka, E. (2008) *Water Law and Policy; Governance without Frontiers*, Oxford University Press, New York.

Lyster, R. (2012) 'Coal seam gas in the context of global energy and climate change scenarios', *Environmental and Planning Law Journal*, 29, 2, 91–100.

Lyster, R. (2012) 'Coal seam gas in the context of global energy and climate change scenarios', *Environmental and Planning Law Journal*, 29, 2, 91–100.

McCarthy, J. (2011) 'Coal Seam Gas Miners Buy-Up 140,000 hectares of Prime Queensland Farmland, *Courier Mail*, 4 July, www.couriermail.com.au/business/miners-snap-up-prime-farming-land/story-e6freqmx-1226086674502, accessed 11 September 2015.

McHugh, B. (2013) 'Industry minister dismisses coal seam gas protests', *ABC Rural*, 17 September, http://abc.net.au/news/2013-09-17/ian-macfarlane-support-forcsg-nsw/4963084, accessed 1 November 2013.

National Competition Council (1993) 'National Competition Policy Review', National Competition Council, Melbourne, Australia, http://ncp.ncc.gov.au/docs/National%20Competition%20Policy%20Review%20report,%20The%20Hilmer%20Report,%20August%201993.pdf, accessed 5 December 2013.

Natural Resources Management Act 2004 (SA).

NSW Department of Industry, Resources & Energy (undated) 'Coal seam gas locations map', Department of Industry, Resources & Energy, Sydney, Australia, http://cdn.digitalservicesnsw.com/csg/mapv2.1/index.html, accessed 31 November 2014.

NSW Department of Primary Industries (2012) 'Aquifer Interference Policy', Department of Primary Industries, Sydney, Australia, http://water.nsw.gov.au/__data/assets/pdf_file/0004/549175/nsw_aquifer_interference_policy.pdf, accessed 8 December 2013.

NSW Department of Primary Industries (2013a) 'Water and Coal Seam Gas, Fact Sheet 3, Regulation and Monitoring', Department of Primary Industries, Sydney, Australia, http://water.nsw.gov.au/__data/assets/pdf_file/0012/548769/groundwater_coal_seam_gas_regulation_monitor.pdf, accessed 2 June 2013.

NSW Department of Primary Industries (2013b) 'NSW Aquifer Interference Policy, Fact Sheet 5, Risk Mitigation', Department of Primary Industries, Sydney, Australia, http://water.nsw.gov.au/__data/assets/pdf_file/0005/549374/law_key_aquifer_interference_factsheet_5_risk_mitigation.pdf, accessed 28 September 2015.

NSW Government (2013) *Strategic Regional Land Use Policy*, Department of Planning and Infrastructure, Sydney, Australia, http://nsw.gov.au/sites/default/files/initiatives/strategicregionallandusepolicy-brochure_sd_v01.pdf, accessed 2 June 2013.

NSW Government (2014) 'NSW Chief Scientist and Engineer's Final Report into Coal Seam Gas', Chief Scientist & Engineer, Sydney, Australia, http://chiefscientist.nsw.gov.au/__data/assets/pdf_file/0005/56912/140930-CSG-Final-Report.pdf, accessed 3 March 2015.

NSW Industry (2014) 'NSW Gas Plan', Department of Industry, Resources & Energy, Sydney, Australia, http://resourcesandenergy.nsw.gov.au/__data/assets/pdf_file/0005/534830/NSW-Gas-Plan.pdf, accessed 12 July 2014.

NSW Industry (2015) 'Petroleum titles', Department of Industry, Resources & Energy, Sydney, Australia, http://resourcesandenergy.nsw.gov.au/miners-and-explorers/applications-and-approvals/current-titles-reports-applications-and-services/title-status-reports/Petroleum-TitlesCurrent-2015-09.pdf, accessed 3 September 2015.

NSW Industry (undated) 'Land access agreement', Department of Industry, Resources & Energy, Sydney, Australia, http://resourcesandenergy.nsw.gov.au/landholders-and-community/coal-seam-gas/community/land-access-agreement, accessed 28 January 2015.
NSW Legislative Council (2012) 'Coal Seam Gas', General Purpose Standing Committee No 5 Report, Sydney, Australia, http://parliament.nsw.gov.au/prod/parlment/committee.nsf/0/318a94f2301a0b2fca2579f1001419e5/$FILE/Report%2035%20-%20Coal%20seam%20gas.pdf, accessed 7 December 2013.
NSW Ministry of Health (2013) 'Drinking Water Guidelines', North Sydney, Australia, http://health.nsw.gov.au/resources/publichealth/environment/water/pdf/nsw_guidelines_for_dwms.pdf, accessed 7 December 2013.
Owens, K. (2012) 'Strategic regional land use plans: Presenting the future for coal seam gas projects in New South Wales?', *Environmental and Planning Law Journal*, 29, 2, 113–28.
Petroleum (Onshore) Act 1991 (NSW).
Petroleum (Onshore) Amendment (NSW Gas Plan) Act 2014 (NSW).
Poisel, T. (2012) 'Coal seam gas exploration and production in New South Wales: The case for better strategic planning and more stringent regulation', *Environmental and Planning Law Journal*, 29, 2, 129–51.
Protection of the Environment Operations Act 1999 (NSW).
Public Health Act 1991 (NSW).
Rights in Water and Irrigation Act 1914 (WA).
Rothwell, D. (2012) 'International law and the Murray Darling Basin Plan', *Environmental and Planning Law Journal*, 29, 4, 268–80.
Smith, D. I. (1998) *Water in Australia: Resources and Management*, Oxford University Press, Melbourne, Australia.
St John, A. (undated) 'The Coal Seam Gas Debate', *Australian Parliament* Canberra, Australia, http://aph.gov.au/About_Parliament/Parliamentary_Departments/Parliamentary_Library/pubs/BriefingBook44p/GasDebate, accessed 31 November 2014.
State Environmental Planning Policy Amendment (Gas Exploration and Mining) Act 2014 (NSW).
Stoeckel, K., Webb, R., Woodward, L. and Hankinson, A. (2012) *Australian Water Law*, Lawbook Co, Pyrmont, Australia.
Stop CSG Sydney (undated), Stop CSG Sydney, Sydney, Australia, www.stopcsgsydney.org.au, accessed 8 May 2015.
Townsend, D. (2012) 'The NSW Strategic Regional Land Use Policy and Aquifer Interference Policy in Context', *Australian Environment Review*, 28, 1, 415–20.
Turnbull, S. (2015) 'Coal seam gas protestors prepare for second Bentley blockage', *ABC News*, 27 April, http://abc.net.au/local/stories/2015/04/27/4224393.htm, accessed 28 April 2015.
United States Energy Information Administration (2013) 'Technically recoverable shale oil and shale gas resources: An assessment of 137 shale formations in 41 countries outside the United States', United States Department of Energy, Washington, DC, http://eia.gov/analysis/studies/worldshalegas/archive/2013/pdf/fullreport_2013.pdf, accessed 31 December 2014.
United States Environmental Protection Agency (2015) 'Assessment of potential impacts of hydraulic fracturing for oil and gas on drinking water resources (external review draft) executive summary', Office of Research and Development, Washington, DC, http://epa.gov/sites/production/files/2015-07/documents/hf_es_erd_jun2015.pdf, accessed 29 September 2015.

Vidas, H. and Hugman, B. (2008) 'Availability, Economics and Production Potential of North American Unconventional Natural Gas Supplies', Interstate Natural Gas Association of America, Washington, DC, http://ingaa.org/File.aspx?id=7878, accessed 23 May 2013.

Waddell, S., Cornwall, A. and Gray, J. (2014) 'Submission to the House Standing Committee on the Environment Inquiry into the impact of "green tape" and issues related to environmental regulation and deregulation', National Environmental Law Association, East Melbourne, Australia, http://nela.org.au/NELA/Documents/NELA_Submissions_140430.pdf, accessed 19 September 2015.

Wallis, P. and Ison, R. (2011) 'Institutional change in multi-scalar water governance regimes: a case from Victoria, Australia', *Journal of Water Law*, 22, 85–94.

Water Act 1989 (Vic)

Water Act 1992 (NT).

Water Act 2000 (Qld).

Water Act 2007 (Cth).

Water Management Act 1999 (Tas).

Water Management Act 2000 (NSW).

Water Management (General) Regulation 2011 (NSW).

Water Resources Act 2007 (ACT).

13 Protecting coastal wetlands in a changing climate

Reinvigorating integrated coastal zone governance

Jan McDonald and Anita Foerster

Introduction

Wetlands are a critical part of the coastal environment. Encompassing estuaries, saltmarsh, mangroves, coastal swamps, lakes and lagoons, saline wetlands occupy the biophysical interface between terrestrial and marine, and between marine and freshwater environments (Australian State of the Environment Committee (ASEC), 2011, p. 849). They have values of local, regional, state, national and, occasionally, international significance. Coastal wetlands perform ecosystem functions important for commercially valuable industries such as fisheries and tourism, including nutrient and pollutant filtering, erosion control to protect coastal infrastructure and aid the adaptation of coastal communities and carbon sequestration (Millennium Ecosystem Assessment (MEA), 2005; Australian State of the Environment Committee (ASEC), 2011, p. 858; McLeod *et al.*, 2011). They also have cultural significance for indigenous people, as well as scenic, research, conservation and recreational values (Rogers *et al.*, 2014, pp. 67–8). Their protection affects both public and private property rights. Accordingly, decisions over their use, protection, destruction and management potentially span multiple tiers of government, public and private actors and a wide range of sector-specific agencies and managers. The need for integrated governance approaches is clear.

Despite the importance of coastal wetlands, like many estuaries around the world the heavy concentration of Australia's population around the estuaries of major river systems has transformed our coasts (Millennium Ecosystem Assessment (MEA), 2005, pp. 41–2). The combined pressures of urban development, catchment land and water use and flood mitigation measures have radically altered and, in many places, destroyed, these coastal ecosystems (Lee *et al.*, 2006; Australian State of the Environment Committee (ASEC), 2011, pp. 858–61; Rogers *et al.*, 2014, p. 68). Climate change will further threaten coastal ecosystems and the land on which they are located (Nicholls *et al.*, 2007, pp. 507, 540; Australian State of the Environment Committee (ASEC), 2011, pp. 878–81; Burley *et al.*, 2012; Wong *et al.*, 2014, p. 18). Sea level rise and more intense storm activity will exacerbate coastal erosion and shoreline recession in many areas (Gilman *et al.*, 2008). The changed frequency and extent of coastal flooding from both storm surge and catchment rainfall will affect the health and composition of these ecosystems. Saltwater intrusion into groundwater will have impacts on both water supply and ecosystem dynamics in some places (ASEC, 2011, p. 854). Warmer conditions will also affect the diversity of wetland ecosystems, and increase the risks of pests and disease.

Under appropriate conditions, many coastal wetlands have some capacity to adapt to higher sea levels *in situ*, by increasing their root mass and hence their elevation. Critical factors in this process include overall sea level rise and sedimentation, which is affected by flood barriers and other structures that alter natural flows (Rogers *et al.*, 2014). Wetlands that are unable to adapt *in situ* could potentially move landward (ASEC, 2011, p. 880), but in countries like Australia, extensive development in the coastal zone restricts the potential for this natural migration (Blankespoor *et al.*, 2012; Rogers *et al.*, 2014). Australia's population is already highly concentrated in coastal areas. Projected population increases are expected to follow this trend and further restrict adaptation space for coastal wetlands. Furthermore, as sea levels rise and coastal hazards worsen, there is likely to be considerable pressure to protect existing settlements and infrastructure with hard coastal and flood defences, such as sea walls, groynes and levies which fundamentally disrupt natural coastal processes (Department of Climate Change (DCC), 2009; ASEC, 2011).

In many situations, planned retreat policies (involving the avoidance or removal of hard coastal defences) will be critical to maintaining coastal ecosystems and providing room for them to migrate inland as sea levels rise. Indeed, 'living shorelines' (the deliberate construction or rehabilitation of coastal wetlands, such as mangroves, as an alternative to traditional engineering structures such as seawalls) are increasingly promoted as an adaptation approach to increase the resilience of shorelines to erosion and inundation (Siders, 2013). Planning for such measures raises difficult trade-offs between the short-term socio-economic costs associated with limiting land use and development, and losing buildings and infrastructure before the end of their economic life; and longer-term goals of public interest nature conservation (Abel *et al.*, 2011). Trans-jurisdictional tensions therefore arise between tiers of government, between public and private legal interests and actors, and across regulatory regimes for different land uses or policy domains. The uncertainties and long timeframes that pervade projected climate change impacts raise new issues of 'temporal trans-jurisdictionality', which compound existing governance challenges.

In light of these current and future pressures, this chapter explores contemporary approaches to the management of coastal wetlands in Australia and argues for urgent reinvigoration of Integrated Coastal Zone Management (ICZM). We begin in Part Two by briefly reviewing the history and current status of ICZM in Australia as a strategic framework for improved coastal management in a climate change context. Part Three then maps and critiques the current regulatory framework, focusing specifically on legal and policy regimes for coastal land use planning, coastal protection works, catchment water and land use, and protected areas. Despite Australia's formal policy commitments to ICZM, our analysis reveals a complex, fragmented regulatory landscape that has hitherto failed to protect coastal wetlands and which is likely to be vastly insufficient to support their adaptation to climate change. Our research also suggests that current approaches reinforce static patterns of land use and environmental protection that are ill-equipped to respond to the impacts of climate change and directly threaten many coastal aquatic ecosystems over the long term. These deficiencies point to the need for a fresh approach to coastal governance, which gives greater prominence to the protection of coastal ecosystems in a changing climate. To this end, in Part Four we suggest some priority reforms for refreshing and reinvigorating the ICZM agenda in Australia.

Integrated Coastal Zone Management

Internationally, ICZM has been the dominant model for improved coastal governance for over 30 years. It encompasses the interactions between marine and terrestrial environments, seeking to transcend the land–marine divide and specifically acknowledging the complexity and cross-sectoral nature of coastal management issues (Sorenson, 1993; Falaleeva *et al.*, 2011; Harvey *et al.*, 2012). The major emphasis of ICZM is on integration and coordination within coastal governance arrangements, both vertically, between tiers of government influencing coastal management and horizontally, between policy domains (Harvey *et al.*, 2012). Consistent with general directions in environmental governance, the ICZM model emphasizes strategic planning and adaptive management and embraces collaborative, participatory approaches to decision making (Sorenson, 1993; Falaleeva *et al.*, 2011).

Implementation of ICZM differs from country to country. Approaches are highly influenced by existing governance arrangements, available resources, institutional capacities and political support. Case studies suggest that effective implementation depends significantly on the establishment of formal institutions for ICZM, especially laws and policies that define the powers and duties of public and private actors, mandate that government undertake and facilitate planning and management actions, and provide effective tools for the implementation of ICZM. Establishing and resourcing dedicated coastal management organizations to perform ICZM functions has also been influential (Celliers *et al.*, 2013). Barriers to effective implementation in practice have included the absence of effective multi-scalar and trans-jurisdictional governance frameworks, particularly in a federal context (Falaleeva *et al.*, 2011; Wong *et al.*, 2014). Incorporating collaborative and participatory processes has proved difficult because of limited resourcing and community capacity constraints (Celliers *et al.*, 2013).

Issues of regulatory fragmentation and the need for coordinated national policy leadership have featured in numerous inquiries into coastal zone management in Australia since 1980. These have all called for greater national leadership in coastal management to prevent 'creeping degradation' of nationally important coastal values (Harvey *et al.*, 2012; Australian State of the Environment Committee (ASEC), 2011, p. 871). Despite this overwhelmingly consistent message, successive national governments have failed to act on many recommendations relating to coastal governance frameworks (George *et al.*, 2009, p. 12). The gamut of national plans and policies potentially affecting the coastal zone highlights the fragmentation and duplication that characterizes this policy space. These conditions are exacerbated by the addition of an equally diverse array of policies and laws at the state and local levels (Norman, 2009). While federal, state and territory governments have given high-level endorsement to a Framework for a National Cooperative Approach to ICZM, the framework has never been properly funded nor has a specific agency been given responsibility for its implementation or delivery (George *et al.*, 2009, p. 251).

With the emergence of climate change as a key issue affecting coastal management, ICZM has been widely promoted as a tool for coastal climate change adaptation (Falaleeva *et al.*, 2011; Harvey *et al.*, 2012; Celliers *et al.*, 2013). Yet the issues of integration, coordination and institutional scale that are central to the ICZM model remain critical unresolved challenges. The next section reviews the regulatory framework governing coastal wetlands and their adaptation needs, providing a context within which to consider progress in addressing these pervasive trans-jurisdictional governance challenges.

Current approaches to coastal planning and wetland protection

Activities affecting coastal wetlands in Australia are governed by state and local laws pertaining to land use planning, environmental impact assessment, coastal protection, catchment management, fisheries, land tenure, protected areas and biodiversity conservation. A range of non-government initiatives including catchment management, landcare and coastcare groups complement regulatory arrangements. This discussion focuses on four interrelated domains within this regulatory matrix where a more strategic, integrated, coordinated and adaptive approach is critical: coastal management, land-use planning, catchment and water management, and land tenure.

Coastal management and protection

Coastal management regimes differ considerably between state jurisdictions. In some states, coastal planning and protection is left primarily to planning and resource legislation, with detail found in policies, manuals and guidelines. The status of, and relationship between, these policies are often unclear and implementation is inconsistent due to limited resources and data at the local level (George *et al.*, 2009, p, 254). New South Wales (NSW), Queensland, Victoria and South Australia have specific coastal management and protection legislation that sets out specific regulatory requirements for activities that affect coastal wetlands, such as the assessment and approval of coastal protection works or the rehabilitation of coastal areas.

Victoria's coastal protection regime provides a good model for coordinated coastal governance that addresses strategic needs and informs local development assessment processes. The Coastal Management Act 1995 (Vic) establishes the Victorian Coastal Council and tasks it with preparation of the Victorian Coastal Strategy (VCS). The detail of the VCS is further elaborated in regional coastal action plans (CAPs), prepared and overseen by the relevant regional Coastal Board and implemented through local planning provisions. This approach provides a solid framework for controlling coastal land-use planning, but has not had a strong focus on environmental conditions or been well integrated into broader river and wetland management arrangements (State of Victoria, 2013). The 2013 Victorian Waterway Strategy purports to provide a rubric that integrates the environmental management issues for estuaries, wetlands and rivers, but with eight-year planning horizons, the success of these strategies as overarching integrative mechanisms remains to be seen.

Like the Victorian model, the Coastal Protection Act 1979 (NSW) combines state-level statutory and policy frameworks with local level planning and management measures. Under the Act, all coastal councils are required to prepare coastal zone management plans (CZMPs) to address the risks from coastal hazards – including those from climate change – and manage threats to estuary health (Coastal Protection Act 1979 (NSW) Part 4A). These plans can address a particular locality within the local government area, such as a beachfront prone to erosion or an estuary area, and can therefore be tailored to an ecologically meaningful scale. Implementation generally relies on integration with other regulatory regimes, for example via the substantive rules and standards of land use planning regimes that restrict or condition development in identified hazard zones. The NSW State Government has not provided clear policy guidance in relation to when planned retreat is recommended to preserve beach environments and amenity, and in which situations coastal armouring is warranted. Accordingly, implementation by local government has been slow and, in some cases, controversial (Corkill, 2012a; Macintosh, 2012; Thom, 2012). Councils have struggled to balance competing interests and take into consideration long-term implications of climate change. Tensions over the impact of erosion control and flood mitigation infrastructure on beach amenity and coastal ecosystem health are particularly fraught (Abel *et al.*, 2011; Measham *et al.*, 2011; Rogers *et al.*, 2014, p. 78).

The CZMP can provide a strategic context for the assessment and approval of coastal protection works, to minimize cumulative environmental impacts on coastal ecosystems. Yet, even where this strategic planning mechanism is available, there are several gaps and loopholes that potentially undermine strategic planning efforts and may allow construction of these works without full consideration of their environmental impacts. These include exemptions from the need to obtain approval for 'temporary' coastal protection works on private land (e.g. geotextile sand bags), the timeframes for which are not restricted in any way (Coastal Protection Act 1995 (NSW) s 55O), even though their effect may be comparable to prohibited hard stabilization measures (Kittinger and Ayers, 2010). For larger-scale public coastal protection works, proponents such as local councils may be able to bypass safeguards aimed at minimizing coastal erosion or loss of public beaches and beach access, in order to protect public infrastructure (Thom, 2012).

The historic focus of the Coastal Protection Act 1979 (NSW) on public access and public safety also limits its application to wetland protection. Of the 40 councils within the NSW coastal zone that are expected to prepare CZMPs, at the time of writing, only four plans contained any reference to the impacts of climate change for wetlands. The CZMP for the Ramsar-listed Hunter Estuary, an area that has benefitted from research into its capacity to adapt to sea level rise (Rogers *et al.*, 2014), provides that '[p]lanning mechanisms to adapt to the environmental impacts of climate change will generally focus on ensuring migration space for estuarine habitat (e.g. saltmarsh) in response to rising sea levels' (Newcastle City Council, 2009, p. 47). The accompanying management study contemplates that future governments may need to consider compulsory acquisition and planned retreat in order to enable endangered wetlands to respond to climate change (Newcastle City Council, 2009, p. 111). The Wallis Lake Wetland Strategy contemplates the development of buffer zones for inland migration of wetlands as a result of sea level rise (Bell, 2010), but both it and the Hunter Estuary Plan illustrate the limits of the CZMP as an overarching strategic plan. It can identify strategic directions but relies heavily on integration with other regulatory regimes, particularly land-use planning regimes, to provide the tools to implement the strategies. These too offer weak protections for wetlands. Consent authorities may approve new coastal protection works if they are satisfied that the works will not unreasonably limit public access to a beach or headland or pose a threat to public safety (Coastal Protection Act 1979 (NSW), s. 55M (1)(a) and (b)). However, they are not required to consider the impacts on adjacent coastal ecosystems. The circumstances in which the coastline may be protected and rehabilitated is also limited to beach protection, and does not explicitly extend to other coastal vegetation and ecosystems (Coastal Protection Act 1979 (NSW), ss. 55ZA and 56A).

Coastal land-use planning

Land-use planning (LUP) regimes are a critical mechanism for implementing ICZM and wetland conservation in a climate change context. Regional and local policies and plans can set a strategic context for implementing coastal adaptation measures, while specific planning tools can restrict private adaptation measures that have negative externalities, such as the construction of hard coastal defences that prevent natural coastal recession and cause erosion on neighbouring properties.

A range of planning instruments is available to protect coastal wetlands. In particular, strategic planning instruments such as restrictive zoning and buffers can be used to limit new development, so as to provide space for potential wetland migration. Many local governments are developing coastal adaptation plans and strategies that include protection of coastal ecosystems. The goals of such strategic plans are typically implemented by either prohibiting or restricting new developments in areas identified as necessary for future conservation. At the statutory planning level, some councils are experimenting with instruments that offer more flexible solutions in situations where land should be earmarked for potential wetland migration in the longer term. These include: information instruments such as planning certificates; the use of conditions revoking consent when a pre-specified event occurs, such as erosion or shoreline recession to a certain point; or covenants on title which oblige property owners to remove or relocate structures when particular conditions, such as shoreline recession, arise (Macintosh *et al.*, 2013, pp. 56–8). These allow land to be used productively until climate change impacts materialize, thereby managing the uncertainties, long timeframes and distributional issues associated with future climate impacts. To date, there has been minimal experimentation with these more flexible instruments, which are exclusively reserved for new developments. There remains a general reluctance to adopt approaches that are perceived to undermine the legal certainty that traditionally attaches to development approvals (Macintosh *et al.*, 2013, pp. 83–4).

For existing development, the role of land-use planning instruments is more limited. Strategic plans can signal long-term intentions with respect to conservation and designation of areas to be protected as future wetland migration areas, but policies of planned retreat to facilitate wetland migration will generally involve a combination of limits on new development and shoreline armouring, and programmes of land acquisition (through voluntary or mandatory schemes). In many parts of the country, low levels of investment in coastal infrastructure will result in hard protection works being the preferred adaptation strategy and retreat options being strongly opposed.

Catchment water and land use

Catchment-scale water and land-management practices are a major influence on water quality and quantity parameters in coastal wetlands. These impacts are generally regulated via a combination of environmental protection, catchment management and water allocation legislation at a state government level. Despite the established institutional framework at a catchment scale, there remain ongoing implementation challenges in tackling the complex and difficult water quality and allocation issues affecting coastal wetlands.

Diffuse source nutrient and sediment pollution from agricultural management practices is a major water quality problem for coastal wetlands. In urban areas, the quality of runoff and storm water is also a major issue (Lee *et al.*, 2006). Solutions to the regulatory challenges posed by diffuse pollution include clearly defining roles and responsibilities and employing a judicious mix of policy instruments, including traditional regulation, planning, incentives and market mechanisms (Gunningham and Sinclair, 2005). Yet diffuse pollution typically experiences fragmented governance arrangements and an ongoing reliance on largely ineffective tools, such as education and voluntary initiatives (Nelson, 2005; Cary and Roberts, 2011).

In addition to water quality issues, many coastal systems are experiencing reduced overall flows and other changes to key elements of the flow regime, as a consequence of diversions for urban and agricultural uses and increased coverage of urban areas by hard surfaces. These changes have implications for the complex freshwater/saltwater regimes of estuaries and dependent ecosystems (Pierson *et al.*, 2002). Extensive recent reforms to water allocation governance have generally resulted in significant improvements (Council of Australian Governments (COAG), 2004) to environmental flows, important for coastal floodplain systems and estuaries. The Water Act 2007 (Cth, s. 3 and 4) makes clear that the Act applies to the water resources of the Murray-Darling Basin only. The Basin is Australia's major inland river system. As a result of major problems of over-allocation and ecological decline, much of the recent emphasis in water law reform and water planning has been in this area (Gardner *et al.*, 2009).

Land tenure and protected areas

The tenure of wetlands and their protected status have important implications for their protection and management under climate change. These factors dictate the actors involved, the resources available and the applicable statutory and policy frameworks, all of which can operate as barriers to their long-term protection. The highest formal level of protection for coastal wetlands is their inclusion in the reserve network. Twenty of Australia's coastal wetlands are listed under the Ramsar Convention on Wetlands of International Importance (Department of Climate Change (DCC), 2009, p. 51).

Conservation planning processes for listed wetlands focus on on-site controls, but do not deal extensively with off-site activities, such as water abstraction and catchment land use, that heavily influence the ecological character of wetlands (Bennet and Gardner, 2013). There have been major improvements in allocating water to environmental assets, such as Ramsar wetlands, within the Murray-Darling Basin, but not all states and territories have strong statutory regimes requiring environmental water allocation for other wetlands.

The majority of formal protected areas in Australia – Ramsar-listed or otherwise – are designated and managed under state legislation which provides that reserves should be managed primarily for nature conservation objectives (Environmental Protection and Biodiversity Conservation Act 1999 (Cth) ss. 328–33; McCormack and McDonald, 2014, p. 120). Climate change challenges these objectives and their focus on preserving the status quo (Dunlop *et al.*, 2012). Seldom do current approaches to protected area management in Australia provide for management of dynamic landscapes (Rogers *et al.*, 2014, p. 68), although climate change will see significant changes to the biodiversity characteristics of land located within reserve boundaries. In some cases, this will mean that protected features will not be adequately protected and that applicable tenure and management arrangements that provide future buffers, corridors and migration pathways to adjoining areas will be as important as those for the wetland itself. Modelling by Rogers *et al.* (2014, p. 72) suggests that on a high-sea level rise scenario, the potential coastal wetland occurring within the Hunter Estuary Ramsar site will decrease from 1393 to only 202 ha, and the area located within the Hunter Wetlands National Park would fall from 1610 to only 618 ha. Narrow protected areas along the coastal strip may therefore require the protection and rehabilitation of substantial adjoining land to allow space for the migration of wetlands. For this to happen, conservation reserve policy must be integrated with broader coastal management and land-use planning.

The rules governing property boundaries in a coastal context may also have a significant bearing on our ability to protect coastal wetlands under climate change. As sea levels rise, particularly in situations of coastal erosion, the public–private property boundary must respond and move landward to maintain public beach access and provide for retreat pathways for coastal ecosystems. Most state-level property regimes provide that boundaries are determined by reference to the mean high-water mark (MHWM), in which case boundaries will shift landwards with coastal recession (Corkill, 2012b, p. 56; 2013, pp. 53–4). However, in rare instances, property titles are determined using fixed or 'right line' boundaries. There are concerns that this may mean that the boundary remains unchanged even when coastal erosion causes beach recession, with the result that beaches and coastal wetlands may end up being located wholly on private land (Corkill, 2012b).

A related issue is whether or not private landholders then have a 'right' to defend their property against the sea. The prevailing view appears to be that statutory regimes for coastal management, such as the Coastal Protection Act 1979 (NSW), have clearly supplanted any ancient common law 'rights' to protect private property from the encroaching sea (Corkill, 2013, pp. 53–4). But given the contentious situations (including litigation) that are likely to arise as coastal erosion begins to affect existing private property, enhanced clarity and consistency through legislation may be worthwhile, to complement the land-use planning tools that can be used allow space for the future adaptation of coastal ecosystems.

Wetland conservation in an integrated, adaptive coastal governance framework

The pervasive trans-jurisdictional and multi-scalar governance challenges associated with coastal management outlined above illustrate that the core tenets of ICZM have not been fully realized in Australia and that current arrangements do not effectively support the conservation of coastal wetlands. Climate change exacerbates existing stressors and poses new challenges for coastal management, underscoring the importance and urgency of achieving a more integrated and adaptive trans-jurisdictional governance regime for the coastal zone. Attention to a few key reforms aimed at enhancing coordination and integration of coastal governance could meet some of these climate change challenges and support the long-term conservation of coastal wetlands.

Multi-scale and multi-sector strategic coastal adaptation planning

Strategic, long-term planning is a core component of all four of the regulatory domains considered in this chapter, and there are already some mechanisms available to achieve a certain level of coordination between these regimes. Yet the discussion above has highlighted how these integrative mechanisms are not effectively used in practice and that the long-term conservation of coastal wetlands demands enhanced integration and coordination across these regimes. There is a clear need for stronger policy leadership from the upper levels of government to guide strategic planning at the local or catchment scale, particularly to advance environmental objectives. Strategic planning instruments have the potential to map out broad adaptation strategies at a meaningful institutional and geographical scale. They can transcend case-by-case decision-making and account for cumulative and longer-term impacts of individual decisions, thereby providing links between temporal and spatial/jurisdictional scales of governance. Strategic planning can also provide a mechanism to link otherwise separate regulatory domains such as coastal management, land-use planning, biodiversity conservation and protected area management (Bates *et al.*, 2013; Siders, 2013).

A nested set of instruments, from the federal to the local or catchment scale, could provide an overarching mechanism for setting the strategic context for coastal management, under which more targeted sector-based strategic planning is undertaken. Current arrangements operate largely at the state and local government scale, so the most obvious remaining gap in the coastal governance framework in Australia is national leadership. Common national policy positions on key issues in coastal adaptation are critical and long overdue, including prioritizing the protection of coastal ecosystems (George *et al.*, 2010, p. xxiv; Coasts and Climate Change Council, 2011). The typical model for such coordination is to develop common policy directions under the auspices of the Council of Australian Governments (COAG) and tie federal funding to the implementation of policy by state governments (e.g. COAG, 2004), but the 2013 abolition of COAG's Standing Council on Environment and Water suggests that such an initiative is unlikely to be a priority in the near future.

At a state government level, mechanisms exist for articulating policy and strategy and implementing these through local government (or catchment-scale) planning instruments. For example, the Victorian Waterway Management Strategy 2013 states that it provides the 'leading statewide strategic direction on the management of the environmental condition of wetlands' and attempts to coordinate provisions contained in the Water Act 1989 (Vic), the State

Planning Policy Framework of the Victorian Planning Provisions and the Environmental Effects Act 1989 (Vic). In the past, coordination across these regimes has been limited and the trans-temporal dimensions of climate change are unrepresented. It is therefore worth considering options to enhance strategic integration.

Properly implemented, consistent statutory coastal management regimes in each state could provide a valuable mechanism for setting overarching statutory objectives and developing strategic coastal management plans and adaptation strategies for particular areas. Local-to-regional scale CZMPs can help identify areas that warrant coastal defences in the short-to-medium term and those where planning strategies and other intersecting mechanisms should be tailored to support efforts to maintain coastal ecosystems, amenity and public access. The experience with Coastal Zone Management Plans under the Coastal Protection Act 1979 (NSW) suggests a considerable implementation deficit, so the reforms outlined in the following sections seek to address this deficit.

Clear statutory and policy guidance on environmental objectives

Statutory objectives and policy positions on how environmental objectives are to be considered by decision makers are typically part of a range of competing considerations, and provide insufficient practical guidance on managing conflicting objectives. Achieving a balance between competing environmental, economic and social values is widely used as an overarching statutory or policy principle, consistent with the principle of ecologically sustainable development. Yet the act of 'balancing' environmental, economic and social impacts in practice requires detailed direction on how these often-competing impacts should be treated in different situations, so as to prevent environmental values being overridden by short term economic considerations. A lack of clear policy direction on contentious coastal adaptation issues has translated to very tentative adoption of planning measures that provide for long-term protection of coastal ecosystems, especially in situations where these measures conflict with the protection of existing property and infrastructure.

Recent experience in a water allocation planning context in Australia highlights that statutory duties and associated environmental standards play a critical role in directing decision makers to consider and provide for public interest environmental values; and in forcing the development of substantive policy guidance on how to achieve more sustainable levels of water allocation (Gentle and Olszak, 2007; Foerster, 2011). Such tools could be used to better effect in conserving coastal wetlands under climate change. State planning and coastal protection laws should be clarified to impose a clear obligation on public authorities to maintain and protect coastal ecosystems and to protect public goods over private interests (Thom, 2012). California's Coastal Act 1976, for example, provides that conflicts between provisions of the Act should be 'resolved in a manner which on balance is the most protective of significant coastal resources' (California Public Resources Code, s 30007.5). Where there are compelling reasons for permitting shoreline protection or armouring, there should be clear statutory requirements for defensive structures to apply 'living shoreline' design principles (Lutz, 2005). To protect priority natural values, strategic planning processes will require good-quality targeted information on these values. The emphasis of information gathering on climate change impacts to date has been on mapping hazard exposure for settlements and infrastructure. Adaptation planning processes have generally not had access to quality information inputs on associated natural values, such as comprehensive biodiversity and landscape function mapping, valuing of ecosystem services, or identification of retreat pathways for coastal ecosystems and measures to facilitate successful retreat, such as revegetation or rehabilitation.

Develop implementation tools for risk and cost-sharing

One of the most critical gaps in the current governance framework for coastal management in Australia is the availability of implementation tools for sharing the risks and costs associated with climate change adaptation, particularly where the adaptation strategy impinges on existing uses and private property rights. This is especially relevant to strategies involving planned retreat, which seek to minimize risks to infrastructure and provide for the long-term conservation and migration of coastal ecosystems. Tools are available under current land-use planning regimes to manage the long timeframes and uncertainties associated with climate change, but these remain largely untested in practice and have been met with opposition in particular localities where coastal erosion is already threatening valuable beachfront property. It is increasingly apparent that local governments will struggle to implement such difficult policy choices without clear guidance from state government.

Along with clear policy guidance on the situations in which a planned retreat policy may be preferable, there is a pressing need to invest in the development of implementation tools that address distributional impacts (McDonald and Styles, 2014). The United States' experience with so-called 'rolling easements' may offer an equitable and effective solution (Titus, 1998; Caldwell and Segall, 2007; Kittinger and Ayers, 2010). Rolling easements typically provide for ongoing beach amenity and public access in cases of coastal erosion, yet they can also provide for the potential migration of important coastal wetlands. For example, an easement may be purchased from property owners adjoining an estuary or other wetland that is likely to move landward as sea levels rise. This could involve a prohibition on engineered coastal defences and a requirement to relocate once water levels reach a certain distance from the dwelling. Importantly, the rolling easement concept involves some level of public investment in securing environmental values into the future, in recognition of the fact that past planning decisions have resulted in legacy development, the protection of which would undermine these environmental values (Caldwell and Segall, 2007, p. 567).

Complementary reforms to intersecting regimes

Despite the importance of coordination and integration across policy regimes, it is not always possible to bring a wide range of intersecting regimes within one coordinating strategic plan. In some cases, formal inclusion may be overkill; in others, it may be inadequate without a corresponding change in the focus or operation of that distinct regulatory regime. For example, water allocation regimes will often be a significant determinant of coastal wetland health and a critical consideration in devising adaptation strategies, but it would be impractical to consider the detail of flow regimes, levels of allocation and operational practices in a coastal management strategy. As well as institutional reform in the coastal management domain, there will also undoubtedly be a range of complementary reforms required *within* intersecting regulatory regimes. For example, more adaptive and long-term approaches to the identification and designation of critical habitat areas would enable the designation of 'future critical habitat', improving the formal protection afforded to areas, which may become important. Ensuring consistent and supportive rules relating to property boundaries, as sea levels rise, would also minimize conflicts with private property rights.

Dedicated institutions and processes for coordination and stakeholder engagement

One of the key governance questions for improved coastal management and climate change adaptation in Australia is whether we need to establish dedicated institutions for strategic planning and to coordinate implementation. A related issue is the extent to which we should embrace more participatory governance models in developing and selecting adaptation strategies (Pahl-Wostl, 2009), a core principle of the ICZM model.

Most of the coastal adaptation planning that has been undertaken to date has been led by local government under the statutory parameters of land use planning or coastal management regimes, with minimal and reactive opportunities for public comment on planning policies. The capacity of local government planners to cover the breadth of coastal management issues, including wetland conservation is limited. Dedicated institutions can overcome these limitations, develop requisite expertise, and apply consistent policies and priorities. For example, stakeholder committees were central to the preparation of water sharing plans in the state of New South Wales under the Water Management Act 2000 (NSW) and also in the subsequent management of environmental water allocations. Similar arrangements have been used in other state jurisdictions (Foerster 2008, 2012). Institutions such as catchment management authorities operate at a biophysical scale that matches the environmental management issues at hand. Further consideration of the best institutional model is warranted. For example, the applicability of models like the Californian Coastal Commission and its partner organization, the Coastal Conservancy, as well as the umbrella coordinating body the Ocean Protection Council, is worth more detailed evaluation (Caldwell and Segall, 2007, p. 545).

Detailed evaluation of the Victorian model under the Coastal Management Act 1995 (Vic), involving regional coastal protection boards with specific planning and management functions, and a state-level Coastal Council, may also reveal opportunities for enhanced institutional design.

Stakeholder involvement in adaptation planning processes is undoubtedly important to ensure that there is adequate consideration of the full range of relevant interests and values (including environmental interests). Engagement can also help integration across otherwise disparate sectors of government, and contribute to more equitable and legitimate outcomes (Burch, 2010; Fünfgeld, 2010). More participatory processes can also help to establish a social licence to support implementation of adaptation responses by government. The practical experience of collaborative governance in adaptation planning (and other natural resource management contexts), however, suggests that considerable care should be exercised in determining the extent, nature and timing of public participation (Holley and Sinclair, 2013). Locally-based deliberative processes face major challenges in moving beyond place-specific, short-term responses in favour of long-term strategic coordination and adequate consideration of the full range of relevant interests and values (Nelson *et al.*, 2007). There has been some experimentation in Australia with broader stakeholder involvement in water and adaptation planning. These planning processes have not generally been undertaken as part of existing statutory or policy regimes and, as a consequence, have varied quite considerably (Macintosh *et al.*, 2013, pp. 116–23). These experiences suggest that while broad community engagement helps build trust and strengthen relationships, without clear policy direction, information on protecting environmental values, implementation guidance and specific representation of public environmental interests, such processes will struggle to transcend local, short-term, private interests (Gentle and Olszak and Gentle, 2007; Foerster, 2011).

Implementing effective participatory processes that also provide for public environmental interests will be resource- and time-intensive. This will add substantially to the costs associated with adaptation planning, so comprehensive community-based processes may not be feasible in all localities. In light of the practical experience of collaborative governance to date, there remains a very real possibility that community processes may prove counterproductive in efforts to ensure that longer-term cumulative impacts on environmental values are considered in adaptation planning.

Conclusions

As sea levels rise and storm weather becomes more intense, the impacts of climate change on coastlines and coastal wetlands are likely to be profound. These impacts will be made worse by the efforts of coastal communities to address their own adaptation needs, through engineered protection of public and private infrastructure and coastal armouring. Some loss of wetlands is inevitable and, like other biodiversity adaptation challenges, we may also need to reassess and redefine our conservation goals – a focus on ecosystem resilience and the protection of ecosystem services may replace preservation of the status quo as the overarching goal of many conservation efforts (Craig, 2010; McCormack and McDonald, 2014). Nonetheless, the critical importance of coastal wetlands in providing erosion control, nutrient filtering, fisheries habitat and carbon sequestration, as well as their recreational, tourism and cultural values, means that their protection should be a matter of local, state, national and indeed international priority. In particular, the role played by these ecosystems in building the resilience of shorelines to erosion and inundation highlights their central importance to the formulation of adaptation strategies for coastal settlements.

This chapter has demonstrated that, despite their importance and the threats they face, the governance of coastal wetlands in Australia, like the coastal zone more generally, is complex, fragmented and tends to undermine environmental values. Integrated coastal zone management may offer the best model by which to address these governance failures, because it explicitly contemplates the need to work across jurisdictions and sectors, both vertically and horizontally. Australian states, territories and the Commonwealth have made hortatory commitments to the principles and goals of ICZM, but without a coordinating mechanism, clear designation of roles and responsibilities, and proper funding, ICZM remains merely a policy aspiration.

The imperative of climate change adaptation presents an additional challenge for ICZM, because the long timeframes, uncertainties and trans-temporal dimensions are greater (Wong *et al.*, 2014). Yet it presents perhaps the best opportunity to embrace the legal, institutional and attitudinal shifts required to reinvigorate the goals of integrated coastal zone management with a strong focus on wetlands protection (Falaleeva *et al.*, 2011). This will require a range of reforms. Consistent state coastal planning legislation would go a long way towards enhancing the strategic framework for wetland conservation under climate change, preferably with clear guidance to decision makers about the way in which environmental objectives should be furthered. Tools and instruments are needed that promote equitable risk and cost sharing, and that recognize the long timeframes involved with climate change and the uncertainties these produce. Many such tools are already available, but require testing, refinement and the will to use them. Along with greater attention to strategic multi-sectoral coastal planning, the ways in which individual intersecting regimes consider and value coastal wetlands must also be reviewed, to ensure that each of them promotes long-term protection. Finally, stronger institutional arrangements and collaborative decision-making techniques are needed.

With concerted efforts towards enhanced trans-jurisdictional governance and ICZM, coastal wetlands may have the opportunity to adjust and adapt naturally to the pressures of the coastal squeeze that sea-level rise portends. Pursuing existing governance pathways destine coastal wetlands to become another casualty of anthropogenic climate change.

The law in this chapter is current to December 2014.

References

Abel, N., Gorddard, R., Harman, B., Leitch, A., Langridge, J., Ryan, A. and Heyenga, S. (2011) 'Sea level rise, coastal development and planned retreat: analytical framework, governance principles and an Australian case study', *Environmental Science and Policy*, 14, 3, 279–88.

Australian State of the Environment Committee (ASEC) (2011) 'State of the Environment 2011: Independent report to the Australian Government Minister for Sustainability, Environment, Water, the Arts and Communities', Department of Sustainability, Environment, Water, Population and Communities, Canberra, Australia.

Bates, L. E., Green, M., Leonard, R. and Walker, I. (2013) 'The influence of fora and multilevel governance on the climate adaptation practices of Australian Organizations', *Ecology and Society*, 18, 4, art. 62.

Bell, M.(2010) 'Wallis Lake Wetland Strategy: Vision, Guiding Principles and Actions', Great Lakes Council, Natural Systems Branch, Forester, Australia.

Bennet, M. and Gardner, A. (2013) 'How do environmental conservation laws interact with the environmental aspects of water laws?', Paper presented to Mainstreaming Environmental Water Law and Practice, Melbourne Law School, Melbourne, Australia.

Blankespoor, B., Dasgupta, S. and Laplante, B. (2012) 'Sea-Level Rise and Coastal Wetlands: Impacts and Costs', Policy Research Working Paper no 6277), World Bank, Washington, DC, http://ssrn.com/abstract=2181286, accessed 25 March 2015.

Burch, S. (2010) Transforming barriers into enablers of action on climate change: Insights from three municipal case studies in British Columbia, Canada', *Global Environmental Change*, 20, 2, 287–97.

Burley, J. G., McAllister, R. R., Collins, K. A. and Lovelock, C. E. (2012) 'Integration, synthesis and climate change adaptation: a narrative based on coastal wetlands at the regional scale', *Regional Environmental Change*, 12, 3, 581–93.

Caldwell, M. and Segall, C. H. (2007) 'No day at the beach: Sea level rise, ecosystem loss and public access along the Californian coast', *Ecology Law Quarterly*, 34, 2, 533–78.

Cary, J. and Roberts, A. (2011) 'The limitations of environmental management systems in Australian agriculture', *Journal of Environmental Management*, 92, 3, 878–85.

Celliers, L., Rosendo, S., Coetzee, I. and Daniels, G. (2013) 'Pathways of integrated coastal management from national policy to local implementation: Enabling climate change adaptation', *Marine Policy*, 39, 72–86.

Clarke, B., Harvey, N. and Nursey-Bray, M. (2012) 'Australian coastal management and climate change', *Geographical Research*, 50, 4, 356–67.

Coastal Act of 1976, Cal. Stat. §§ 30000–30900. (West 1977 & Supp. 1980).

Coastal Management Act 1995 (Vic).

Coastal Protection Act 1979 (NSW).

Coasts and Climate Change Council (2011) 'Coasts and Climate Change Council Advice to Minister Combet', Department of the Environment, Canberra, Australia, http://environment.gov.au/climate-change/adaptation/australias-coasts/coasts-climate-change-council-advice, accessed 25 March 2015.

Corkill, J. (2012a) 'Low Tide for Non-Government Interests in NSW Coastal Management', Paper presented at the 21st NSW Coastal Conference, Kiama,

Australia, http://coastalconference.com/2012/papers2012/John%20Corkill%20Full%20Paper.pdf, accessed 25 March 2015.

Corkill, J. (2012b) 'Principles and problems of shoreline law', National Climate Change Adaptation Research Facility, Gold Coast, Australia, www.nccarf.edu.au/settlements-infrastructure/sites/www.nccarf.edu.au.settlements-infrastructure/files/DPG_ACCA RNSI_Node_1_Principles%26ProblemsOfShorelineLaw_FINAL.pdf, accessed 25 March 2015.

Corkill, J. (2013) 'Claimed property right does not hold water', *Australian Law Journal*, 87, 1, 49–58.

Council of Australian Governments (COAG) (2004) 'Intergovernmental Agreement on a National Water Initiative 2004', COAG, Barton, Australia, http://nwc.gov.au/nwi, accessed 25 March 2015.

Craig, R. K. (2010) 'Stationarity is dead – Long live transformation: Five principles for climate change adaptation law', *Harvard Environmental Law Review*, 34, 1, 9–75.

Department of Climate Change (DCC) (2009) 'Climate change risks to Australia's coast: a first pass national assessment', Department of Climate Change, Canberra, Australia, http://environment.gov.au/system/files/resources/fa553e97–2ead-47bb-ac80-c12adffea944/files/cc-risks-full-report.pdf, accessed 25 March 2015.

Dunlop, M., Hilbert, D., Ferrier, S., House, A., Liedloff, A., Prober, S., Smyth, A., Fletcher, C., Harwood, T., Martin, T., Murphy, H. and Williams, K. (2012) 'The Implications of Climate Change for Biodiversity Conservation and the National Reserve System: Final Synthesis', CSIRO Climate Adaptation Flagship, Canberra, Australia, https://publications.csiro.au/rpr/download?pid=csiro:EP105380&dsid=DS4, accessed 25 March 2015.

Environmental Effects Act 1989 (Vic).

Environmental Protection and Biodiversity Conservation Act 1999 (Cth).

Falaleeva, M., O'Mahony, C., Gray, S., Desmond, M., Gault, J. and Cummins, V. (2011) 'Towards climate adaptation and coastal governance in Ireland: Integrated architecture for effective management?', *Marine Policy*, 35, 6, 784–93.

Fletcher, M. and Haines, P. (2009) 'Hunter Estuary Coastal Zone Management Plan'Final report, prepared for Newcastle City Council, Port Stephens Council, Maitland City Council and NSW Department of Environment and Climate Change. Broadmeadow: BMT WBM, Broadmeadow, Australia, http://maitland.nsw.gov.au/UserFiles/File/Environment/Estuary%20Plan.pdf, accessed 25 March 2015.

Foerster, A. (2008) 'Managing and protecting environmental water: Lessons from the Gwydir for ecologically sustainable water management in the Murray Darling Basin,' *Environmental and Planning Law Journal*, 25, 1, 130–53.

Foerster, A. (2011) 'Developing purposeful and adaptive institutions for effective environmental water governance', *Journal of Water Resource Management*, 25, 15, 4005–18.

Foerster, A. (2012) 'Emerging environmental water governance in Australia,' *The Journal of Water Law*, 22, 2–3, 58–72.

Foerster, A., Macintosh A. and McDonald, J. (2013) 'Transferable lessons for climate change adaptation planning? Managing bushfire and coastal climate hazards in Australia', *Environmental and Planning Law Journal*, 30, 6, 469–90.

Fünfgeld, H. (2010) 'Institutional challenges to climate risk management in cities', *Current Opinion in Environmental Sustainability*, 2, 3, 156–60.

Gardner, A. and Bennet, M., (2014) 'How do environmental conservation laws interact with the environmental aspects of water laws?', *Environmental and Planning Law Journal*, 31, 1, 3–10.

Gardner A, Bartlett R. and Gray, J. (2009) *Water Resources Law*, LexisNexis Butterworths, Chatswood, Australia.

Gentle, G. and Olszak, C. (2007) 'Water planning, principles, practice and evaluation', in K. Hussey and S. Dovers (eds) *Managing Water for Australia: The Social and Institutional Challenges*, CSIRO Publishing, Collingwood, Australia, pp. 59–73.

George, J., Washer, M., Cobb, J., D'Ath, Y., Dreyfus, M., Irwin, J., Livermore, K., Marino, N., Murphy, J., Scott, B., Wood, J. and Zappia, T. (2009) 'Managing our coastal zone in a changing climate: The time to act is now', House of Representatives Standing Committee on Climate Change, Water, Environment and the Arts, Canberra, Australia.

Gilman, E. L., Ellison, J., Duke, N. C. and Field, C. (2008) 'Threats to mangroves from climate change and adaptation options: A review', *Aquatic Botany*, 89, 2, 237–50.

Gunningham, N. and Sinclair, D. (2005) 'Policy instrument choice and diffuse source pollution', *Journal of Environmental Law*, 17, 1, 51–81.

Hamstead, M., Baldwin, C. and O'Keefe, V. (2008) '*Water Allocation Planning in Australia – Current Practices and Lessons Learned*,' Waterlines Occasional Paper no 6, National Water Commission, Canberra, Australia.

Harvey, N., Clarke, B. and Nursey-Bray, M. (2012) 'Australian coastal management and climate change', *Geographical Research*, 50, 4, 356–67.

Holley, C. and Sinclair, D. (2013) 'Deliberative participation, environmental law and collaborative governance: insights from surface and groundwater studies', *Environmental and Planning Law Journal*, 30, 1, 32–55.

Kittinger, J. N. and Ayers, A. L. (2010) 'Shoreline armoring, risk management, and coastal resilience under rising seas', *Coastal Management*, 38, 6, 634–53.

Lee, S. Y., Dunn, R. J. K., Young, R. A., Connolly, R. M., Dale, P. E. E., Dehayr, R., Lee, S. Y., Lemckert, C. J., McKinnon, S., Powell, B., Teasdale, P.R. and Welsh, D. T. (2006) 'Impact of urbanization on coastal wetland structure and function', *Austral Ecology*, 31, 2, 149–63.

Lutz, L. (2005) 'Shoring up coasts against erosion', *Chesapeake Bay Journal*, http://bayjournal.com/article/shoring_up_coasts_against_erosion, accessed 20 April 2013.

Macintosh, A. (2012) 'Coastal climate hazards and urban planning: how planning responses can lead to maladaptation', *Mitigation & Adaptation Strategies for Global Change*, 18, 7, 1035–55.

Macintosh, A., Foerster, A. and McDonald, J. (2013) 'Limp, leap or learn? Developing legal frameworks for climate change adaptation planning in Australia', National Climate Change Adaptation Research Facility, Gold Coast, Australia, www.nccarf.edu.au/publications/limp-leap-or-learn, accessed 20 April 2013.

McCormack, P. and McDonald, J. (2014) 'Adaptation strategies for biodiversity conservation: Has Australian law got what it takes?', *Environmental and Planning Law Journal*, 31, 2, 114–36.

McDonald, J. and Styles, J. (2014) 'Legal strategies for adaptive management under climate change', *Journal of Environmental Law*, 26, 1, 25–53.

McLeod, E., Chmura, G., Bouillon, R., Salm, S., Björk, M., Duarte, C. M., Lovelock, C. E., Schlesinger, W. H. and Silliman, B. R. (2011) 'A blueprint for blue carbon: Toward an improved understanding of the role of vegetated coastal habitats in sequestering CO2', *Frontiers in Ecology and the Environment*, 9, 10, 552–60.

Measham, T. G., Preston, B. L., Smith, T. F., Brooke, C., Gorddard, R., Withycombe, G. and Morrison, C. (2011) 'Adapting to climate change through local municipal planning: Barriers and challenges', *Mitigation and Adaptation Strategies for Global Change*, 16, 8, 889–909.

Millennium Ecosystem Assessment (MEA) (2005) 'Ecosystems and human well-being: wetlands and water: Synthesis', World Resources Institute, Washington, DC.

Nelson, R. (2005) 'Legislation for ICM: Advancing water resources sustainability?', *Environmental and Planning Law Journal*, 22, 96–129.

Nelson, R., Adger, W. N. and Brown, K. (2007) 'Adaptation to environmental change: Contributions of a resilience framework', *Annual Review of Environment & Resources*, 32, 395–419.

Newcastle City Council (2009) 'Hunter Estuary Management Plan', Newcastle City Council, Newcastle, Australia, http://newcastle.nsw.gov.au/__data/assets/pdf_file/0009/87057/HEMP_Adopted_Final_October_2009.pdf, accessed 15 April 2015.

Newton v Great Lakes Council [2013] NSWLEC 1248.

Nicholls, R. J., Wong, P. P., Burkett, V. R., Codignotto, J., Hay, J., McLean, R., Ragoonaden, S. and Woodroffe, C. D. (2007) 'Coastal systems and low-lying areas', in: M. L. Parry, O. F. Canziani, J. P. Palutikof, P. J. van der Linden and C. E. Hanson (eds) *Climate Change 2007: Impacts, Adaptation and Vulnerability. Contribution of Working Group II to the Fourth Assessment Report of the Intergovernmental Panel on Climate Change*, Cambridge University Press, Cambridge, UK, pp. 315–56.

Norman, B. (2009) 'Planning for Coastal Climate Change: an Insight into International and National Approaches', Department of Planning and Community Development and Department of Sustainability and Environment, East Melbourne, Australia.

Pahl-Wostl, C. (2009) 'A conceptual framework for analyzing adaptive capacity and multi-level learning processes in resource governance regimes', *Global Environmental Change*, 19, 3, 354–65.

Pierson, W. L., Bishop, K., Van Senden, D., Horton, P. R. and Adamantidis, C. A. (2002) 'Environmental Water Requirements to Maintain Estuarine Processes. Environmental Flows Initiative', Technical report no 3, Environment Australia, Canberra, Australia.

Rogers, K., Saintilan, N. and Copeland, C. (2014) 'Managed retreat of saline coastal wetlands: Challenges and opportunities identified from the Hunter River Estuary, Australia', *Estuaries and Coasts*, 37, 1, 67–78.

Shoo, L. P., O'Mara, J., Perhans, K., Rhodes, J. R., Runting, R. K., Schmidt, S., Trail, L.W., Weber, L.C., Wilson, K.A. and Lovelock, C.E. (2014) 'Moving beyond the conceptual: specificity in regional climate change adaptation actions for biodiversity in South East Queensland, Australia', *Regional Environmental Change*, 14, 2, 435–47.

Siders, A. (2013) 'Managed Coastal Retreat: A Legal Handbook on Shifting Development Away from Vulnerable Areas', Columbia Public Law Research Paper, no. 14–365, Columbia Center for Climate Change Law, New York.

Sorenson, J. (1993) 'The international proliferation of integrated coastal zone management efforts', *Oceans and Coastal Management*, 21, 1, 45–80.

State of Victoria (2013) 'Improving our waterways: Victorian waterway management strategy', Department of Environment and Primary Industries, East Melbourne, Australia, http://depi.vic.gov.au/__data/assets/pdf_file/0015/200373/VWMS_Part1.pdf, accessed 25 March 2015.

Thom, B. (2012) 'Climate change, coastal hazards and the public trust doctrine', *Macquarie Journal of International and Comparative Environmental Law*, 8, 2, 21–41.

Titus, J. G. (1998) 'Rising seas, coastal erosion, and the takings clause: How to save wetlands and beaches without hurting property owners', *Maryland Law Review*, 57, 4, 1279–399.

Water Act 1989 (Vic).

Water Management Act 2000 (NSW).

Wong, P. P., Losada, I. J., Gattuso, J. P., Hinkel, J., Khattabi, A., McInnes, K. L., Saito, Y. and Sallenger, A. (2014) 'Coastal systems and low-lying areas', in C. B. Field *et al.* (eds) *Climate Change 2014: Impacts, Adaptation, and Vulnerability. Part A: Global and Sectoral Aspects. Contribution of Working Group II to the Fifth Assessment Report of the Intergovernmental Panel on Climate Change*, Cambridge University Press, Cambridge, UK and New York, pp. 361–409.

14 Contests over 'wild' rivers in Queensland

Implications for trans-jurisdictional water governance

Poh Ling Tan

Introduction

> Water is the life for us all. It's the main part. If we are going to lose that, I don't know where we gonna stand. If that water go away, everything will die. That's the power of water. He connect with the land. Pukarrikarra (the dreaming) put 'em all together. One life.
>
> (John 'Dudu' Nangkiryn, Bidyadanga, in Yu, 1999)

With those powerful words, one Aboriginal Australian[1] tells us what water means for him. Because water is core to their identity, Aboriginal people continue to assert shared values of caring for their land and water despite an unhappy history of dispossession (Queensland Aboriginal Forum, 2013). Three prime ministers have now acknowledged that history: Keating in a historic 1992 Redfern address; Rudd in his 2008 apology to the Stolen Generations; and Abbott in 2013, stating that he wanted a new engagement with Aboriginal people to be one of the hallmarks of his government.

Contests over land and water between the state of Queensland and Indigenous communities have resulted in several landmark decisions relating to human rights and native title.[2] These rights, later statutorily confirmed under the Native Title Act 1993 (Cth), were subsequently limited by amendments, with reading down of its provisions also providing barriers for Indigenous claims (Tehan, 2003; Bartlett, 2004; Young, 2008; McAvoy, 2009). For many Indigenous Australians, historical exclusion from legal rights to their traditional estates left them without claims to sovereignty, limiting their day-to-day land and water use decisions, leaving some embittered about native title, seeing it as a 'white-man's' construct that pits groups against each other (Queensland Aboriginal Forum, 2013).

Over the past six decades, various international conventions and protocols have expressed global concern over Indigenous rights. However, within Australia, despite water reforms, the interests and values of Indigenous people in regard to freshwater have not been adequately addressed (Behrendt and Thompson, 2004; Durette, 2010; Tan and Jackson, 2013).

This chapter examines the question of what the Aboriginal struggle for water in Queensland offers, in the context of the management of near-natural rivers, to the development of trans-jurisdictional water governance. This chapter considers access to water outside of the native title framework. It begins with a brief discussion of the rights-based approach that Indigenous people have used to frame their arguments in claiming a right to access and protect land and water, and the examines the wild rivers debate in two vast regions in Queensland, the Cape York Peninsula and the Channel Country. It then draws together the main messages relating to Indigenous access to water from the two case studies while questioning whether there are lessons for improved trans-jurisdictional water governance. Data on which the analyses are based were collected from interviews, focus groups and general discussion at a Queensland Aboriginal Land and Water Management Forum attended by 30 Aboriginal participants (Queensland Aboriginal Forum, 2013).

Indigenous water rights and the rights-based approach

Since 'the federal [US] government finally articulated a policy of self-determination for Indian nations in the 1970s (Tsosie, 2006, p. 21), the rights-based approach has underpinned the implementation of the regime for access to water for the Indigenous peoples in the United States of America (see, e.g. Thorson *et al.* 2006; Tarlock, 2010). These legal rights are based on treaties that allow the Indian nations to form tribal governments and give them important rights to land, water, hunting and fishing. However, in the absence of a clearly articulated treaty-protected and prior right for Indigenous Australians, the Australian position is different.

In Australia, a rights-based approach relies on a different reading of the term *sovereignty*. In brief, while sovereignty may be read as an exclusive authority exercised by an absolute state power, more generally it is about authority and authority to govern (Brennan *et al.*, 2005, p. 70). Contemporary interpretations consider sovereignty as 'divisible and capable of being shared or pooled across different entities or locations' (Brennan *et al.*, 2004, p. 312; 2005). In this context, Australian Indigenous calls for sovereignty do not extend to secession or to the creation of an independent nation state, or even self-government within geographical boundaries, but, rather, uniquely seek a recognition of cultural distinctiveness, self-determination and a sense of shared authority in a political system (Behrendt, 2003; Robbins, 2010). This argument for sovereignty continues to hold currency, not as an end in itself, or as a justification to separate from the Australian state, but as a basis for the recognition of rights and transference of power from the state.

However, this contextualized interpretation is contested. Arzey and McNamara (2011) find that claims disputing the application of Australian law to Indigenous groups and persons on the basis of Indigenous sovereignty have been consistently rejected by the courts. Nevertheless, Robbins (2010, p. 261) finds:

> . . . a significant body of scholarly literature attests to the growing acceptance of the language of rights to address the inequality of Indigenous peoples and the need for a just solution for their desire to control their own future and claim their own identity. Self determination as a form of shared sovereignty is a crucial concept in such a theory of justice.

The concept of self-determination has become less focused on political gains than on practical ones – on finding innovative means to support decision-making and governance (Anderson, 2007, p. 152). Thus a practical self-determination approach informs Australian Indigenous people's claims to land and water. Adopting elements of this view, a recent national report calls for self-determination; full free and informed consent; the right to maintain and protect culture and environment, including water; and a right to social and economic development (First Peoples Water Engagement Council (FPWEC), 2012). This report draws from the discourse of international human rights, in particular the 2007 *United Nations Declaration on the Rights of Indigenous People* (UNDRIP) signed by Australia in 2009.

The Australian legislative context for Indigenous water rights

The Australian Constitution provides for states' rights and responsibility over land and water management (Gardner *et al.*, 2009). However certain Commonwealth powers such as the corporations power, the external affairs power, the defence power and possibly the implied nationhood power, are used to extend the Commonwealth's jurisdiction over the management of interstate water basins and river systems.[3] This federal law recognizes Indigenous rights to water through various statutes including the Indigenous Land Rights (Northern Territory) Act 1976 (Cth), the Native Title Act 1993 (Cth) and the Water Act 2007 (Cth). In relation to freshwater, native title allows a limited, non-exclusive and non-commercial right to use water without the need for a licence (Gardner *et al.*, 2009).

Outside of native title policy and legislative context, two iterations of national water documents have slowly recognized Indigenous peoples' interests in water. The 1994 Council of Australian Governments (COAG) water reforms mainly addressed economic and environmental implications of water use, with no mention of Indigenous access to water. However, the 2004 National Water Initiative (NWI) (COAG, 2004) expressly recognized the interests of Indigenous peoples in water while concentrating on providing for sustainable use of water, increasing the security of water access entitlements and ensuring the economically efficient use of water.

Nevertheless, only two states, New South Wales and Queensland, reformed their water statutes in the early 2000s to expressly recognize Indigenous water interests, arguably the most underdeveloped of any area covered by the NWI (Tan, 2009).

Responding to its own biennial evaluations, in 2011 the National Water Commission (NWC), an independent statutory body overseeing implementation of water reform, established the First People's Water Engagement Council (FPWEC), a national body for the articulation of interests (National Water Commission (NWC), 2011, p. 45), and commenced funding projects to further Indigenous engagement. In its 2011 audit of water reforms, the NWC stressed that despite improved consultations with Indigenous communities in water planning and management by most jurisdictions, there was a general failure to incorporate effective strategies for achieving Indigenous social, spiritual and customary objectives into water plans as envisaged under the NWI (NWC, 2011, p. 46). With the termination of the NWC in 2014, the progress made through focused support and development of Indigenous water rights was disrupted.

Parallel to the actions of the federal government, Queensland has recognized Indigenous interests in water but, as discussed below, Queensland's recent actions have also been seen by some as a retrograde step in terms of Indigenous self-determination in water management.

Water management of wild rivers in Queensland

Water planning and Indigenous engagement in Queensland

Queensland has a third of Australia's freshwater resources. Its diverse waterways reflect the large state's highly seasonal and variable climate (Queensland Department of Environment and Heritage Protection, 2012). In the north, tropical monsoons bring heavy rainfall whereas in the interior south-west, rainfall is patchy. Parts of the state have river catchments relatively untouched by development, for example in Cape York Peninsula, the Channel Country in the west, and the Gulf Country in the northern and eastern coastal areas.

The Water Act 2000 (Qld) provides an overall planning framework for allocating and managing water. It introduces a water market through plans that 'grandfather' water licences into tradeable water entitlements while allocating secure rights to water for the environment. Community consultation is an important part of the water planning process. This element of water planning in Queensland has been rated highly (NWC, 2009) and formal legal provisions for engagement with Indigenous peoples are clear (Tan, 2009).

In 2012, the state government adopted five principal objectives for water management, two of which relate to Indigenous people: to provide a coherent institutional and policy framework that promotes integrated, adaptive and inclusive water management (Objective 1); and the wise management and use of water by an informed and engaged community (Objective 5). Both involve effective consultation and engagement with community and stakeholders, including Indigenous people.

Planning and managing wild rivers

In spite of the general success in water reform, management of rivers in natural conditions through the Wild Rivers Act 2005 (Qld) (WRA) has been exceptionally contentious. Upon the introduction of the WRA, the concept of 'wildness' immediately came under criticism (Iles and Johns, 2010; Neale, 2012) for colonial connotations ascribed to places occupied by Indigenous people. At the heart of the objections to wild rivers legislation are issues of informed consent by Indigenous people to declarations, and Indigenous peoples' concern that economic activities of benefit to their communities are curtailed (Holmes, 2011; Neale, 2012). The WRA was repealed at the end of 2014.

Under the WRA the responsible minister could propose any part of Queensland for consideration as a wild river area: §7. A Wild River Declaration (WRD) was a statutory instrument that describes the wild river area, wild river requirements and circumstances in which parts of the Wild Rivers Code (WRC) or other development assessment codes apply. A declaration was able to protect natural values including unimpeded runoff, stream flow, aquifer and spring recharge; geomorphic processes, meaning free movement of sediments along the river system to allow for stable beds and banks and sediment delivery to estuaries and floodplains; and intact riparian function along stream banks and for food and habitat for native animals. It was the only statute that took precedence over mining legislation in designated areas (Tan, 2013).

WR Declarations did not affect native title, cultural heritage or the functioning and operation of the Aboriginal Cultural Heritage Act 2003 (Qld). Thus Indigenous activities such as fishing, hunting and conducting ceremonies and traditional fire management were not subject to wild river requirements and did not constitute development. A variety of management areas were defined by a declaration, and the areas could overlap with each other. Typically these included high-value or preservation areas (HPAs) which acted as buffer zones, normally 1 km either side of the river. They could also include major tributaries and features such as floodplain wetlands. Within these areas, strip mining, intensive irrigation and dams was prohibited. Also protected were floodplain management areas with strong hydrologic connection to the river system and aquifer recharge areas with a strong hydrologic connection to the river system. Each declaration was accompanied by a map of the wild rivers catchment area, essential for understanding the location and extent of the management areas.

WR Declarations were to be read together with Water Resource Plans (WRPs), which are established through a statutory process under the Water Act 2000 (Qld). In some catchments, lengthy political negotiations resulted in WRPs providing for Indigenous water reserves which were set aside for economic development, although there is no evidence of Indigenous input into the volume of water set aside in the reserve (Cranney and Tan, 2011). Indeed, this lack of input resulted, in some cases, in Indigenous people's groups contesting the declaration of a number of river catchments. Chronologically, the contest in the Cape York Peninsula region preceded that in the Channel Country. However, as there was support for Wild River declarations in the Channel Country, and a better-informed process was followed, it is that example which is considered next.

Widespread support of protection in the Channel Country

The Channel Country lies in the Queensland portion of the Lake Eyre Basin (LEB), which sprawls across three other jurisdictions, NSW, South Australia and the Northern Territory. With an ephemeral large salt lake, the LEB is one of the world's largest internally draining river systems. Three main rivers of the LEB, the Cooper, the Georgina and the Diamantina, have large floodplains, low topographic relief and occasional floods characteristic of the Channel Country. Boom-and-bust ecology in the arid and semi-arid climate is driven by episodic weather events that generally do not follow predictable annual cycles. About 60% of the water that enters South Australia flows from the Georgina and Diamantina, with the remainder from the Cooper. Reliance on rain events and floods, and widely scattered springs and waterholes, is an integral part of Aboriginal culture. Although there are a number of registered native title claims, only two have been finalized. Reportedly, there are 25 Aboriginal people groups in the Channel Country.

Current economic activities in the LEB continue to rely on natural values of wet-and-dry cycles. Rain-fed grazing is the major land use in the LEB and the region has major organic beef production. Tourism and mining are the other two important natural resource-based industries in the LEB.

Tourism based on natural and cultural heritage features of the vast landscape has potential to significantly contribute to the economic viability and livelihoods of the LEB (Measham *et al.*, 2009).

The rivers of the Channel Country are presently subject to minor diversions for stock and domestic use after the Cooper Creek community warded off a 1994 proposal to take 42,000 ML from Cooper Creek for cotton farming. Vigorous public action by a coalition of local graziers, concerned scientists and NGOs (Walker *et al.*, 1997) led to the adoption of a Wate Management Plan (Coopers Creek) to restrict irrigation development. This was rushed through the Queensland state Parliament and deemed to be compliant with procedures under the new Water Act 2000 (Qld).[4] There remain unutilized, currently suspended, licences for irrigation subject to the water planning framework.

Notwithstanding controversy over wild rivers elsewhere, local graziers in the Channel Country and environmental groups formed an alliance in 2009 to lobby for legislative protection of their rivers (Desert Channels Queensland, undated). This became an election issue and, after its return to power, the state Labor government began the process of WR declarations over the three rivers in the region.

Two-year consultation over proposed wild rivers declaration

At the time there were no specific directions as to consultation mechanisms in the WRA.[5] Nevertheless, the Department of Environment and Resource Management (DERM) held a stakeholder forum in 2009, and after release of a consultation paper in March 2010, formed a LEB Wild Rivers Advisory Panel (LEB WRAP). The Panel met six times over a 12-month period. Of the 15 members of LEB WRAP, four were Aboriginal, four represented local councils, with one member each from peak farming, mining and natural resource management groups, and an environmental NGO, and one member from the LEB Scientific Advisory Panel. Minutes taken and interviews with members of the LEB WRAP confirm that all members supported the WRA declaration.[6] Rather than a consultation, Scott Gorringe, a Mithaka member of the LEB WRAP, considers that the process was more akin to negotiation (Gorringe, 2013):

> It was us having a bit of a say and gaining benefits . . . the process was solid. We were committed to long-term *protection of the rivers* (emphasis added).

After 20 months of consultation, WR Declarations were made over the Cooper and Georgina-Diamantina Rivers in December 2011.

Capacity building through LEB fora

In addition to the LEB WRAP, Aboriginal groups had significant input into the WR declaration process through sustained capacity building and engagement in biennial LEB Aboriginal fora. These fora discussed a wide range of topics including:

- increasing Aboriginal cultural awareness for non-Aboriginal people;
- addressing the need for a social and community profile of Aboriginal people in the Basin;
- strong concern over the potential changes to the natural flows of the rivers in the Lake Eyre Basin that sustain their livelihoods and culture.

Thus the 3rd forum saw a broadening of alliance building between Aboriginal and non-Aboriginal participants (Norris, 2013). Significantly at the 2011 4th LEB Aboriginal Forum, attended by 40 Aboriginal participants and 35 non-Aboriginal invitees (scientists, historians, government and NGOs), the Tibooburra Declaration endorsed wild river declarations for the Cooper and Georgina-Diamantina (LEB 4th Indigenous Forum, 2011).

Post Declaration – Western Rivers Review

In 2011, the Newman Liberal-National Party (LNP) government was elected, replacing the Labor (ALP) government. The new Premier, Newman, had made abolition of the WRA an election promise (Queensland Department of Natural Resources and Mines, 2012). To that end, a Western Rivers Advisory Panel was formed in late 2012 to give stakeholder advice on alternative strategies in respect of the three main rivers of the LEB. While it had the same range of stakeholders as the LEB WRAP, its membership was different. Members of the LEB WRAP who were strong advocates of Wild Rivers Declaration were left off the Western Rivers Advisory Panel (Gorringe, 2013) and Aboriginal members were reduced from four to two. Under the terms of reference, the Western Rivers Advisory Panel was directed not to consider an option of retaining the Wild Rivers Declaration. Instead the new panel were to primarily identify 'values or assets of the Basin which were the most important, where protection of these values should be focused and the level of protection that is required' (Western Rivers Advisory Panel (WRAP), 2013, p. 4).

Unsurprisingly, this panel reported that the natural assets identified through this process were 'nearly identical to the Wild Rivers values' previously identified (WRAP, 2013, p, 4). The report presented views from foursectors, resources, natural resource management (NRM), local government and agricultural producers, with no record of an Aboriginal view. One of the panel's Aboriginal members issued a public statement soon after the publication of the report to the effect that Aboriginal people in the Channel Country had not been fully consulted, and that all Aboriginal people had strong views on the protection of natural river systems and supported the WR declaration (Fogarty, 2013). While a robust and effective stakeholder process had been successful in putting into place wild rivers declarations in the Channel Country, rather than being held up as an example of effective Indigenous consultation and water management operating effectively within a trans-jurisdictional context, the process was dismantled.

Contestation over Cape York's wild rivers

The Cape York Peninsula region (the Cape) has exceptional conservation assets, including relatively intact and extensive coastal dune-fields, wetlands, rainforest, heathlands and river systems. Eastern Cape rivers have stable summer base flows while those in the west have massive flood events during the monsoons. Declared Wild Rivers are located in the Western Cape.

Protracted struggle over futures in the Cape

Prior to the 1970s, the Cape region relied on pastoralism and mining (Holmes, 2011, 2012). From the 1970s to 90s, the strategies of both Conservative and Labor state governments set a backdrop for the struggle over wild rivers in the Cape. At that time Aboriginal people in the Cape, as elsewhere in Queensland, held land under a variety of legal instruments.[7] In the 1970s, the Conservative state government adopted a pro-development rule with the sudden, unanticipated declaration of seven national parks in the 1977–79 period interpreted as preventing the purchase of land by the federal government for transfer to Aboriginal owners (Holmes, 2011, 2012). Faced with a failing pastoral industry, and unable to find alternative commodity outputs other than mining, the Labor government in the 1990s explored two potentially complementary futures for the Cape – the expansion of the conservation estate while supporting native title, and self-management by Aboriginal people (Holmes, 2011).

Several key events took place in the 1990s into which the wild rivers debate played. First, the Native Title Act 1993 (Cth) meant that Aboriginal people had the prospect of owning their own land, albeit without freehold rights. The Wik peoples, who live in the Western part of the Cape, established through a High Court decision that pastoral leases do not by themselves extinguish native title (*The Wik Peoples v. Queensland* (1996)). Coexistence between the two types of title is possible and requires each particular lease to be examined for inconsistency with native rights.

Second, in what seemed a rare meeting of hearts and minds, an alliance between the state of Queensland, the Cattlemen's Union, Cape York Land Council, conservation groups and others signed a Cape York Heads of Agreement (CPHoA) in 1996, endorsing the Cape York Peninsula Land Use Strategy 1990–96 (Australian Government Department of the Environment, undated(a)). In the lead-up to the federal elections of 1996, Prime Minister Keating hailed this as a 'water shed in the reconciliation process'. It received bipartisan support and more importantly, a commitment of $40 million funding (Stevenson, 1998, p. 4). However, the mining industry had not been consulted about the Agreement and initially expressed concern that it referred to World Heritage status, which it understood as a 'no-go area for miners' (Stevenson, 1998, p. 9). CPHoA opened a space for a debate on the benefits of regional agreements for Indigenous title to land and resources (Stevenson, 1998), one of the many steps that led to the adoption of Indigenous Land Use Agreements (ILUAs) under the Native Title Act 1993 (Cth).

In 2001, Queensland formally adopted the Heads of Agreement. In a short time, however, the alliance crumbled. Explicit about protecting World Heritage values in the Cape, the CYHoA gave an 'impression of consensus' while 'in reality no issue of substance was resolved, nor was there any procedure towards resolution' (Holmes, 2011, p. 63). Others attribute the collapse of the agreement to schisms between Indigenous and conservation groups, and powerful Cape leader Noel Pearson's dissembling opinions of wild rivers legislation (Neale, 2011, p. 318).

Despite the CYHoA failing, the Cape York Peninsula Heritage Act 2007 (Qld) provided the first Indigenous reserves over water in Queensland, if not all of Australia.[8] Nevertheless, while these and other positive outcomes in relation to negotiated settlements under the Native Title Act's ILUA's provisions were achieved, Hafner *et al.* (2006, p. 123) point to the obvious distrust of state parties arising from the history of dispossession in the region, and the need to overcome divisiveness among Aboriginal parties using principles of 'practical self-determination'. The desire for self-determination was also the basis for raising the issue of consent in the federal legislative context, which is next discussed.

Federal debate over wild rivers

This distrust of the state was further illustrated when the then leader of the federal opposition, Tony Abbott, introduced a Private Members Bill to protect the rights of Traditional Owners (TOs) by requiring that wild rivers declarations could proceed only with their consent. In his speech introducing the *Wild Rivers (Environmental Management) Bill 2011* (Cth), Abbott referred to UNDRIP and called on the Rudd government to give practical effect to Aboriginal peoples' rights to 'own, use, develop and control' their own land and resources (Commonwealth, 2011, p. 9950). By doing so, Abbott highlighted that whether a declaration of wild rivers was acceptable to Indigenous people was an issue of consent and not of consultation; in other words, that the Aboriginal owners of the land subject to the declaration had not consented to the perceived freezing of development rights over affected land (Neale, 2012). This view echoes the objections of a number of powerful leaders in the Cape – that the WRA was the result of political deal struck between ALP politicians and city environmentalists, and that Indigenous communities had not been asked for their views, and indeed were in opposition to the Act and the declaration of Cape York rivers (Pearson, 2010).

Neale's analysis of the submissions and public hearings into the Abbott Bill finds that critics of the WRA point to the effect of the Act on the Cape's Aboriginal population as 'choking', 'killing' and 'suffocating' (Neale, 2012). Yet he notes that repeatedly in the Parliamentary inquiries, proponents and opponents were asked what environmentally sustainable projects had the WRA rendered impossible? None were named, and the absence of specific answers to these questions rendered the Act more and more marginal, as the proximate cause of the controversy was shifted to other legislative and bureaucratic factors (Neale, 2012).

Moreover, the Bill's requirement of the agreement of 'relevant titleholders' (s. 5) to development or use of wild rivers areas was criticized by Cape traditional owners who, 'suspicious of Abbott's motives . . . openly criticized this right to consent' (Neale, 2012). In these circumstances, the Bill was defeated in May 2011.

Resolution of Cape York's wild rivers debate?

In the meantime, Aboriginal groups applied to annul WR declarations over three of the four wild rivers in the Cape, namely the Archer, Lockhart and Stewart. The three WR declarations were held invalid on the grounds that the Minister of the Environment had failed to satisfy the statutory pre-conditions to the exercise of power as required under the WRA (*Koowarta v. State of Queensland* [2014] FCA 627). Under § 13 of the WRA, matters that were mandatory for consideration *before* Ministerial exercise of power include the results of community consultation and all properly made submissions. There was evidence that the Minister had not sighted the required material accompanying the Ministerial Brief on the declarations until *after* he had made a decision to declare these three rivers (para. 214). There was also evidence that the Ministerial Briefing note and the accompanying material did not incorporate any maps showing the boundaries of the Wild River areas (para. 50).

There was no appeal of this judgement because Queensland repealed the WRA in 2014.[9] The new *Regional Planning Interests Act 2014* (RPI) provides that the river systems in the Cape and other regions subject to WR declarations are rolled into the RPI framework as Strategic Environmental Areas (SEAs). Environmental groups have expressed grave concerns over the repeal of the WRA, on the basis that the RPI does not provide for a similar level of protection for the natural values of the regions (Environmental Defenders' Office Queensland, 2014).

With the exception of the Wenlock region where Aboriginal communities have consistently supported the WR declaration (Queensland Aboriginal Forum, 2013), the Wild Rivers debate appears to have been resolved over much of the Cape. The Cape York Regional Plan 2014 provides regulated areas that mirror the four Cape basins declared under the repealed WRA. In response to Aboriginal feedback, a decision was taken by the Minister to review the plan annually. The resolution has not satisfied groups that have previously called for the repeal of the WRA and who maintain that the 2014 regional plan still blocks economic development. While conservationists argue that safeguards over natural rivers and floodplains have been compromised, it appears that the Queensland Government has alienated both Aboriginal communities and conservationists with the Cape's regional plan (Walker, 2014).

Implications for trans-jurisdictional water governance

Aboriginal responses to wild rivers legislation in Cape York and the Channel Country have been sharply divided. The differences cannot be easily contrasted or explained, as much of the explanation lies in the history of division in the Cape over its vision of development. One explanation is intense localism, which Sutton considers as precluding regional consensus on any critical issues (Sutton, 1990, cited in Holmes, 2012). Holmes (2011, 2012) suggests a modernist–traditionalist divide built on differentiation of land tenure, as the Aboriginal people on Cape York generally retained occupation of land and now have native title over vast tracts of the region and the potential capacity to develop their land, if not for regulatory measures such as the WRA. This is contrasted with the Channel Country where many native title claims are still outstanding and may not succeed, thus cutting off this potential for the Indigenous people in that region. Yet land tenure *in itself* does not provide a coherent explanation – some native title-holders in Cape York are strong supporters of WR declarations, and there is also similar support from the few native title-holders in the Channel Country (Queensland Aboriginal Forum, 2013). The following sections seek to draw out some of the consistent messages about Aboriginal access to land and water.

Free, prior and informed consent

During the Queensland Aboriginal Forum (2013), participants noted that arts 10, 25 and 26 of the 2007 UNDRIP require governments to seek free, prior and informed consent (FPIC) of Indigenous peoples to decisions being made by governments *or* private developers. The Cape study suggests that some Aboriginal groups believed that the wild rivers legislation was a political compact made without sufficient Aboriginal engagement. They thus considered themselves to have been locked out of economic development of the floodplains of the four rivers. However, while Indigenous people are fully supportive of the FPIC principle, (First Peoples Water Engagement Council (FPWEC), 2012), few political leaders formally endorse it, although there have been notable exceptions such as the assurance by the Federal Minister for the Environment that the World Heritage nomination under the Cape York Peninsula Heritage Act 2007 (Qld) would proceed with FPIC (Logan, 2013, p. 169). FPIC was also expressed as the core rationale behind the ultimately unsuccessful attempt to introduce Federal legislation to override Queensland's wild river legislation.

The implementation of the FPIC principle may provide a coherent explanation for the different responses to WR consultation in the two case studies. While there was a large amount of resistance to WR from Cape York Aboriginal communities and leaders, consultation in the Channel Country found a high degree of support for WR declarations.

Moreover, the consultations in the Channel Country started two years before the declarations were made and were described as 'negotiations' rather than mere information giving. The fact that amendments were made to the WRA to provide for an Indigenous Reference Group for wild rivers consultations was interpreted by Aboriginal groups in the Channel Country as evidence of good faith. FPIC clearly has implications for a transactional model of trans-jurisdictional governance.

Groups speak for Country at multiple levels

However, even if FPIC is endorsed, given the spiritual significance of land and water for people at clan and group scales,[10] its implementation is compromised if governments fail to recognize the diversity of groups and the importance of multi-scaled participatory processes. Hillman and Howitt (2008) posit that adopting a homogenous set of management objectives and outcomes across a region produces 'substantial injustices particularly where the scales at which injustice originates and is experienced are disparate' (p. 60). Where dominance of certain scientific discipline or stakeholder interests becomes embedded in dominant political and cultural values about what counts and who has been involved, then informal or unstated characteristics confound change, resulting in disadvantaged groups being neglected (Hillman and Howitt, 2008). Willis *et al.* (2008) found that Cape York Institute's spokesperson, Pearson's, sophisticated, intellectual, 'centrist' reasoning appealed to a Western 'modern' engagement model. Referring to this, Aboriginal leaders said 'there is an Aboriginal voice being heard, but it is not ours' (Queensland Aboriginal Forum, 2013). Participants in the Forum were unanimous in asserting their right to speak for their own Country.

Moreover, differing understandings of what is sustainable and what is needed for their own future also underpin the responses to wild rivers. There are two views of opportunity (thus affecting comprehensions of sustainability) held by Aboriginal communities. On the first view, preservation of a broad legal space will allow vast opportunities to develop at the discretion of Aboriginal landholders. Thus, shifting capital to Aboriginal landholders – even if it be for a short-term venture such as a 15-year-old mine – is considered sustainable. On the second view, 'the opportunity to continue management practices on country through associated funding – whether governmental or through ILUAs with resource companies – and market economic activity (carbon sequestration, ecotourism, etc.)' is considered sustainable (Neale, 2012). These perceptions of the future largely shape responses of communities and the differences within the Cape itself, and also for the Channel Country.

While the Queensland Aboriginal Forum (2013) called for the retention of WR declarations over their Country, participants were aware that there was a need to achieve appropriate and traditional models that could speak at a regional scale and called for establishing their own hybrid model, combining Western and traditional institutions.

Participants aspired to a 'kitchen table' approach at a local scale not dependent on government funding, with a loose-knit federation of interests for representation at regional scale. One particular model is the Northern Basin Aboriginal Nations, a body representing 22 Aboriginal groups, not based on native title. This was flagged as important for further development.

Innovative collaborative structures for whole-of-basin approach

A whole-of-basin approach informed the Channel Country debate for Aboriginal and non-Aboriginal participants (Norris, 2013). This integration of scientific and technical understanding of the ecology of 'boom-and-bust' desert rivers, and the social and institutional aspects of the related dryland management systems in the LEB, was the outcome of Federal government investment (Bellamy *et al.*, 2012). Indeed, the LEB Intergovernmental Agreement 2000 signed by South Australia, Queensland, Northern Territory and the Australian Commonwealth established a Ministerial Forum with a supporting secretariat, a Scientific Advisory Panel, a Community Advisory Committee and a Senior

Officer's Group (Australian Government Department of the Environment, undated(b)). The LEB Ministerial Forum's support of Aboriginal engagement created an appropriate 'space' in biennial 2–3-day Indigenous fora, allowing a diversity of Indigenous people to collaborate in the LEB governance processes, thereby increasing capacity to understand NRM issues from a 'Western' perspective.

Channel Country WRAP meetings also developed NRM capacity. Hosted by a local organization and supported by the Queensland water agency, they were a 'solid process' (Gorringe, 2013). Over ninety public meetings were held with communities from Longreach to Barcaldine (Queensland Department of Environment and Resource Management, 2011). Over the process, non-Aboriginal individuals on the LEB WRAP shifted to recognizing the values and interests of Aboriginal peoples. However, this realization was personal to the individual and was not translated to the culture of the organization that they represent. Gorringe feels that there is a long way to go to building 'a joint culture that values different ways of being and thinking' (Gorringe, 2013).

Bellamy *et al.* (2012) identify another key enduring feature from LEB institutional development, namely a collaborative vision of keeping the whole basin in good ecological condition, building a collaborative 'self-organizing' community-based cross-border process at the whole of basin level. When developed, this approach can inform federal, state and local politicians and policy makers enabling institutional change.

Science enabling an evidence-based debate on governance

Indigenous communities and the larger LEB community also appreciated the role of independent scientists in reshaping the policy debate (Kingsford *et al.*, 1998; Queensland Aboriginal Forum, 2013). Organized and funded by the local community and visiting scientists, workshops were held in 1996 in the heart of the communities affected by proposals to irrigate the Cooper Creek. Attended by more than one hundred people, the workshops presented up-to-date information on the state of ecological and hydrological knowledge of the catchment, with insights into the ecological impact downstream from the diversion of waters from the river systems of the Channel Country. These events allowed locals to freely question scientists while allowing two-way conversations on local history and observations (Kingsford *et al.*, 1998, p. 372).

Building strategic alliances

Along the lines of developing a collaborative ecological vision, the 2013 Queensland Forum identified that a key action was to build strategic alliances principally with scientists, conservationists and universities. Participants also recognized sharing similar values with others in the Channel Country community:

> Them graziers are more black than what we (traditional owners) are, so we can have people on our side which we never thought were on our side.
>
> (Queensland Aboriginal Forum, 2013)

This resonates with findings elsewhere that such alliances have resulted in significant gains for Indigenous communities, for example in stopping the construction of a hydroelectric dam in Brazil (Conklin and Graham, 1995) and construction of logging roads in Canada (Willow, 2012). Strong alliances ultimately empowered indigenous communities to participate in environmental protection on terms closer to their own, such as political empowerment, self-determination and economic development. Alliances were conceived as *middle ground*, an 'arena of intercultural communication, exchange and joint political action' (Conklin and Graham, 1995, pp. 695–6 as cited in Willow, 2012, p. 372).

Nevertheless, Measham *et al.* (2008) remind us that equal alliances are far from easy to achieve, while Willow (2012) observes that the ultimate goals of Indigenous activists are far wider than the environmental protection sought by conservation groups. There is a danger that alliances will collapse where environmentalists expect Indigenous people goals to match their own. Thus, others consider that 'equitable' is a more useful concept than equal, 'since partners could well have unequal powers and yet contribute in different ways' (Zurba *et al.*, 2012, p. 1131). Co-management models may go a long way to counter the risk of different aspirations identified by Willow (Carlsson and Berkes, 2005, Ross *et al.*, 2009; Hoverman *et al.*, 2012).

Conclusion

At both federal and state levels, the past 10 years have seen a strengthening of policy providing Indigenous access to water. Indigenous people have consistently articulated a desire to both effectively participate in water planning so that Indigenous values inform water management and, reversing past injustices, to own special Indigenous water allocations for self-determined purposes.

Queensland's wild rivers debate illustrates that Aboriginal people seek self-determination in their struggle for access to water. In Cape York, this struggle has largely been shaped by a vision of being able to determine their own economic future – for some this has meant rejecting the constraints of regulations imposed by wild rivers on development of floodplains. Conversely, this vision of self-determination has been viewed through an ecological lens in the Channel Country. For any rights-based, self-determination approach to deliver outcomes satisfactory to Indigenous peoples, there must be recognition first that each group speaks for its own Country, and second that 'space' for consensus building within Indigenous groups at regional scales – such as the biennial Lake Eyre Basin Aboriginal for a – is needed and that negotiation may extend to an examination of co-management models. All of these measures need sustained investment by governments.

The state-centric jurisdictional model adopted by the Queensland government in planning and managing wild rivers posed difficulties for Aboriginal peoples, particularly in Cape York. Whether or not the reader agrees that natural values of 'wild' rivers need to be protected and managed, the state-centric model largely failed in the Cape because of the failure to consider the principle of Aboriginal groups' free, prior and informed consent, especially in the Archer, Lockhart and Stewart catchments. Thus, in 2011, the new Liberal-National Party government promised wide-sweeping reforms, including repealing wild rivers legislation and seeking alternative management strategies for the Georgina-Diamantina and Cooper Rivers. Ironically, *after* its own 2013 Western Rivers review (which was criticized for not presenting Aboriginal views) confirmed that the majority opinion was to maintain a high level of protection of natural values in the Channel Country, the government abolished the protection given by Wild Rivers Declarations, thus opening up fragile floodplains to mining. It is suggested that the intent behind this reversal of wild rivers legislation was not that of honouring free, prior and informed consent of Aboriginal peoples.

Environmental justice is emerging as a discourse in natural resource management. FWPEC (2012), as do others in Cape York, refers to environmental justice in relation to accessing water for economic development. While the rights-based approach has been predicated on rights issuing from or pursued against a nation state, in a post-colonial world, transnational corporations and globalized institutions threaten Indigenous communities and their well-being (Hosmer, 2012, p. 265). In the Channel Country there has been a positioning of powerful large corporations offering to take advantage of the vast mining resources of the Cooper Basin, with the potential for State interests to become intertwined with those of multinational corporations. Ultimately what is needed are whole-of-basin approaches that go beyond a state-centric, jurisdictional governance model. The Lake Eyre Basin Agreement provides a trans-jurisdictional model for the management of river system across four jurisdictions, with the objective to keep the LEB in a good ecological condition. These approaches require strategic alliances that go beyond the formal community advisory group. Relationships are based on meeting the communities on their own land, using authentic communication and negotiation instead of mere consultation, and governments willing to go back to the drawing board to accommodate the wishes of Aboriginal peoples. Models such as these need resourcing, and new relationships which build a coherent whole of basin understanding may not survive when administrative and financial support is withdrawn.

The last word should be reserved for Indigenous peoples. For implementing trans-jurisdictional water governance:

> . . . we need a new umbrella structure, a hybrid model based on clan and family. The best institutions are those that merge Western and Indigenous practices.
>
> (Queensland Aboriginal Forum, 2013)

Notes

1 The term 'Indigenous' refers to both Aboriginal and Torres Strait Islander peoples, whereas 'Aboriginal' is intended to refer to mainland people specifically. These terms, which may be confusing for readers, honour Indigenous preferences.
2 See *Koowarta v. Bjelke-Petersen* (1982) 153 CLR 168; *Mabo v. Queensland* (No. 2) (1992) 175 CLR 1; *The Wik Peoples v. Queensland* (1996) 187 CLR 1.
3 See Water Act 2007 (Cth) ss. 9A, 9(a) and (b), 119(3).
4 There is no public record of Indigenous involvement at this point of public protest although George Gorringe, a Mithaka man, is Vice-Chair of the Coopers Creek Action Group.
5 By amendments in 2011 to the WRA, provision was made for an Indigenous reference group to advise on matters relating to the declaration, or the management of the wild river area. The Minister is to consider the results of community consultation, all properly made public submissions and, if an Indigenous reference group has been established, advice from that group.
6 Half of the meetings have online minutes/communiqué (Remote Area Planning and Development Board, undated). Minutes and interviews of LEB WRAP record that members found process beneficial (minutes of meeting, 14 December 2011) and signed off on WR declarations for CC and GD.
7 Land was granted in trust (commonly referred to as DOGIT land) under the Land Act 1962 (Qld); or reserves under that Act, or land successfully claimed under the Aboriginal Land Act 1991 (Qld).
8 The Mitchell Water Resources Plan 2008 provides an Indigenous reserve of 5,000 ML, and 1,000 ML is reserved in the Gulf Water Resource Plan 2008. Although these volumes of water available are small (in comparison to the general reserve that comprises 175,000 ML per annum in the Gulf and 55,000 ML per annum in the Mitchell), they are a precedent for other Indigenous water reserves.
9 The Queensland government passed the State Development, Infrastructure and Planning (Red Tape Reduction) and Other Legislation Amendment Act 2014 to repeal the WRA in its entirety.
10 Classifications may be on broad linguistic lines within which there are sub-linguistic groups, or by political or social groups delineated by totemic or religious geography. There are reportedly over 50 language groups in Cape York and 25 in the Channel Country.

References

Aboriginal Cultural Heritage Act 2003 (Qld).

Altman, J. (2010) 'Wild Rivers and Informed Consent on Cape York', CAEPR Topic issue no 02/2010, Australian National University, Canberra.

Anderson, I. (2007) 'The end of Aboriginal self-determination', *Futures*, 39, 2–3, 137–54.

Arzey, S. and McNamara L. (2011) 'Invoking international human rights law in a rights-free zone: Indigenous justice campaigns in Australia', *Human Rights Quarterly*, 33, 3, 733–66.

Australian Government Department of the Environment (undated(a)) 'Cape York Peninsula land use strategy', Department of the Environment, Canberra, Australia, http://environment.gov.au/archive/erin/cyplus/index.html, accessed 17 April 2015.

Australian Government Department of the Environment (undated(b)) 'Lake Eyre Basin Intergovernmental Agreement', Department of the Environment, Canberra, Australia, http://environment.gov.au/resource/lake-eyre-basin-intergovernmental-agreement, accessed, accessed 17 April 2015.

Australian Human Rights Commission (2009) 'Inquiry into the Native Title Amendment Bill 2009: Submission by the Aboriginal and Torres Strait Islander Social Justice Commissioner to the Senate Standing Committee on Legal and Constitutional Affairs, Australian Human Rights Commission, Sydney, Australia, https://humanrights.gov.au/inquiry-native-title-amendment-bill-2009.

Bartlett, R. (2004) *Native Title in Australia*, LexisNexis Butterworths, Chatswood, Australia.

Behrendt, J. and Thompson, P. (2004) 'The recognition and protection of Aboriginal interests in New South Wales rivers', *Journal of Indigenous Policy*, 3, 37–140.

Behrendt, L. (2003) *Achieving Social Justice: Indigenous Rights and Australia's Future*, Federation Press, Annandale, Australia.

Bellamy, J. Head, B. and Ross, H. (2012) 'Blurring boundaries and building bridges: Challenges of managing across borders in the Lake Eyre Basin', in J. R. Grove and I. D. Rutherford (eds) *Proceedings of the 6th Australian Stream Management Conference, Managing for Extremes, 6 – 8 February, 2012*, River Basin Management Society, Canberra, pp. 1–11.

Brennan, B. Gunn, B. and Williams, G. (2004) '"Sovereignty" and its relevance to treaty making between indigenous peoples and Australian governments', *Sydney Law Review*, 26, 3, 307–52.

Brennan S., Behrendt L., Strelein L. and Williams G. (2005) *Treaty*, Federation Press, Annandale, Australia.

Cape York Peninsula Heritage Act 2007 (Qld).

Carlsson, L. and Berkes, F. (2005) 'Co-management: concepts and methodological implications', *Journal of Environmental Management*, 75, 1, 65–76.

Commonwealth, Parliamentary Debates, House of Representatives (2011) 12 September 2011, 9950 (Tony Abbott), http://parlinfo.aph.gov.au/parlInfo/download/chamber/hansardr/24fe5c5f-1391-455b-808d-7943b3d1ab3e/toc_pdf/House of Representatives_2011_09_12_411_Official.pdf;fileType=application%2Fpdf - search ="r4646 first reading

Conklin, B. A. and Graham, L. R. (1995) 'The shifting middle ground: Amazonian Indians and eco-politics', *American Anthropologist*, 97, 4, 695–710.

Council of Australian Governments (COAG) (2004) 'Intergovernmental Agreement on a National Water Initiative 2004', COAG, Barton, Australia, http://nwc.gov.au/nwi, accessed 17 April 2015.

Cranney K. and Tan, P. L. (2011) 'Old knowledge in freshwater: Why traditional ecological knowledge is essential for determining environmental flows in water plans', *The Australasian Journal of Natural Resources Law and Policy*, 14, 2, 71–113.

Desert Channels Queensland (undated) 'Rivers Protection Alliance formed', http://dcq.org.au/rivers-protection-alliance-formed, Desert Channels Queensland, Longreach, Queensland, accessed 17 April 2015.

Durette, M. (2010) 'A comparative approach to indigenous legal rights to freshwater: key lessons for Australia from the United States, Canada and New Zealand', *Environmental and Planning Law Journal*, 27, 296–315.

Environmental Defenders' Office Queensland (2014) 'EDO Qld submission on the State Development, Infrastructure and Planning (Red Tape Reduction) and Other Legislation Amendment Bill 2014, Environmental Defender's Office, West End, Australia, https://parliament.qld.gov.au/documents/committees/SDIIC/2014/22-StatDevIPRTROLAB/submissions/082.pdf, accessed 17 April 2015.

First Peoples Water Engagement Council (FPWEC) (2012) 'Advice to the National Water Commission', Canberra, Australia, http://.nwc.gov.au/__data/assets/pdf_file/0004/22576/FPWEC-Advice-to-NWC-May-2012.pdf, accessed 17 April 2015.

Fogarty, G. (2013), Media release, 27 July.

Gardner, A. W., Bartlett, R. H. and Gray, J. (2009) *Water Resources Law*, LexisNexis, Chatswood, Australia.

Gorringe, S. (2013) Personal communication, 23 July.

Hafner, D., Epworth, D., and Salmon, M. (2006) 'Planning for Agreement in Cape York Peninsula: The Pastoral Properties Planning Project,' in M. Langton et al. (eds.), *Settling with Indigenous People: Modern Treaty and Agreement-Making*, The Federation Press, Annandale, NSW, pp. 114–133.

Hillman, M. and Howitt, R. (2008) 'Institutional change in natural resource management in New South Wales, Australia: Sustaining capacity and justice, local environment', *The International Journal of Justice and Sustainability*, 13, 1, 55–66.

Holmes, J. (2011) 'Contesting the future of Cape York Peninsula', *Australian Geographer*, 42, 1, 53–68.

Holmes, J. (2012) 'Cape York Peninsula, Australia: A frontier region undergoing a multifunctional transition with Indigenous engagement', *Journal of Rural Studies*, 28, 3, 252–65.

Hosmer, B. (2012) 'Indigenous communities, nation states, extranational sovereignties and the challenge of environmental justice in the age of globalization', *Environmental Justice*, 5, 5, 264–9.

Hoverman, S., Delfau, K. and Tan, P.L. (2012) 'Developing alternative management models for indigenous water plans and strategies in Australia's north: final report', TRaCK NAWFA Social and Cultural Values Project: Sub-project 3, Griffith University, Nathan, Australia.

Illes, S. and Johns, G. (2010) 'The economic unravelling of the precautionary principle: The Queensland Wild Rivers Act 2005', *Agenda*, 17, 73, 73–92.

Kingsford, R. T., Bolton, A. J. and Puckridge, J. T. (1998) 'Challenges in managing dryland rivers crossing political boundaries: Lessons from Cooper Creek and the Paroo River, Central Australia', *Aquatic Conservation: Marine and Freshwater Ecosystems*, 8, 361–78.

Koowarta v. State of Queensland [2014] FCA 627.

Lake, R. W. (1996) 'Volunteers, NIMBYs and environmental justice: Dilemmas of democratic practice', *Antipode*, 28, 2, 160–74.

Lake Eyre Basin Aboriginal Forum (2011) 'Event summary: 4th Indigenous Forum – Water, land and connections across the Lake Eyre Basin – Sharing the journey and passing on knowledge', http://lebmf.gov.au/publications/pubs/ aboriginal-forum-event-summary-2011.pdf, accessed 17 April 2015.

Langton M., Teehan M., Palmer L. and Shain K. (eds) (2004) *Honour Among Nations? Treaties and Agreements with Indigenous People*, Melbourne University Press, Melbourne, pp. 220–36.

Langton, M., Mazel, O., Palmer, L. and Tehan, M. (2006) 'Sharing land and resources: Modern agreements and treaties with indigenous people in settler states', in M. Langton, M. Tehan, L. Palmer, K. Shain and O. Mazel (eds) *Settling With Indigenous People. Modern Treaty and Agreement-Making*, Federation Press, Sydney, Australia, pp. 1–18.

Lee, C. (1995) 'The origins of environmental justice: Developing the vision of environmental justice: A paradigm for achieving healthy and sustainable communities', *Virginia Environmental Law Journal*, 14, 571–769.

Logan, W. (2013) 'Australia, Indigenous peoples and World Heritage from Kakadu to Cape York: State Party behaviour under the World Heritage Convention', *Journal of Social Archaeology*, 13, 2, 153–76.
McAvoy, T. (2009) 'Native title litigation reform', *Australian Law Reform Commission Reform Journal*, 93, 30–2.
McRae, H. and Nettheim, G. (2009) *Indigenous Legal Issues: Commentary and Materials*', Lawbook Co, Pyrmont, Australia.
Measham, T.G., Robinson, C., Larson, S., Richards, C., Herr, A., Williams, L., Brake, L. and Smith, T. (2009) 'Synthesis of the "People, communities and economies of the Lake Eyre Basin" project', in T. G. Measham, L. Brake (eds) *People, Communities and Economies of the Lake Eyre Basin*, DKCRC Research Report 45, Desert Knowledge Cooperative Research Centre, Alice Springs, Australia, pp. 1–40.
National Native Title Tribunal (2014) 'Fact Sheet about indigenous land use agreements', Native Title Tribunal, Sydney, Australia, http://nntt.gov.au/Information%20Publications/1.About%20Indigenous%20Land%20Use%20Agreements.pdf, accessed 17 April 2015.
National Water Commission (NWC) (2009) Australian Water Reform 2009: Second Biennial Assessment of Progress in Implementation of the National Water Initiative', NWC, Canberra, Australia.
National Water Commission (NWC) (2011) The National Water Initiative – Securing Australia's Water Future: 2011 Assessment', NWC, Canberra, Australia.
National Water Commission (NWC) (2014) 'Closure in 2014', NWC, Canberra, Australia, http://nwc.gov.au/organisation/closure-in-2014, accessed 17 April 2015.
Native Title Act 1993 (Cth).
Neale, T. (2011) 'Duplicity of meaning: Wildness, indigeneity and recognition in the Wild Rivers Act debate', *Griffith Law Review*, 20, 2, 310–32.
Neale, T. (2012) '"A substantial piece in life": Viabilities, realities and given futures at the Wild Rivers inquiries', *Australian Humanities Review*, 53, http://australianhumanitiesreview.org/archive/Issue-November-2012/neale.html, accessed 17 April 2015.
Norris, V. (2013) Lake Eyre Basin Facilitator, Personal communication, 14 October.
O'Sullivan, D. (2012) 'Justice, culture and the political determinants of Indigenous Australian health', *Ethnicities*, 12, 6, 687–705.
Pearson, N. (2010) 'Decision is in: Wild Rivers laws stink', *The Weekend Australian*, 2–3 October, http://cyi.org.au/wp-content/uploads/2014/files/Decision-is-in_Wild-Rivers-laws-stink.pdf, accessed 17 April 2015.
Productivity Commission (2003) 'Water Rights Arrangements in Australia and Overseas', Productivity Commission, Melbourne, Australia, http://pc.gov.au/__data/assets/pdf_file/0006/8457/waterrights.pdf, accessed 17 April 2015.
Queensland Aboriginal Forum (2013) 'Report of Proceedings for Queensland Aboriginal Forum held 15–16 November 2013, Brisbane', Queensland Aboriginal Forum, Brisbane, Australia, available from author.
Queensland Department of Environment and Heritage Protection (2012) State of the Environment Queensland 2011 Report, Part B – State', Department of Environment and Heritage Protection, Brisbane, Australia, http://ehp.qld.gov.au/state-of-the-environment/report-2011/pdf/state.pdf, accessed 17 April 2015.
Queensland Department of Environment and Resource Management (2011) 'Cooper Creek Water Resource Plan 2011: Consultation Report', Department of Environment and Heritage Protection, Brisbane, Australia, https://dnrm.qld.gov.au/__

data/assets/pdf_file/0011/105212/cooper-wrp-consreport.pdf, accessed 17 April 2015.

Queensland Department of Natural Resources and Mines (2012) '2011–12 Annual Report', Department of Natural Resources and Mines, Brisbane, Australia, https://dnrm.qld.gov.au/__data/assets/pdf_file/0009/29385/dnrm-annual-report-2011–12.pdf, accessed 17 April 2015.

Queensland Department of State Development, Infrastructure and Planning (2014) 'Cape York Regional Plan', Brisbane, Australia, http://dsdip.qld.gov.au/regional-planning/cape-york-regional-plan.html, accessed 17 April 2015.

Regional Planning Interests Act 2014 (Qld).

Remote Area Planning and Development Board (RAPAD) (undated) 'LEBWRAP documents', RAPAD, Longreach, Australia, http://rapad.com.au/lebwrap, accessed 17 April 2015.

Robbins, J. (2010) 'A nation within? Indigenous peoples, representation and sovereignty in Australia', *Ethnicities*, 10, 2, 257–74.

Ross, H., Grant, C., Robinson, C.J., Izurieta, A., Smyth, D. and Rist, P. (2009) 'Co-management and Indigenous protected areas in Australia: Achievements and ways forward', *Australasian Journal of Environmental Management*, 16, 242–52.

Rural Solutions (2008) *Aboriginal Access to Water Across Australia*, Department of Water, Land and Biodiversity Conservation, Adelaide, Australia.

State Development, Infrastructure and Planning (Red Tape Reduction) and Other Legislation Amendment Act 2014 (Qld).

Stevenson, B. (1998) 'Cape York Land Use Heads of Agreement', Research Bulletin no. 3, Queensland Parliamentary Library, Brisbane, Australia.

Sutton, P. (1990) 'Planning and decision-making in Cape York Peninsula: Whose principles, for whose benefits', Paper presented at Remote Aboriginal Community Futures Conference, Australian Institute of Aboriginal and Torres Strait Studies, Cairns.

Tan, P. L. (2009) 'National Indigenous Water Planning Forum: A Review of the legal basis for indigenous access to water', NWC, Canberra, Australia.

Tan, P. L. (2013) 'Queensland Government's proposal to replace the Wild Rivers Act 2005', *Australian Environment Review*, 28, 10, 789–93.

Tan P. L. and Jackson, S. (2013) 'Impossible dreaming – does Australia's water law and policy fulfil Indigenous aspirations?', *Environmental Planning and Law Journal*, 30, 132–49.

Tarlock, D. (2010) 'Tribal justice and property rights: the evolution of Winters v. United States', *Natural Resources Journal*, 50, 2, 471–99.

Tehan, M. (2003) 'A hope disillusioned, an opportunity lost? Reflections on common law native title and ten years of the Native Title Act', *Melbourne University Law Review*, 27, 2, 523–70.

Thorson J. E. Birtton, S. and Colby B. G. (eds) (2006) *Tribal Water Rights: Essays in Contemporary Law, Policy and Economics*, University of Arizona Press, Tucson, AZ.

Tsosie, R. (2006) 'Tribal sovereignty and intergovernmental cooperation', in J. E. Thorson, S. Birtton, and B. G. Colby (eds) *Tribal Water Rights: Essays in Contemporary Law, Policy and Economics*, University of Arizona Press, Tucson, AZ, pp. 13–22.

United Nations General Assembly (2007a) 'Declaration on the rights of indigenous peoples' (UNDRIP), GA Res 61/295, UN GAOR, 61st sess., UN Doc A/RES/61/295 (13 September 2007).

United Nations General Assembly (2007b) 'Declaration on the rights of indigenous peoples', GA Res 61/295, UN GAOR, 61st sess., UN Doc A/RES/47/1 (13 September 2007).

Walker, J. (2014) 'Noel Pearson loses will to fight Wild Rivers II as hopes swept away', *The Australian*, August 23, www.theaustralian.com.au/national-affairs/indigenous/noel-pearson-loses-will-to-fight-wild-rivers-ii-as-hopes-swept-away/story-fn9hm1pm-1227033848752, accessed 26 June 2015.

Walker, K.F., Puckridge, J.T. and Blanch, S.J. (1997) 'Irrigation development on Cooper Creek Central Australia: Prospects for a regulated economy in a boom and bust ecology', *Aquatic Conservation: Marine and Freshwater Ecosystems*, 7, 63–73.

Water Act 2000 (Qld).

Western Rivers Advisory Panel (WRAP) (2013) 'Report to Minister Cripps', Office of Heritage and Environment, Griffith, Australia, https://dnrm.qld.gov.au/__data/assets/pdf_file/0015/82500/wrap-report.pdf, accessed 17 April 2015.

The Wik Peoples v. Queensland (1996) 1987 CLR 1.

Wild Rivers Act 2005 (Qld).

Wild Rivers (Environmental Management) Bill 2011 (Cth).

Willis, E., Pearce, M., McCarthy, C., Ryan, F. and Wadham, B. (2008) 'Indigenous responses to water policymaking in Australia', *Development*, 51, 3, 418–24.

Willow, A. J. (2012) 'Re(con)figuring alliances: Place membership, environmental justice, and the remaking of indigenous-environmental relationships in Canada's Boreal Forest', *Human Organization*, 71, 4, 371–82.

Young, S. (2008) *The Trouble with Tradition: Native Title and Cultural Change*, Federation Press, Annandale, Australia.

Yu, S. (1999) 'Ngapa kunangkul: Living water – Report on the indigenous cultural values of groundwater in the La Grange sub-basin', The Centre for Anthropological Research, Perth, Australia, http://water.wa.gov.au/PublicationStore/first/11504.pdf, accessed 17 April 2015.

Zurba, M., Ross, H., Izurieta, A., Rist, P., Bock, E. and Berkes, F. (2012) 'Building co-management as a process: Problem solving through partnerships in country, Australia', *Environmental Management*, 49, 6, 1130–42.

15 Trans-jurisdictional water governance and implementing the human right to water in South Africa

Cristy Clark

Introduction

In April 1994 South Africa held its first full, democratic elections and the African National Congress (ANC) won a 62.6% majority (Harvey, R., 2003, p. 242). This signalled the beginning of a new era of democracy in South Africa and carried with it the promise of equality and dignity for black South Africans. This promise was then enshrined in the new Constitution, which contained a Bill of Rights and protected participation, equality and a wide range of socio-economic rights, including the human right to water (Constitution of the Republic of South Africa 1996, ss. 7(1), 27, 33(1)). As the Constitution allocates responsibility for the delivery of water services to local government (Constitution of the Republic of South Africa 1996, sch. 4, pt. B), it binds all levels of government to take 'reasonable legislative and other measures' to progressively realize the right to water. This chapter situates this division of constitutional responsibilities as a trans-jurisdictional water governance arrangement.

In theory, the decentralized delivery of water services should serve to increase community participation in water governance, and encourage greater responsiveness and accountability from service providers. However, as this chapter will discuss, the realization of these anticipated benefits of decentralization, and of the constitutional right to water, has been compromised by the cost recovery imperatives imposed on municipalities by the National Government, the limited capacity of many of the poorer municipalities and the challenges of overcoming entrenched class (and racial) divisions (or 'boundaries') within South African society. These challenges have hobbled what could have been an effective trans-jurisdictional system.

This chapter is divided into four parts. Part one sets out the rights to both participation and water that exist in the South African Constitution and the cross-jurisdictional obligations that these rights impose on both the national and municipal governments. Part two examines the system of water governance that has been established under this trans-jurisdictional arrangement. Part three details the shift in water policy which took place in 2001 when the National Government introduced a policy of Free Basic Water (FBW). Part four highlights the current move towards a more participatory style of water governance and the way that this serves to enhance the trans-jurisdictional approach.

Constitutional rights to participation and water

The end of apartheid and the ascendance of the ANC to government were the culmination of a long struggle on behalf of South Africa's majority population for equality and full citizenship. Apartheid laws had denied the black population access not only to civil and political rights, but also to socio-economic rights in a manner that had underscored the interdependent nature of these rights (Liebenberg, 2010, p. xxi). A lack of democratic rights had disempowered the population from demanding access to socio-economic rights through the democratic process, while their everyday struggle for survival in the face of systematic deprivation had limited their capacity to demand political freedom more generally.

In response to this history, a dual commitment to democracy and socio-economic rights was enshrined in South Africa's Constitution (Mandela, 1991). Section 7(1), for example, describes the Bill of Rights as 'a cornerstone of democracy in South Africa. It enshrines the rights of all people in [South Africa] and affirms the democratic values of human dignity, equality and freedom' (Constitution of the Republic of South Africa 1996, s. 7(1)). Scholars have argued that the concept of democracy that is enshrined in the South African Constitution is a deep version of democracy – one that envisions opportunities for citizen participation well above and beyond voting in periodic elections (Roux, 2006, p. 80; Brand, 2008, p. 97; Liebenberg, 2010, pp. 29–32).

Liebenberg argues that democratic deliberation is central to the transformative agenda of South Africa's Constitution (Liebenberg, 2010, p. 29). Here she cites Sachs J. in *Minister of Health v. Treatment Action Campaign*, where he argues that in a society like South Africa, 'where the majority were for centuries denied the right to influence those who ruled over them, the right "to be present" when laws are being made has a deep significance'. As will be discussed in Part four, below, this right 'to be present when laws are being made' is also significant to the realization of the right to water for the poor and marginalized in South Africa, particularly under the trans-jurisdictional governance arrangement that has been established.

The Constitution also recognizes a right of participation in governance, and the ANC has recognized the interdependence of participatory rights to both the realization of socio-economic rights and the process of social and economic transformation envisioned for post-apartheid South Africa (Liebenberg, 2010, p. 9). Section 33(1) entitles everyone to administrative action that is lawful, reasonable and procedurally fair. This right is given legislative effect through the Promotion of Administrative Justice Act 2000 (PAJA), which states that administrative action that materially and adversely affects the rights or legitimate expectations of any person must be procedurally fair (s 3(1)).

South Africa was also the first country in the world whose Constitution included explicit recognition of the right to water in its Constitution, and it is contained in s27, which states that '[e]veryone has the right to have access to . . . sufficient food and water'. Section 27 also lays out the state's responsibilities, stating that it 'must take reasonable legislative and other measures, within its available resources, to achieve the progressive realisation of each of these rights'. Accordingly, the South African Government has given the right to water legislative force (Water Services Act 1997, s. 3), and implemented a series of water governance reforms, including decentralization and the introduction of a policy of providing FBW (DWA, 2012).

The ANC had formally embraced socio-economic rights in its Reconstruction and Development Plan (RDP), which it adopted in the lead-up to the April 1994 elections (ANC, 1994). The RDP declared that the state was going to provide equal access to essential services, which had long been denied to black South Africans under apartheid, and committed the Government to providing 20–30 litres of water per capita per day (Lpcpd) as a matter of priority, with a medium-term goal of 50–60 Lpcpd (ANC, 1994, para. 2.6.6–2.6.7).

Water governance in South Africa

Flowing from this dual commitment to participatory governance and the realization of a human right to water, South Africa could have been expected to adopt a highly participatory water governance process that focused on the rapid expansion of affordable and accessible water services. However, working against these constitutional commitments to participation and access was South Africa's legacy of extreme economic polarization and the constrained economic circumstances facing both the national and municipal governments.

Economic background

Two centuries of colonial rule, followed by 50 years of formal apartheid, left South Africa with a legacy of extreme polarization. In 1993, the Gini coefficient (an economic measure of equality where '1' represents perfect equality) was estimated to be 0.61, placing South Africa in a tie with Brazil as the most unequal country in the world (World Bank, 2012). The ongoing legacy of apartheid polices has meant that although black South Africans were given formal equality in 1994, they remain significantly disadvantaged in comparison to white South Africans in terms of income, employment, and access to resources and essential services, and this is even more marked in respect of those living in rural areas (UNDP, 2012). This polarization has created several kinds of 'jurisdictional' divisions that water governance must straddle – between white and black populations, rural and urban communities, and poor and wealthy municipalities.

Apartheid inequalities were starkly reflected in water services. In 1993, only 43% of black South Africans had on-site access to running water as compared to 99% of white South Africans (Klasen, 2002). When the ANC came to power it had to deliver on the promise of universal access to water services through a water system that had been specifically designed to service only a small minority of the country. The ANC was also faced with the reality that all of its promises had to be delivered under constrained economic conditions. It was in this context that the ANC began to implement a new vision for water governance reform.

The ANC was committed to improving the lives of black South Africans, but it was limited in its capacity to redistribute resources (Bond, 2000). It was because of this that the ANC moved to investigate options for rapidly expanding the economy. Adopting a neoliberal economic approach came with a range of incentives, including World Bank and IMF approval, increased access to finance and the promise of increased foreign investment (Bond, 2000, pp. 58–62). The ANC has been criticized for these compromises (Bond, 2000, p. 63; Peet, 2002), but others argue that such an approach was critical to the success of the relatively peaceful transition process (Francis, 2005, p. 159).

In 1996, the ANC made a distinct policy shift towards neoliberalism and replaced the RDP with the Growth, Employment and Redistribution (GEAR) strategy (Department of Finance, 1996). GEAR is a macroeconomic policy framework that has been described as 'South Africa's own, "home-grown" structural adjustment programme' (Coalition Against Water Privatisation (CAWP) *et al.*, 2004, p. 5). It covers a range of policy areas, including trade liberalization, tax breaks for big business, labour market adjustments and cuts in social spending – all designed (according to the wisdom of the time) to expand the South African economy and make it competitive in the international market (Department of Finance, 1996).

In relation to the provision of essential services, GEAR adopted policies of 'fiscal responsibility', including cost recovery, private sector participation (PSP) and corporatization (Department of Finance, 1996). These were the kind of market-driven policies that were encouraged globally by the development banks and others (Gutierrez *et al.*, 2003, pp. 10–13; Gray, 2013, pp. 158–9) and specifically in South Africa by the World Bank (1994, p. 44), which has stated that only a 'fee reflecting the costs will encourage users to correctly value the service they receive, [. . . and] help reverse the "entitlement mentality" that has been the historical result of subsidising public services'.

The result of these economic policies is that despite a transition to democracy, the dramatic economic polarization of South African society remains entrenched and gaining the vote has not significantly improved the real political power or the standard of living for poor communities (World Bank, 2006, p. 201). An illustration of this effect is that between 1995 and 2005, South Africa actually went backwards on UNDP's Human Development Index, despite a global trend of progress (UNDP, 2012).[1]

This process of entrenching inequality has compounded the challenges posed by the prevailing water scarcity in South Africa (DWA, 1997a). For example, the decision to grandfather the water rights of large commercial farms, by exempting them from the new water licence rules, exacerbated existing levels of inequality (Klasen, 2002, p. 618; Francis, 2005, p. 155). There is not enough water left for the majority of South Africa's population to enjoy the same levels of water use that are currently the norm in white South African households (Muller, 2008, p. 69). Despite accounting for only 10% of the population, white urban households accounted for more than 50% of residential water use in 2000 (Francis, 2005, p. 150). In Johannesburg, for example, Klasen (2002, p. 609) reports that white households used an average of 3,000 litres per household per day, while 27% of black households did not have access to piped water within the guaranteed radius of 200 metres of their homes.

The Government acknowledged the imperative of addressing historic water inequality in the White Paper on National Water Policy (DWA, 1999, para. 6.5.2):

> It is important that the introduction of realistic pricing for water does not further penalize disadvantaged communities who were already penalized during the apartheid era. . . . The price to be levied for water reserved to meet basic needs must merit particular attention.

Despite this acknowledgement, the Government instituted a policy of marginal cost recovery for basic water (high enough to cover operation and maintenance) without first investigating whether the resulting price would be affordable for poor households (Muller, 2008, p. 72).

Full or marginal cost recovery policies are central to a market-driven approach to water governance, and this approach was reflected in the ANC's new water policies (DWA, 1999). The 1997 Water Policy document (DWA, 1997b, para. 6.5) also states that 'water is valued as an economic resource . . . it is widely agreed that the setting of the appropriate price for a natural resource such as water can be an effective mechanism to achieve its efficient and productive use'. In line with this approach, bulk water charges (the price charged to the water provider for water taken from dams or other sources) were steadily increased in order to raise money to increase infrastructure investment (DWA, 1999).

Devolution of services

In 1996, the provision of water and sanitation services became the constitutional responsibility of local government in South Africa, and the role of delivering water services was gradually devolved to municipal governments from 2000 (Constitution of the Republic of South Africa 1996, sch. 4, pt. B). In theory, decentralized delivery ought to bring government decision-making processes closer to the local community. However, the trans-jurisdictional nature of revenue-raising capacity in South Africa meant that decentralization was compromised by the National Government's decision to simultaneously withdraw financial and technical support from the water sector (Gowlland-Gualtieri, 2007, p. 6; Tissington *et al.*, 2008, pp. 57–9).

Local governments in South Africa receive funding from the National Government through several mechanisms outlined in the Constitution (s. 214(I)). Among other considerations, this funding allocation is based on each municipality's need to provide basic services, fiscal capacity and development needs, and on 'the economic disparities within and among the provinces' (ss. 214(2)(d)-(g)). National funding and local tariffs are the only finance available for local governments to run their water and sanitation services (Tissington *et al.*, 2008, pp. 18–19), but from 1991 to 1998, funding to local governments was reduced by 85%, while an additional 55% was withheld from 1997 to 2000 (McDonald, 2002; Smith and Hanson, 2003, p. 1522).

As national funding to the water sector decreased, local officials were left without the resources to manage their water systems effectively (Smith, 2006, p. 5; Tissington *et al.*, 2008, pp. 57–9; Bond and Dugard, 2008a, pp. 16–21). Some efforts have been made since 2000 to increase the level of funding provided to local governments for water management (Francis, 2005, pp. 168–9), but as recently as 2011 funding remained inadequate, particularly for uniformly poor municipalities (Water and Sanitation Program, 2011, p. 2).

When responsibility for water and sanitation services was devolved to municipalities, staff had little relevant technical experience. Since insufficient funds were made available for capacity building, or the wage increases necessary to attract skilled staff, many municipalities lacked the technical capacity to adequately carry out the responsibilities of water sector management (Muller, 2008, p. 37; Tissington *et al.*, 2008, p. 60).

In 2008, eight years after the devolution of service delivery under South Africa's trans-jurisdiction water governance model, Tissington *et al.* (2008, p. 60) documented a lack of understanding of roles, responsibilities and local government legislation among municipal staff, in addition to high levels of illiteracy – finding that one in three municipal councillors could not read or write. This lack of basic skills hampered the capacity of municipal workers to facilitate public participation in water governance – a central goal of decentralized delivery – while community members historically lacked the experience of genuine participation (Goldberg, 2009, p. 47). Until recently (Smith, 2009), little has been done to provide the relevant training, and researchers report that where capacity building of local authorities and community members has been attempted it has proven challenging due to widespread illiteracy, language barriers and reality of the 'digital divide', which separates those who are familiar with technology from those are not (Schreiner *et al.*, 2004, p. 176). This capacity building could have been a key role for the National Government under the trans-jurisdiction division of responsibility, but was overlooked in the devolution process.

In response to the pressure of trying to meet an increasing demand with severely limited funds and capacity, local governments elected to commercialize their water and sanitation services (Smith and Vawda, 2003). The imperative to raise funds brought an increased focus on cost recovery, and tariffs were increased in order to ensure the financial sustainability of water and sanitation services and to enable providers to expand services (McDonald and Pape, 2002). For households who could not afford these rising costs, commercialization had a devastating effect on their ability to maintain access to water (Smith and Hanson, 2003). Significant water tariff increases between 1996 and 2000 had already led to unprecedented levels of household debt and to millions of people having their water cut off for non-payment (Bond and Dugard, 2008a, p. 17).

Women, in particular, were forced to bear the burden of increased cost recovery measures (Dugard and Mohlakoana, 2009, p. 546). As the traditional caregivers for the elderly and the sick (a significant issue with the high prevalence of HIV and AIDS), it fell to women to source enough water to meet these responsibilities and to meet the needs of the household (CAWP *et al.*, 2004, p. 10).

Some government officials argued that the payment boycotts used against the apartheid Government had created a 'culture of non-payment' in South Africa (Tomlinson *et al.*, 2003, p. 9). A related criticism is the argument that black South Africans refused to pay for services due to a 'culture of entitlement' (World Bank, 1994, p. 44). However, the application of these criticisms to South Africa's poor failed to account for the issue of affordability. In a 2003 study, the number one reason given for non-payment of bills was 'incapacity to pay due to insufficient income' (Smith and Hanson, 2003, pp. 1540–1).

Somewhere between 25 and 40% of South Africans are unemployed (Statistics South Africa, 2013). The rate of unemployment is also considerably higher among black South Africans (Hazelhurst, 2012), with some townships, like Soweto, having rates of 53% and higher (City of Johannesburg, 2008, p. 85; Future Directions International, 2012, p. 3). In a 2000 study of household affordability (Harvey, R., 2003, p. 7), 90% of low-income households gave unemployment, no income or too low an income as the main reason for the non-payment of water tariffs. Although providers were required to make arrangements with households to manage their repayments, many of these arrangements left households permanently in debt, while others imposed unrealistic payment schedules and then imposed disconnections or more serious action such as eviction proceedings when they were unmet (Xali, 2002, pp. 101, 113; Smith and Hanson, 2003, p. 1533).

Water disconnections are permissible under the 2003 Strategic Framework for Water Services, but are only intended as a last resort (DWA, 2003, para. 4.5.8). Additionally, s. 4(3) of the Water Services Act 1997 states that procedures for the limitation or discontinuation of water services must not result in a person who is unable to pay being denied access to basic water services. In practice, however, researchers found that water service disconnections were the most common response to a poor household's failure to pay their water bills, even when this failure was a clear result of an inability to pay (Tissington *et al.*, 2008, pp. 50–3). This uncompromising approach to disconnections was a direct result of the preoccupation with cost-recovery imposed on municipalities by national government policies (Bond and Dugard, 2008a, pp. 16–21).

In 1994, when the ANC first came to power, 12 million South Africans did not have access to clean drinking water (ANC, 1994). While the ANC was able to provide an additional seven million people with access to water by 2002, the imposition of disconnection as a means of credit control meant that at least half that number (if not many more) had their water cut off for non-payment (Bond and Dugard, 2008a, p. 17; Dugard, 2010, p. 178).[2]

These early policies produced a number of negative effects, including mass service delivery protests, but the most significant issue was the impact on public health. The corporatization of water services, as recommended under a market-driven approach, involves 'ring-fencing' the budgets of water utilities, meaning that they are separated from that of other government departments and cannot be subsidized from other budgets or have their profits diverted for other purposes (Cortez *et al.*, 2010, p. 1). This artificial boundary can serve to mask the inherently interconnected nature of the different jurisdictions of public service provision, and creates the potential for irrational externalities, particularly in relation to health care (Bond and Dugard, 2008, pp. 7–8).

In South Africa's case, households that could not afford to purchase sufficient water, or whose water services were cut off for non-payment, were forced to use water from untreated natural sources like streams, ponds or muddy puddles (Hemson, 2006). Often these natural sources were polluted and unsafe, and households experienced negative health consequences as a result of consuming contaminated water (Deedat and Cottle, 2002). One consequence was cholera, which saw an upsurge in cases after the introduction of cost recovery and credit control policies (Hemson 2006; Bond and Dugard 2008b, pp. 7–8). In one area of KwaZulu-Natal, a cholera outbreak became one of the worst in Africa, spreading across the country and infecting 117,147 people and killing 265 of them (Hemson, 2006, p. 4). Ultimately it was this cholera outbreak that demonstrated the need to ensure economic, as well as physical, access to water, and forced authorities to reconsider their approach (Muller, 2008, p. 72).

Free Basic Water

Despite its early adoption of a marginal cost-recovery approach for all water services, by 2001 the National Government realized that, in the context of the apartheid legacies of widespread unemployment, geographic clusters of poverty and low levels of education, affordability for the poor meant that a lifeline amount needed to be given away for free (Muller, 2008). The Government's vision for a *lifeline tariff* to provide free basic water (DWA, 1997b, para. 6.5.3) was realized in February 2001 when then President Mbeki announced a new national policy of FBW (DWA, 2001). The FBW allocation was set at six kilolitres per household per month and was designed to provide a household of eight people with 25 Lpcpd (DWA, 2002, para3.3).

The need for water affordability has widespread support among international water specialists (see, e.g. Winpenny, 2003, pp. 18–19), but providing free water is more contentious (Foster *et al.*, 2000, p. 4; Cardone and Fonseca, 2003, pp. 31, 47, 88). Muller (2008, p. 67) argues that the FBW policy was controversial because it challenged the conventional wisdom that water being an economic good should be sold and not given away. In this respect, the introduction of a free lifeline amount was a progressive policy in that it acknowledged affordability as a central element in South Africa's constitutionally enshrined right to water. It could also be read as an acknowledgement that water governance crosses over into other jurisdictions, such as health, and must be viewed holistically as a de-commodified 'public good'.

Due to the trans-jurisdictional division of responsibilities for water governance, this new FBW policy placed a new fiscal burden on the municipal governments that were tasked with providing free water. This was acknowledged by the National Government, along with the need to provide more support (both technical and financial) to poorer municipalities (DWA, 2002, pp. 37–45). However, the National Government only partly acted on this acknowledgement through a small increase in the equitable share (national funding that is directed primarily to poor municipalities) (DWA, 2002, pp. 17–18), meaning that the implementation of the FBW policy has been compromised by inadequate national funding (Gowlland-Gualtieri, 2007, p. 6), which has led to a slow roll-out in many areas and the adoption of a number of regressive policies (Hemson, 2006, p. 7; DWA, 2012).

In response to funding constraints, most municipalities restrict access to FBW to those who are registered as indigent (DPLG, 2005, para. 2.5). This is problematic, because the various indigent registers are highly under-representative of the true numbers of poor households in South Africa (Tissington *et al.*, 2008, p. 34). There are no national criteria for indigent status (DPLG, 2005, para. 2.7), but the criteria set by each municipality are typically onerous and exclusionary (Department of Provision and Local Government (DPLG), 2005, para. 2.7; Calfucoy *et al.*, 2009, p. 9). Requirements of proof of income and account-holder status exclude many poor tenants, all informal occupiers and many undocumented poor people, creating a boundary to access for many of the most needy (DPLG, 2005, para. 2.2).

A second regressive feature of the FBW policy is the inadequacy of the lifeline amount, which provides for 25 Lpcpd in households of eight people, but less in the larger households that are common among South Africa's poor black communities. While around 20–25 Lpcpd is sufficient for people to have enough water for consumption, it is inadequate for them to maintain personal hygiene, to wash their clothes, clean their homes or maintain adequate sanitation (Howard and Bartram, 2003, pp. 22–3). In order to undertake these activities and lead a healthy life, a minimum of 50–70 Lpcpd is required – and more for people with higher than average needs (Howard and Bartram, 2003, pp. 22–4; Reed, 2005). To put these figures into context, it takes around 12.5 litres to flush an ordinary toilet and 200 to run a bath.

The lack of national funding provided to support the roll-out of FBW has also meant that municipalities, who have the jurisdictional responsibility to deliver water services, have had to cross-subsidize from poor communities themselves – meaning that they have imposed significant tariff increases in the second block of consumption (above the FBW amount) in order to recover the cost of providing the first six kilolitres free (Bond and Dugard 2008a, p. 21). In some areas, such as Durban, this has resulted in water tariffs actually doubling for poor consumers after the introduction of FBW (Bond and Dugard 2008a, pp. 20–1). As a result, affordability remains a significant barrier to water access.

In South Africa's first Constitutional Court case on the right to water, the applicants in *Mazibuko v. City of Johannesburg* argued that the definition of 'sufficient water' under s. 27(1)(b) of the Constitution should be the amount of water that is necessary for dignified human existence, which for those households with waterborne sanitation (flush toilets) is at least 50 Lpcpd. As the applicants pointed out, context is highly relevant to the determination of sufficiency. For example, unlike in many rural households, urban users do not have access to alternative sources of water for washing and sanitation, which increases the amount of water they need to obtain from improved sources (Gleick, 2005, para.22). There is also a range of factors that affect individual requirements for water, including membership of high-needs population groups, including young children, pregnant and lactating women, the elderly and the terminally ill (Howard and Bartram, 2003, pp. 5–6).

Muller (2007, p. 69) (former Director-General of DWA), however, has defended the low allocation of FBW, describing the policy as a demand management device that was necessary to prevent the poor from adopting the 'water-intensive suburban lifestyle of South Africa's minority population at the upper end of the household income scale'. He has also rejected arguments for a larger FBW allocation because this would provide 'more water than is required for basic health needs and . . . [benefit] predominately urban consumers rather than the poorer rural communities, which have limited infrastructure capacity' (Muller, 2007, p. 76). These arguments seem to be an attempt to justify several divisions or 'jurisdictions' within South Africa's water governance policies – the continuing divide between the wealthy, well-serviced white population and poorer black communities, along with a new division between the 'deserving' rural poor, for whom funding must be set aside, and the 'undeserving' urban poor, who ought to be grateful for what little they have.

The central weakness of Muller's argument lies in the fact that more than 25 Lpcpd is required to meet basic needs, especially for urban users due to the absence of alternative sources of water and the fact that their houses often have waterborne sanitation. Differentiating on the basis of need is not the same as privileging poor urban users. A further obvious weakness is the entrenchment of the ongoing privilege of 'water-intensive suburban lifestyle of South Africa's minority population at the upper end of the household income scale' and the refusal to consider policies that would enable genuine redistribution.

The final regressive feature of the FBW policy has been its relationship to punitive and discriminatory cost recovery and demand management measures (DWA, 2003, para. 4.5.8). These have included water restriction devices (a metal disc with a small hole in the centre, which reduces the diameter of the pipe and restricts the flow of water) (Loftus, 2004, p. 195), prepaid water meters (PPMs – devices that are programmed to dispense the monthly FBW amount and then to cut off the water supply unless additional water credit tokens are purchased) (*Mazibuko v. City of Johannesburg*) and the delivery of lower levels of water service, such as a yardpipe or standpipe to poor households rather than a household water connection (City of Johannesburg, 2008, s. 3).

All of these measures are designed to restrict poor households to the FBW allocation. This is a highly regressive policy, given that this allocation is inadequate for many households and highlights the problems with the trans-jurisdictional implementation of FBW. While the initial policy was designed to ensure access, the municipal implementation process has undermined the adequacy of that access.

The negative effects of these cost-recovery policies on poor households and, particularly, on uniformly poor municipalities, raise the question of whether a single national water governance approach for all jurisdictions is sufficiently responsive to local conditions. Trans-jurisdiction water governance needs to be more adaptable to local contexts, and it may be that South Africa's constitutional right to participation could play a positive role in this process.

Participatory governance

An alternative to the market-driven water governance model imposed on municipalities through National Government policies is community participation in water governance. Participatory water governance has been shown to increase equity and access (Malzbender *et al.*, 2009, pp. 29–31). One theory as to why participation has this effect is that it empowers marginalized groups to voice their rights and needs (UNESCO, 2006, pp. 372, 389). Beyond just voicing these rights and needs, participation can also help marginalized sections of the community to identify themselves as rights-bearers, increasing their expectation of having their rights fulfilled and the likelihood that they will organize among themselves to demand responsiveness from government and other non-state actors (Osmani, 2008, p. 7). UNESCO (2006, p. 389) reports that this empowering effect of participation can assist in combating the tendency for policy and planning to reflect predominantly the needs of elite sections of the community. This process is particularly important in relation to gender equity, as recognized by the World Water Council, which has argued that 'democratic processes give women better opportunities to benefit equitably from the use of water resources and to take full part in decision-making' (Cosgrove and Rijsberman, 2000, p. 3).

Community participation can also serve to increase accountability and to bridge the divide between policy and implementation in water governance, but this requires transparency and access to accountability-enforcing mechanisms through which community members can assert power over more powerful actors. This can take the form of an effective regulator empowered to act in the public interest, or arbitration and judicial review processes. It is notable that the National Government has only recently assumed the regulatory role allocated to it under the constitutional division of water governance responsibilities (DWA, 2008). DWA has declared that its regulatory priorities will include social regulation, including affordability and access to services (DWA, 2008, p. 55). However, the Department's reporting thus far has been focused on drinking water quality and economic efficiency rather than on social issues (DWA, 2011a, b, 2013).

Despite the constitutional commitment to participatory governance, South Africa has struggled to develop a truly participatory approach to water governance within its trans-jurisdictional framework. Smith (2011a, p. 147) has suggested one reason for this is the tension between the cost recovery imperatives imposed by the National Government and the low capacity of many poor South Africans to pay for water services. This tension has created a climate in which municipalities have resorted to a paternalistic and authoritarian style of water governance, particularly in relation to the introduction of credit control and demand management devices (Tissington *et al.*, 2008, p. 72). This top-down approach to water governance has served to exclude the interests and experiences of the poor and marginalized, and to disempower them in the process (Tissington *et al.*, 2008, p. 69).

As discussed above, the Promotion of Administrative Justice Act 2000 (PAJA) does provide a range of accountability-enforcing mechanisms that are designed to empower the community to request that the government provide reasons for its decisions or reconsider them entirely. However, these mechanisms are only effective if they are adhered to and if the policy in question is deemed to be 'administrative' rather than 'executive' in character. In the *Mazibuko* case (paras 127–31), this distinction meant that the city's policy of imposing PPMs on a poor community in Soweto was not required to adhere to the procedural justice requirements of the PAJA because it was deemed to be an executive action.

Another accountability-enforcing mechanism in South Africa is the requirement under s. 4(3) of the Water Service Act 1997 that water disconnections will not be carried out without first providing an opportunity for households to challenge the decision or to provide evidence that they are unable to pay their water bill. This right to procedural justice has been undermined by the imposition of PPMs in some areas of South Africa, because they can discontinue a household's water service without providing any opportunity for that household to make representations regarding their (lack of) capacity to afford additional water.

Considering the challenges that many municipalities have faced in attempting to improve payment rates and the highly contested nature of their efforts to enforce debt collection, it is easy to see the appeal of this hands-off approach to credit control (McDonald and Pape, 2002; Smith, 2011a). When households are restricted by default to the FBW amount and there is a mechanical boundary between them and the service provider, municipalities do not have to spend any time or money on reading meters, billing or debt collection. However, this automatic disconnection serves to shield service providers and the government from the consequences of cost recovery on the poor (Bond and Dugard, 2008a). By closing off the important channel of communication provided by procedural justice rights, this technological approach also risks damaging the relationship between the service providers and the community, and reducing the community's perception of the legitimacy of rules of service and willingness to comply with them.

In theory, the recognition of constitutional rights to both participation and water should provide an additional mechanism for community members to ensure participation and accountability in water governance. However, South Africa's landmark water rights case, *Mazibuko*, was unsuccessful and resulted in a judgement that was highly deferential to the government's right to determine its own water policy priorities. Furthermore, the Court (paras 122–3) specifically considered the issue of PPMs and found that they 'suspended' rather than 'disconnected' water services, thus rendering them immune from the procedural justice requirements contained in s. 4(3) of the Water Services Act 1997.

Nonetheless, there have been some positive developments which indicate a promising new approach to participatory water governance in South Africa. These include the establishment of the *Raising the Citizens' Voices* programme, beginning with a 2006 pilot project in Cape Town, to support the training and empowerment of South Africans to strengthen the role of civil society in water governance and to increase the community's capacity to hold local government to account (Smith, 2009, p. 20).

Smith (2011b, pp. 504–5) reports that the *Raising the Citizens' Voices* initiative was developed in response to the way cost-recovery policies have commodified the relationship between municipalities and service users, and undermined efforts to put public participation mechanisms in place. The initiative seeks to change this approach by fostering active public participation in water governance (Smith, 2011a). Smith (2011a, pp. 509–11) reports that although it has experienced many challenges, including ongoing mistrust between municipalities and the more radical civil society organizations, the initiative has resulted in several positive outcomes. In eThekwini Municipality (Durban) the initiative empowered ward councillors to 'feel better equipped to discuss water service issues when representing their constituencies at ward committee meetings', and provided citizens and civil society representatives with access to 'user platforms', which successfully focused on addressing structural problems with water services (Smith, 2011a, p. 511). In Cape Town, the initiative has also helped to increase communication between the community and the Municipality, and to 'increase community-based capabilities and confidence to engage with councils on individual service delivery concerns and to better articulate structural issues that required higher level lobbying for resolution' (Smith, 2011a' p. 508).

Several South African municipalities have also taken part in the international *Water Dialogues* and, by doing so, opened up their water policies to a multi-stakeholder review process (Galvin, 2009). Both of these initiatives recognize the centrality of participation to improving water governance and equity in South Africa – of bridging the jurisdictional boundary between municipalities and their communities. Both have also demonstrated positive results and the constructive power of meaningful dialogue and participation (Smith, 2011a, pp. 509–11; interview with Kathy Eales cited in Chapman and Mancini, 2009, p. 20), which is promising for the future of water governance in South Africa.

Conclusions

The simultaneous adoption of rights to participation and water in the 1996 Constitution was not a panacea to the entrenched inequality of power and wealth in South Africa society. Indeed, although the devolution of water governance offered an opportunity to bridge boundaries of race and class, one of the effects of the decentralization of water service delivery in South Africa's first decade of democracy was the continued polarization of service levels between wealthy and poor municipalities, and between wealthy and poor communities.

Nonetheless, as this chapter has discussed, the recognition and implementation of the constitutional rights to water and participation are having a positive, albeit gradual, effect on water governance in South Africa. The shift from a national policy of marginal cost recovery to the adoption of FBW is one example of this effect. While the details of the policy do require improvement, it has been a positive first step in increasing access and equality, and a growing number of municipalities have begun adopting more generous allocations (Szabo, 2009, pp. 4, 12, 25).

Recent innovations in participatory water governance, including the *Raising the Citizens' Voices* programme, have also begun to demonstrate promising results (Smith, 2009, 2011a). This is a particularly important development, as participation can enable the poor and marginalized (and their representatives in civil society) to claim their own water governance jurisdictions and to promote policies and processes that better reflect their experiences and better respond to their needs. It can also promote community ownership over the water governance process, potentially enhancing the process of trans-jurisdictional governance by increasing the accountability of the state and the responsible management of water by the community.

The main future challenge in South Africa will be for the National Government to adopt a stronger role in supporting those poorer municipalities that are currently being left behind under the trans-jurisdictional governance arrangement. It is these municipalities that still lack the technical and financial capacity to expand the FBW allocation and to adopt a more participatory style of water governance. This is the current weakness of South Africa's trans-jurisdictional approach to water governance and, without National Government intervention, stark inequalities in access to water services will continue to be a reality in South Africa.

Notes

1 For further discussion of the negative effects of austerity measures see Blanchard, O. and Leigh, D. (2013) 'Growth Forecast Errors and Fiscal Multipliers', Working Paper 13/1, International Monetary Fund, Washington, DC.

2 Precise data on water disconnections in South Africa are not possible to obtain, but research supports a figure of between 1 and 1.5 million disconnections per year.

References

African National Congress (ANC) (1994) 'Reconstruction and Development Program', www.anc.org.za/show.php?id=235, accessed 1 September 2015.

Bond, P. (2000) *Elite Transition: From Apartheid to Neoliberalism in South Africa*, 2nd edn, Pluto Press, London.

Bond, P. (2010) 'South Africa's 'rights culture' of water consumption: Breaking out of the liberal box and into the commons?', in B. Johnston, L. Hiwasaki, I. Klaver and V. Strang (eds) *Water, Cultural Diversity & Global Environmental Change: Emerging Trends, Sustainable Futures?*, UNESCO, Paris and Springer, Dordrecht, Netherlands.

Bond, P. and Dugard, J. (2008a) 'The Case of Johannesburg Water: what really happened at the pre-paid parish pump', *Law, Democracy and Development*, 12, 1, 1–28.

Bond, P. and Dugard, J. (2008b) 'Water, human rights and social conflict: South African experiences', *Law, Social Justice and Global Development*, 10, 1, www.go.warwick.ac.uk/elj/lgd/2008_1/bond_dugard, accessed 3 June 2010.

Brand, D. (2008) 'Writing the law democratically: A reply to Theunis Roux', in S. Woolman and M. Bishop (eds) *Constitutional Conversations*, University Law Press, Pretoria, South Africa.

Calfucoy, P., Cibulka, J., Davison, J., Hinds, T. and Park, M. (2009) *Improving Free Basic Water Provision in South Africa – Prepared for the Financial and Fiscal Commission*, La Follette School of Public Affairs, University of Wisconsin-Madison, Madison, WI.

Cardone, R. and Fonseca, C. (2003) *Financing and Cost Recovery*, International Water and Sanitation Centre, Den Haag, Netherlands.

Chapman, J. and Mancini, A. (2009) *The Story of an International Multistakeholder Process*, The Water Dialogues, London.

City of Johannesburg Metropolitan Muncipality (2008) 'Metropolitan Municipality Water Services By-Laws', Gauteng Provincial Gazette extranordinary no. 162, City of Johannesburg Metropolitan Muncipality, Johannesburg, South Africa.

Coalition Against Water Privatisation (CAWP), Anti-Privatisation Forum (APF) and Public Citizen (2004) 'Nothing for Mahala: The forced installation of prepaid water meters in Stretford, Extension 4, Orange Farm, Johannesburg, South Africa', Research Report no 16, Centre for Civil Society, Glenwood, South Africa.

Constitution of the Republic of South Africa 1996 (South Africa).

Cortez, M., Orda, E., Elvas, L., Mendoza, A. M. and Maris Salas, S. (2010) 'Guide to Ring-Fencing of Local Government-Run Water Utilities', The World Bank, Washington, DC.

Cosgrove, W. and Rijsberman, F. (2000) *World Water Vision Making Water Everybody's Business*, Earthscan, London.
de Albuquerque, C. (2012) *Stigma and the realization of the human rights to water and sanitation*, 21st Session, United Nationas Human Rights Council, A/HRC/21/42, Geneva, Switzerland.
Deedat, H. and Cottle, E. (2002) 'Cost cecovery and prepaid water meters and the cholera outbreak in KwaZulu-Natal: A case study in Madlebe', in D. A. McDonald and J. Pape (eds) *Cost Recovery and the Crisis of Service Delivery in South Africa*, Zed Books, London, pp. 81–100.
Department of Finance (1996) *Growth, Employment and Redistribution: A macroeconomic strategy (GEAR)*, Department of Finance, Republic of South Africa, Pretoria, South Africa.
Department of Provision and Local Government (DPLG) (2005) *Guidelines for the Implementation of the National Indigent Policy by Municipalities*, DPLG, Pretoria, South Africa.
Department of Water Affairs and Forestry (DWA) (1997a) *Overview of Water Resource Availability and Utilisation in South Africa*, DWA, Pretoria, South Africa.
DWA (1997b) *White Paper on National Water Policy for South Africa*, DWA, Pretoria, South Africa.
DWA (1999) *A Pricing Strategy for Raw Water Use Charges*, DWA, Pretoria, South Africa.
DWA (2001) *Free Basic Water Implementation Strategy (Version One)*, DWA, Pretoria, South Africa.
DWA (2002) *Free Basic Water Implementation Strategy (Version Two)*, DWA, Pretoria, South Africa.
DWA (2003) *Strategic framework for water services*, DWA, Pretoria, South Africa.
DWA (2008) *National Water Services Regulation Strategy*, DWA, Pretoria, South Africa.
DWA (2011a) *Blue Drop Report 2010*, DWA, Pretoria, South Africa.
DWA (2011b) *Blue Drop Handbook (Version 1)*, DWA, Pretoria, South Africa.
DWA (2012) *Free Basic Water*, DWA, Pretoria, South Africa, http://dwaf.gov.za/dir_ws/fbw/#, accessed 5 April 2012.
DWA (2013) *Blue Drop and No Drop Handbook 2014*, DWA, Pretoria, South Africa.
Dugard, J. (2010) 'Can human rights transcend the commercialization of water in South Africa? Soweto's legal fight for an equitable water policy', *Review of Radical Political Economics*, 42, 2, 175–94.
Dugard, J. and Mohlakoana, N. (2009) 'More work for women: A rights-based analysis of women's access to basic services in South Africa', *South African Journal on Human Rights*, 25, 3, 546–72.
Foster, V., Gomez-Lobo, A. and Halpern, J. (2000) *Designing Direct Subsidies for the Poor: A Water and Sanitation Case Study*, World Bank, Washington, DC.
Francis, R. (2005) 'Water justice in South Africa: Natural resources policy at the intersection of human rights, economics, and political power', *Georgetown International Environmental Law Review*, 18, 1, 149–202.
Future Directions International (2012) *Strategic Analysis Paper – Food and Water Security in South Africa: Present and Future*, Future Directions International, Perth, Australia.
Galvin, M. (2009) *Straight Talk to Strengthen Delivery in the Water Services Sector: The Water Dialogues – South Africa Synthesis Report*, The Water Dialogues, London.
Gleick, P. (2005) 'Supporting affidavit in *Mazibuko v. City of Johanesburg*', High Court of South Africa, Witwatersrand Local Division.

Goldberg, K. (2009) 'The Water Dialogues: Cape Town Case Study', http://waterdialogues.org/south-africa/documents/CapeTownCaseStudy-FullReport.pdf, accessed 7 May 2011.

Gowlland-Gualtieri, A. (2007) 'South Africa's Water Law and Policy Framework: Implications for the Right to Water', IELRC Working Paper 2007–03, International Environmental Law Research Centre, Geneva.

Gray, J. (2013) 'Dollars and dreams: legal aspirations and report cards in the Murray Darling Basin in Australia', in L. Westra, P. Taylor and A. Michelot (eds) *Confronting Ecological and Economic Collapse*, Earthscan, Abingdon, UK, pp. 156–71.

Gutierrez, E., Calaguas, B., Green, J. and Roaf, V. (2003) *New Rules, New Roles: Does PSP Benefit the Poor?: Synthesis Report*, WaterAid, London.

Harvey, E. (2003) 'Researching the Public's Perceptions, Views and Concerns about the Planned Installation of Pre-Paid Water Meters in Soweto, South Africa', paper presented at an African Studies Association Conference, Boston, MA.

Harvey, R. (2003) *The Fall of Apartheid: The Inside Story from Smuts to Mbeki*, 2nd edn, Palgrave Macmillan, London and New York.

Hazelhurst, E. (2012) 'SA unemployment gets worse', *Independent Online*, 31 October, http://iol.co.za/business/news/sa-unemployment-gets-worse-1.1414454#.VfULNfmqpBc, accessed 24 June 2015.

Hemson, D. (2006) 'Still Paying the Price: Revisiting the Cholera Epidemic of 2000–2001 in South Africa', Occasional paper no 10, Municipal Services Project, Kingston, Canada.

Howard, G. and Bartram, J. (2003) *Domestic Water Quantity, Service Level and Health*, World Health Organisation, Geneva.

Kasrils, R. (2001) 'The value and price of water (the women of Lutsheko)', *Water Science and Technology*, 43, 4, 51–5.

Klasen, S. (2002) 'Social, economic, and environmental limits for the newly enfranchised in South Africa?', *Economic Development and Cultural Change*, 50, 3, 607–42.

Liebenberg, S. (2010) *Socio-Economic Rights: Adjudication under a Transformative Constitution*, Juta, Claremont, South Africa.

Loftus, A. (2004) '"Free water" as commodity: The paradoxes of Durban's water service transformations', in D. A. McDonald and G. Ruiters (eds) *The Age of Commodity: Water privatization in Southern Africa*, Earthscan, London, pp. 189–203.

McDonald, D. A. (2002) 'The Bell Tolls for Thee: Cost Recovery, Cutoffs, and the Affordability of Municipal Services in South Africa', in D. A. McDonald and J. Pape (eds) *Cost Recovery and the Crisis of Service Delivery in South Africa*, HSRC Press, Pretoria, South Africa, pp. 161–82.

McDonald, D. A. and Pape, J. (eds) (2002) *Cost Recovery and the Crisis of Service Delivery in South Africa*, HSRC Press, Pretoria, South Africa.

Malzbender D., Earle, A., Deedat, H., Hollingworth, H. and Mokorosi, P. (2009) *Review of Regulatory Aspects of the Water Services Sector – Report to the Water Research Commission*, African Centre for Water Research, Cape Town, South Africa.

Mandela, N. (1991) 'Address: On the occasion of the ANC's Bill of Rights conference', in *A Bill of Rights for a Democratic South Africa: Papers and Report of a Conference Convened by the ANC Constitutional Committee*, University of Western Cape, Belleville, South Africa.

Mazibuko v. City of Johannesburg CCT 39/09 [2009] ZACC 28.

Minister of Health v. Treatment Action Campaign (2002) 5 SA 721 (ZACC).

Muller, M. (Former Director-General of the Deparmtne ot Water Affairs and Forestry) (2006) Interview, 4 October.

Muller, M. (2007) 'Parish pump politics: the politics of water supply in South Africa', *Progress in Development Studies*, 7, 1, 33–45.

Muller, M. (2008) 'Free basic water – a sustainable instrument for a sustainable future in South Africa', *Environment and Urbanization*, 20, 67–87.

Osmani, S. (2008) 'Participatory governance: An overview of issues and evidence', in United Nations Department of Economic and Social Affairs (UNDESA) (ed.), *Participatory Governance and the Millennium Development Goals (MDGs)*, UNDESA, New York, pp. 1–45.

Peet, R. (2002) 'Ideology, discourse and the geography of hegemony: from socialist to neoliberal development in post apartheid South Africa', *Antipode*, 34, 1, 54–84.

Promotion of Administrative Justice Act 2000 (South Africa) (PAJA).

Reed, B. J. (2005) 'Minimum water quantity needed for domestic uses', WHO technical note for emergencies no. 9, World Health Organisation, Geneva.

Roux, T. (2006) 'Democracy', in S. Woolman and M. Bishop (eds) *Constitutional Law of South Africa*, 2nd edn, Pretoria University Law Press, Pretoria, South Africa.

Schreiner, B., Mohapi, N. and Van Koppen, B. (2004) 'Washing away poverty, democracy and gendered poverty eradication in South Africa', *Natural Resources Forum*, 28, 3, 171–8.

Smith, L. (2006) 'Neither Public nor private: unpacking the johannesburg water corporatisation model', Programme paper no. 27, United Nations Research Institute for Social Development, Geneva, Switzerland.

Smith, L. (2009) 'Raising citizens voice in the regulation of water services: building the foundations for local regulation in South Africa', Paper presented at the Fifth World Water Forum, Istanbul, Turkey.

Smith, L. (2011a) 'Conflict versus cooperation between the state and civil society: A water-demand management comparison between Cape Town and Johannesburg, South Africa', in B. Barraque (ed.) *Urban Water Conflicts*, Taylor & Francis, Paris, pp. 147–68.

Smith, L. (2011b) 'The limits to public participation in strengthening public accountability: a reflection on the "citizens voice" initiative in South Africa', *Journal of Asian and African Studies*, 46, 5, 504–17.

Smith, L. and Hanson, S. (2003) 'Access to Water for the urban poor in Cape Town: Where equity meets cost recovery', *Urban Studies*, 40, 8, 1517–48.

Smith, L. and Vawda, A. (2003) 'Citizen vs. customer: Different approaches to public participation in service delivery in Cape Town', *Urban Forum*, 14, 1, 26–52.

Statistics South Africa (2013) *Quarterly Labour Force Survey: Quarter 4 (October to December), 2012*, Statistics South Africa, Pretoria, South Africa.

Szabo, A. (2009) *The Value of Free Water – Analyzing South Africai's Free Basic Water Policy*, University of Minnesota, Minneapolis, MN.

Tissington, K., Dettmann, M., Langford, M., Dugard, J. and Conteh, S. (2008) *Water service Fault Lines: An Assessment of South Africa's Water and Sanitation Provision Across 15 Municipalities*, Centre for Applied Legal Studies (CALS), Centre on House Rights and Evictions (COHRE) and Norwegian Centre for Human Rights, Johannesburg, South Africa.

Tomlinson, R., Beauregard, R.A., Bremner, L. and Mancgu, X. (2003) 'The post-apartheid struggle for an integrated Johannesburg', in R. Tomlinson, R. A.

Beauregard, L. Bremner and X. Mancgu (eds) *Emerging Johannesburg: perspectives on the post-apartheid city*, Routledge, London, pp. 3–20.
United Nations Development Program (UNDP) (undated) 'National Human Development Reports for South Africa International. Human Development Indicators', http://hdrstats.undp.org/en/countries/profiles/ZAF.html, accessed 10 December 2012.
UNDP (1998) *Human Development Report – Consumption for human development*, UNDP, New York.
UNDP (2012) 'National Human Development Reports for South Africa International Human Development Indicators', UNDP, New York, http://hdrstats.undp.org/en/countries/profiles/ZAF.html, accessed 10 December 2012.
United Nations Economic Social Council (2002) 'General Comment No 15 on the Right to Water', 29th CESCR sess, Agenda item 3, UN Doc E/C/12/2002/11, United Nations Economic and Social Council, Geneva, Switzerland.
United Nations Educational, Scientific and Cultural Organisation (UNESCO) (2006) 'Chapter 11: Sharing Water', in UNESCO and WWAP (eds) *The United Nations World Water Development Report 2: Water Shared Responsibility*, UNESCO, Geneva, Switzerland, pp. 372–97.
Water and Sanitation Program (2011) *Water Supply and Sanitation in South Africa: Turning Finance into Services for 2015 and Beyond*, The World Bank, Nairobi, Kenya.
Water Services Act 1997 (South Africa).
Winpenny, J. (2003) *Financing Water for All: Report of the World Panel on Financing Water Infrastructure*, Global Water Partnership, World Water Council and Third World Water Forum, http://worldwatercouncil.org/fileadmin/world_water_council/documents_old/Library/Publications_and_reports/CamdessusSummary.pdf, accessed 24 June 2015/
World Bank, The (1994) *World Development Report 1994: Infrastructure for Development*, Oxford University Press, New York.
World Bank, The (2006) *World Development Report 2006: Equity and Development*, Oxford University Press, New York.
World Bank, The (2012) 'GINI index', http://data.worldbank.org/indicator/SI.POV.GINI, accessed 3 March 2012.
Xali, M. (2002) '"They are killing us alive": A case study of the impact of cost recovery on service provision in Makhaza Section, Khayelitsha', in A. David, D. A. McDonald and J. Pape (eds) *Cost Recovery and the Crisis of Service Delivery in South Africa*, Zed Books, London, pp. 101–22.
Zitha, D. 'Developing crisis number two', *The Weekender*, 17–18 May.

Part V
Conclusion

16 The future of trans-jurisdictional water law and governance

Janice Gray, Cameron Holley and Rosemary Rayfuse

Introduction

This book's starting premise, or hypothesis, was that achieving sustainable water outcomes requires a trans-jurisdictional approach to water law and governance. We defined this governance approach as including legal and policy arrangements for the conservation, management and use of water resources that cross either functional or geographic administrative borders and either seek to reinforce human-imposed borders, divisions and boundaries or to achieve coherence and institutional coordination both between and across different bodies of law and policy at the local, national, regional and international levels. We acknowledged that this broad concept of trans-jurisdictional water governance might gloss over theoretical and practical differences, particularly between international, national and local tiers. However, we offered this definition as a means of orienting contributions to the book around a structured, albeit limited, understanding and conceptual framework. While there was not necessarily agreement that trans-jurisdictional approaches would resolve all the problems of water governance, the book's chapters unpacked and interrogated different aspects of trans-jurisdictional water governance in order to test the hypothesis as to its utility. In doing so, as discussed in Chapter 1, contributors adopted a range of definitions and theoretical approaches to the concept of trans-jurisdictionality and identified a number of models – some more successful than others – for its practical implementation as a matter of law and policy.

In Chapter 1 we proposed four questions as the foundational architecture for the book and its chapters. Each chapter examined theory or case studies to respond to some or all of these questions and advance our understanding of trans-jurisdictional water governance. They did this by canvassing and probing international, national and sub-national water policy arenas and emerging water problems.

This concluding chapter steps back from these individual arenas and problems to synthesise key generalizable lessons and insights arising from the book. Returning to the four organizing questions set out in the introduction, we compare and contrast the core conclusions and insights from the chapters to offer some possible solutions. As we will see below, the diversity of contexts and issues examined in each chapter meant there were few definitive answers to these questions. However, there were a number of cross-cutting themes, each providing unique insights and implications for how we think about and respond to legal and governance designs for addressing modern trans-jurisdictional water challenges.

First, we turn to the question of why trans-jurisdictional water governance remains such an intractable problem. Second, we consider the key challenges and confounders to successful trans-jurisdictional water governance, before turning to the third issue of opportunities and successes for more effective trans-jurisdictional water governance practices. Fourth, despite having outlined many recommendations for policy and theory of trans-jurisdictional water governance arising from this research, we ask what still remains to be learned and tested regarding trans-jurisdictional water governance in practice and theory. We accordingly conclude by outlining where this research could lead.

Why does trans-jurisdictional water governance remain such an intractable problem?

A key organizing issue for the chapters was the perplexing question: why does trans-jurisdictional water governance remain such an intractable problem? Many of the chapters offered insights on this issue and at least three common causes of obstinacy can be identified. Not all causes arose in all chapters and nor were they given equal weight, but each was a core focus across a range of contexts and problems, suggesting they evidence fundamental obdurate elements of trans-jurisdictional water governance.

The fundamental mismatch between the nature of the problem and the nature of the response

First, and perhaps most obviously, many chapters highlighted the mismatch between water problems on the one hand and traditional governance thinking and arrangements on the other. The latter tended to solidify around established fixed legal boundaries that conflict with the more complex and interconnected social and ecological systems of the former. This concern echoed broader critiques discussed in ecological and wider environmental governance literatures about the ineffectiveness of government attempts to regulate human impacts on ecological systems (Odum, 1953; Holling, 1978). In the context of trans-jurisdictional water governance discussed in this book, Thompson's concept of 'fragmentation' (Ch. 7) offered a nice anchor for conceptualizing and diagnosing this perplexing feature across three levels: geographic fragmentation between governments with little hydrologic logic; hydrologic fragmentation between groundwater and surface-water regimes; and functional fragmentation among different regulatory tasks.

These points were echoed in Godden's chapter (Ch. 2), which acknowledged that governance problems were confounded by long-held divisions in water law, such as that between groundwater and surface water, and the continuation of separate regulatory spheres for freshwater and coastal marine areas. Such hard divides were further scrutinyzed in detail by McDonald and Foerster's analysis (Ch. 13) of coastal management in Australia, which confirmed the challenge of governing across different tiers of government and across regulatory regimes for different land uses or policy domains. Waschka and Gardner (Ch.11), who highlighted the polycentric features of Federal and State relations in water quality management, identified a similar challenge. Van Rijswick's analysis (Ch. 4) of European water law also served to remind us that insufficient integration between water law, regulations and measurements taken in other policy fields continues to hinder the achievement of good water governance.

Inadequate participatory frameworks

A second common (and overlapping) cause of intractability was the continued tendency for legal and regulatory arrangements to privilege statist institutions and lock out the knowledge and capacity of other actors. Yet, it was this very knowledge and capacity, it was argued, that can help to develop the kind of polycentric cross-jurisdictional governance systems necessary to manage water effectively. As with broader theoretical critiques on the limits of state capacity (see, e.g. Hayek, 2007), a number of chapters cast a critical gaze on the hegemony of the state in water governance thinking. Tan's chapter (Ch. 14), for example, highlighted how state-centric jurisdictional models adopted for wild rivers in northern Australia largely fell short because of a failure to consider broader principles of indigenous engagement and consent. Kidd (Ch. 9) similarly identified the authoritarian approach of South African municipalities as excluding the interests and experiences of the poor and marginalized, and undermining water governance models. Paisley *et al.* (Ch. 3) also pinpointed a relative dearth of public participation in the functioning of the Mekong River Commission as an issue of growing concern in Mekong governance. Meanwhile Gray's chapter (Ch. 12) highlighted how state-centric water governance models have sought to lock out the voices of resistance and protest in the related unconventional gas sector.

Certainly, there were positive signs of change toward approaches that were more inclusive of non-state knowledge and capacity. Butterly and Techera (Ch. 5), for instance, shone light upon recent participatory developments in the form of the UNDRIP and the WIN, while Ruru's chapter (Ch. 10) analysed the possibilities of reconciliation and public participatory decision making in New Zealand (see also suggestions made by Tan (Ch. 14)). Nelson and Casey (Ch. 8) and Thompson (Ch. 7) also illustrated more consultative approaches in the USA. Even so, as Godden pointed out, path dependencies can often limit such participatory opportunities and the extent to which decision making about water has been effectively relocated from the conventional 'top-down' models remains debatable. Even in examples where constitutional recognition of rights to participation and water exist, the theory seems to fall short. Such rights in South Africa are gradually having a positive effect on water governance. However, as Clark (Ch. 15) showed in her analysis of the landmark South African water rights case, *Mazibuko v City of Johannesburg* (2009), considerable latitude can still be provided to the government to determine its own water policy priorities.

Lack of clear theoretical and policy focus

The third and final cause of intractability was less a failure of state legal systems to embrace polycentricism and relevant governance scales, and more a matter of contestation and confusion within water governance theory and policy itself. This was a point tackled most directly by Godden, who noted that the concept of water governance and trans-jurisdictional water governance is often not articulated, even if it is alluded to in management principles.

Similar critiques of definitional disagreement in governance literature (Rhodes 2007; Drahos 2016) and water governance (Araral and Wang, 2013) are commonplace. And this lack of consensus was equally evident in this edited collection. Each chapter in this book adopted (but did not always specify) its own nuanced definition of trans-jurisdictionally and water governance, which varied among international, national and local levels. Such diversity is arguably appropriate at this stage of inquiry into trans-jurisdictional water governance, where there are few settled articulations and models precisely identifying the content and function of water governance at national, regional and international levels (see Godden, Ch. 2). However, such diversity also can limit diagnostic and predictive value of insights and ideas for advancing water governance (Araral and Wang, 2013).

Godden's chapter offered a path towards resolving some of these challenges in making the case that trans-jurisdictional water governance is often normative. The challenge, it was argued, lies in articulating fit-for-purpose 'architectures' to cover the diverse demands upon, and contexts in which, trans-jurisdictional water governance comes into being. Each chapter started to do this across a diversity of water problems and policy arenas. In what remains of this conclusion, we will summarize some common core challenges and responses of trans-jurisdictional water governance, and in doing so we sketch out normative architectures that may be generalizable across a number of contexts.

What are the key challenges to successful trans-jurisdictional water law and policy?

The multitude of water problems and legal contexts covered in this book revealed a diverse range of challenges, inhibitors and confounders to successful trans-jurisdictional water law and policy. Some of these overlap with the broader factors of intractability noted above, particularly issues of fragmentation (which can cause problems such as conflicts, distributional disagreements and ineffective management, see, e.g. Thompson (Ch. 7) and Van Rijswick

(Ch. 4)) and state-centrism (that can deliver outcomes unsatisfactory to affected populations and interests – see, e.g. Tan (Ch. 14) and Butterly and Techera (Ch. 5)). However, a range of other specific factors can be unpacked from the rich work completed across the book, and below we highlight four confounders common across a number of contexts.

Lack of political will

The first was that trans-jurisdictional water problems remained in the 'too hard' basket for many national governments, which were reticent to take the necessary steps to implement successful trans-jurisdictional water governance. Kidd's analysis (Ch. 9) of the South African Constitution and legislative framework illustrated this point nicely. He argued it was often a lack of political will, rather than the absence of adequate law, that undermined a satisfactory response to water problems. Gray (Ch. 12) also raised the issue of a lack of political will. She observed that a form of grassroots-level governance had begun to emerge in response to growing frustration over how coal seam gas mining and the related field of water were conventionally being governed. As governments and legislatures shied away from taking a clear stance on coal seam gas governance (and its related effects on water governance), protest movements stepped in to set new directions and fashion alternative governance responses. Waschka and Gardner (Ch. 13) also identified a similar challenge in their analysis of diffuse water pollution, noting that despite the existence of national strategies and proposed models of reform in Australia, governments were unwilling to pursue necessary regulatory and compliance measures. In a similar vein, Thompson (Ch. 7) argued that a generally taciturn US Congress had limited direct intervention in the management of politically heated interstate waterways. While reticence to implement water laws was thus an evident problem, McDonald and Foerster (Ch. 13) identified a different sort of government 'foot dragging'. In their analysis of Integrated Coastal Zone Management, they contended that despite issues of regulatory fragmentation and the need for coordinated national policy leadership consistently featuring in inquiries into coastal zone management over many decades, successive national governments simply failed to act at all, or to provide clear statutory and policy guidance. Kidd (Ch. 9) perhaps mostly aptly sums up this seemingly pervasive challenge: '. . . water is a critical resource, yet one often gains the impression that it is regarded as somewhat peripheral by government instead of a crucial policy issue'.

Inter- and intra-state rivalries

Even where governments saw water as a crucial policy issue, a second major challenge for trans-jurisdictional water law and policy was the rivalries and conflicts between different nation states and tiers of government. Perhaps not surprisingly, this was prominent at the international level, where achieving effective cooperation between nation states remains an ongoing challenge

(McCaffrey, 2007; Boisson de Chazournes, 2013). As Paisley *et al.* (Ch. 3) point out in their analysis of the Mekong, '. . . the most notable challenge [in managing international drainage basins] is the tendency for regional politics to exacerbate the already difficult task of understanding and managing complex natural systems' (see also Martin and Kennedy, Ch. 6). Similar conflicts at the intra-state level were also evidenced across many chapters, and had many causes, including a desire to maintain jurisdictional autonomy (Nelson and Casey, Ch. 8), as well as political allegiances and vested economic interests (Thompson, Ch. 7). Examples of conflicts between governments were also highlighted in Kidd's analysis of water services in South Africa, Gray's investigation of water quality and quantity governance in the coal seam gas-mining context in Australia, and Waschka and Gardner's examination of diffuse pollution in Australia. In all cases, rivalries (and the absence of coordination) between multiple governments (and often multiple agencies) were found to compound the challenge of successful trans-jurisdictional water law and policy. Martin and Kennedy's chapter (Ch. 6) perhaps provides the most detailed analysis of both intra- and interstate conflicts that arise in places such as Australia, China, the United States, the Andes, Patagonia, Iceland, South Africa and the Mekong. Their chapter also extended this inter- and-intra state understanding, to identify not just state-focused rivalries, but also those between corporations and NGOs that can undermine trans-jurisdictional water law and policy. As Martin and Kennedy (Ch. 6) note: '. . . legal arrangements which do not adequately contemplate [water] conflicts, or which marginalise actors in these conflicts, are unlikely to be socially optimal or robust over time'.

Lack of legal compulsion

Conflict, it seems, is a likely challenge for trans-jurisdictional law and policy regardless of which level or which actor is involved in the water problem. The result is often slow progress on achieving agreed, yet enforceable, modes of water governance across political dimensions (see, e.g. Godden, Ch. 2). Indeed, a third major challenge, arising in part from the above rivalries, is that many measures of trans-jurisdictional water law and policy remain voluntary and, as a result, are often insufficient (c.f. Butterly and Techera (Ch. 5), who noted weaknesses in this binding/non-binding debate).

Reliance on voluntary measures alone is often patently unsuccessful in delivering water outcomes. Waschka and Gardner (Ch. 11) illustrated these failures in areas such as diffuse water pollution from agriculture, where water quality continued to decline despite significant promotion of voluntary efforts. Paisley *et al.*'s (Ch. 3) analysis of the Mekong Agreement also highlighted how the unclear legal status of periodic assessments of the Mekong Basin Plan weakened their influence on state behaviour. Nelson and Casey (Ch. 8) similarly cautioned that the voluntary nature of many planning approaches meant that collaboration and commitment remained fundamental to the successful management of groundwater.

Inadequate funding

A fourth and final challenge raised in the chapters related to the funding and financial capacities needed to facilitate effective trans-jurisdictional water law and policy. Often typified by its absence and inadequacies, calls for more funding are commonplace in most water law and environmental contexts. However, there is arguably a special need for more funding in trans-jurisdictional contexts as the type of coordinated efforts needed can be very expensive, and often come on top of existing responsibilities (Holley *et al.*, 2012). Thompson (Ch. 7) nicely illustrated this point in his analysis of US contexts, finding that Congress often inadequately funded commissions and thus constrained the ability to deliver effective comprehensive management. McDonald and Foerster (Ch. 13) made a similar point in the context of Integrated Coastal Zone Management, noting that although federal, state and territory governments had given high-level endorsement to a framework for national cooperation, the framework had never been properly funded and its effectiveness had suffered as a result.

The adverse effects of inadequate funding expose a crucial contradiction evident in trans-jurisdictional water governance. Governments and other actors can find such governance approaches attractive in part because of their perceived promise to reduce expenditure by engaging other actors and their resources to do more together.

However, this preoccupation with minimizing expenditure often results in trans-jurisdictional governance being denied the very resources that are essential to its success (Holley *et al.*, 2012). Clark's chapter (Ch. 15) on water services in South Africa starkly captured this problem: '. . . decentralized delivery ought to bring government decision-making processes closer to the local community. However, the trans-jurisdictional nature of revenue-raising capacity in South Africa meant that decentralization was comprised by the National Government's decision to simultaneously withdraw financial and technical support from the water sector'. Kidd (Ch. 9) points to similar issues, noting that huge amounts of money have to be spent on upgrading or replacing infrastructure and hiring appropriately qualified technical personnel to address problems, but this is money most municipalities do not have, nor are there prospects for them to be able to raise the necessary funds.

What opportunities, successes and learning experiences can be gleaned from existing trans-jurisdictional governance practices?

Despite the challenges documented across the chapters, there were also many experiences gleaned from these challenges as to how trans-jurisdictional water governance could be done better. A number of these recommendations have been covered in the specific chapters. Here we take a step back to identify three sets of lessons shared across a number of the examples discussed in the book's chapters.

The need for adequate funding, flexible conflict management mechanisms and strong policy vision and leadership

The first set of lessons for more successful trans-jurisdictional water governance follows from the above common challenges we have identified. For example, overcoming the challenge of financial capacities needed to facilitate trans-jurisdictional governance suggests that a key lesson on success is to ensure sufficient funding to support these efforts. Indeed, despite what appears to be a consistent problem of underfunding across many contexts, it is important to note that there are some examples where sufficient funding has been a key success factor, such as international aid in the governance of the Mekong River Basin (see Paisley *et al.*, Ch. 3).

Overcoming the abovementioned challenges of rivalry will likely require a flexible conflict management mechanism (see discussion by Martin and Kennedy (Ch. 6), while Thompson (Ch. 7) reminded us of the need to avoid 'clunky, costly and time-consuming' conflict resolution experiences. Furthermore, utilizing sufficient legal powers (whether explicitly or in the background) appears a necessary mechanism to avoid the apparent pitfalls of voluntary trans-jurisdictional approaches.

Finally, and perhaps most problematically, finding ways to overcome the abovementioned resistance of national governments to implement trans-jurisdictional water law and policy remains difficult. Certainty a number of chapters highlighted how a strong role for national governments or intergovernmental organizations like the UNDP was central to the success of trans-jurisdictional water governance. Various arguments were made for the benefits provided by strong states, including steering processes and bringing unique support such as roles in technical studies, perceived impartiality and offering incentives. Nelson and Casey (Ch. 8) demonstrate many of these unique features through their analysis of the Federal Government in the United States. McDonald and Foerster's analysis (Ch. 13) appears consistent with this view, albeit that the absence of federal steering led them to advocate that a common national policy and strong policy leadership from the upper levels of government was critical to successful management and protection of coastal zones in Australia. The same idea is presented in relation to Indigenous marine conservation by Butterly and Techera (Ch. 5), who note that national governments will always have a role to play in creating the context for bottom-up approaches. Paisley *et al.* (Ch. 3) also offer a comparable international example, suggesting that neutral assistance provided by the UNDP was able to provide technical and financial assistance in formulating a new agreement in the Mekong.

Despite this list, it appears often that national governments need to be spurred to take action on trans-jurisdictional water governance. While this is an area that needs further research attention, one option may be for coalitions of interested actors (both lower-level government and non-government actors) to focus their attention on not only building responses from the 'bottom up', but also on actively working to engage national levels (e.g. policy entrepreneurs) to encourage national action. As Nelson and Casey suggest (Ch. 8), this may also involve minimizing political barriers to adopting the governance arrangements by being politically palatable.

The importance of information sharing

A second series of lessons coalesce around the importance of information sharing as a means to lubricate and facilitate the formation of cooperative relationships, effective decision making and informed action within trans-jurisdictional water governance. As Godden pointed out, information sharing can lay the groundwork for collaboration, as well as reduce innovation timelines and improve practice though mutual policy learning and sharing of best practice. Gray's chapter (Ch. 12) demonstrated what can happen when actors are locked out of information-sharing activities and processes. Importantly, she outlined how scientific information could be fed into law and policy in non-water areas to help direct better water outcomes.

A range of other chapters demonstrated how such benefits arise from information sharing among various stakeholders, including citizens, scientists, government, indigenous interests and water managers. Nelson and Casey (Ch. 8), for instance, neatly illustrated the benefits of collaborative science in the case of Washington and Idaho in the United States, where information sharing was useful for building a common scientific understanding of the trans-jurisdictional problems among a broad range of state and non-state actors, and formed the cross-institutional relationships for later cooperative management. Paisley *et al.* (Ch. 3) made a similar point about the successes of collaborative science and information sharing in the Mekong, noting that information exchange among the Lower Mekong River Basin States had been an important element in the lead-up to the Commission.

In the Australian Lake Eyre Basin community, Tan (Ch. 14) also identified the role of independent scientists in reshaping the policy debate. In this case, workshops were conducted in local communities which allowed locals to freely question scientists, in turn facilitating two-way conversations on local history and observations and building strategic alliances principally with scientists, conservationists and universities. Tan's chapter (see also Ruru, Ch. 10) also served to extend the information sharing to include greater sharing of Indigenous and Western practices, suggesting the need for a new umbrella structure based on clan and family that would be able to merge different practices.

Perhaps the best exemplar of this form of information sharing was highlighted by Butterly and Techera's analysis (Ch. 5) of the cross-jurisdictional WIN network approach to local marine management. As they explained, WIN was a knowledge-sharing network that has created dialogues and exchanges between land and sea managers and other organizations and has the potential to be a clearinghouse mechanism whereby best practice models of environmental governance can be shared and communicated across jurisdictions.

Matching organizational arrangements with the physical system

A third and final set of insights on achieving good trans-jurisdictional water governance relates to the importance of organizational arrangements matched to natural water systems via an integrated basin approach. Godden introduces this point in her chapter (Ch. 2), noting that using the river basin, watershed or catchment as an organizing principle may constitute better governance units than municipalities, regions, states or nations. Nevertheless, Godden cautions that although we have spent many years promoting integrated water management, disparate regimes remain for governing water separated across jurisdictions and geospatial and biodiversity regions. Indeed, recent literature suggests that even a watershed-focused approach may be insufficient to address water security and that the politics of scale matter more than previously acknowledged (Norman *et al.*, 2015).

Despite this challenge, there were signs of at least some level of success across the various problems and policy arenas discussed in the book. These included the whole-of-basin governance in the Lake Eyre Basin in Australia discussed by Tan (Ch. 14) where, although there were challenges, innovative collaborative structures allowed a diversity of Indigenous people to collaborate in governance processes. Van Rijswick (Ch. 4) made a similar point about the river basin approach of the Water Framework Directive, noting that the obligation to make appropriate administrative arrangements for every river basin district was key to its success.

Centralized water basin authorities, although not always typical, possibly represent the most successful examples. These included bodies such as the Mekong River Commission (Paisley *et al.*, Ch. 3), whose major strength was its ability to generate shared data and information regarding the Mekong. Nelson and Casey (Ch. 8) also singled out centralized regulation as a key element to successful trans-jurisdictional water governance approaches, noting, in particular, the Susquehanna Compact, where the Basin Committee continually strengthened its powers to regulate groundwater-using projects in a way that superseded state law. Thompson (Ch. 7) also identified the 'hard restructuring' of the Delaware River Basin Compact into an interstate partnership that successfully promoted holistic, basin-wide planning.

What does 'trans-jurisdictionality' suggest for further water governance research?

Throughout this book, numerous research questions were raised that remain important to rethinking and redesigning water governance. Some of these have already been discussed in the various chapters. Godden (Ch. 2) in particular has identified many areas where more remains to be learned and tested regarding trans-jurisdictional water governance. However, we have identified at least four other broad areas for further research that are discussed below.

Theorizing 'trans-jurisdictional water governance'

First, this book has taken a theoretical orientation that approaches water governance from the basis of trans-jurisdictionality rather than through other lenses such as trans-boundary law or other forms of 'joined-up' governance. We argued in the introduction that the trans-jurisdictionality lens offers a number of advantages to analysis, not the least of which is that it enables comparisons among national, sub-national and international contexts; and it is arguably able to recognize that water crosses not just physical state boundaries/jurisdictions, but also administrative, legal, social and political ones (Yu, 2011). Each chapter demonstrated the utility of using the conceptual 'tool' of trans-jurisdictionality to develop a structured understanding of the governance challenges and responses to water's unique nature (from the Mekong to local places and waterways in the unitary island country of New Zealand).

Nevertheless, some chapters also suggest that it is not yet possible to conclude that trans-jurisidictional governance will necessarily generate the desired outcomes. Godden (Ch. 2) observes that '. . . before seizing on new theories and tools, it will be important to examine the expectations of effective governance and to identify when and where trans-jurisdictional governance is the best response'. Although supporting a trans-jurisdictional approach, Gray (Ch. 12) similarly hints at the possibility that a greater, deeper and more all-encompassing change may be required to bring about desired outcomes, by suggesting that water governance regimes need 'to rely on mechanisms and tools which are fundamentally appropriate and effective', whatever they might be.

Thus, an important area for both theoretical and empirical study is to more closely examine and test some of the distinctions and differences among trans-jurisdictional governance and other theories about governing across boundaries, in order to determine which theories may better account for governing water descriptively or provide useful normative visions in view of the realities of practice.

Empirical studies of other manifestations and experiments in trans-jurisdictional water governance

Related to this issue is a second area for further research. This book has conducted an analysis of trans-jurisdictionality in the regions of South Africa, New Zealand, the USA, the European Union, the lower Mekong Basin states and Australia. Issues and developments in those jurisdictions highlighted many interesting trans-jurisdictional legal and governance issues including various state, cooperative and rights-based approaches for diffuse source pollution, the intersections between unconventional gas mining and water, Indigenous and non-Indigenous water relationships and water services. Furthermore, the analysis of multiple regions and issues has allowed the book to reflect on a significant variance in trans-jurisdictional water governance designs and contexts. Even so, research into other manifestations and experiments in trans-jurisdictional governance is still vital to test and confirm what the findings and implications drawn in the conclusion and individual chapters hold for other institutions and contexts.

Continuing analysis of existing attempts at trans-jurisdictional water governance

Third, there is a need for further research on many of the examples of trans-jurisdictional water governance, particularly some of the more novel examples discussed in the book, such as the WIN (discussed in Butterly and Techera, C. 5) and the Delta Stewardship Council (discussed in Thompson, Ch. 7). Both remain works in progress, and lessons on their long-term success can offer important insights into the success or otherwise of these novel trans-jurisdictional water governance structures. Future research into these and other cases could also reflect longitudinal aspects such as changes over time, and to what extent sustainable environmental and water outcomes are achieved.

Analysis of the effect of trans-temporal factors

A fourth and final area for research, which has not been explored in its entirety in this book or indeed elsewhere, is the influence or effect of trans-temporal factors (cumulative impacts, intergenerational equity, enhanced climatic risk) in trans-jurisdictional water governance. As Godden (Ch. 2) noted, change and dynamism in water's ecological cycle inherently transgresses imposed legal, political and social boundaries. Any comprehensive trans-jurisdictional model must therefore be oriented towards proactively seeking to reduce stresses to water systems in advance, rather than being reactive. This is particularly important, as McDonald and Foerster (Ch. 13) remind us, because such temporal challenges are only likely to continue to confound governance thinking due to growing dynamism in water systems caused by climate change.

Conclusion

In conclusion, each of these areas will require a robust and detailed agenda for trans-jurisdictional water law and governance research. It is an agenda that we hope many will join us in pursuing so as to achieve more sustainable water outcomes in the future.

References

Araral, E. and Wang, Y. (2013) 'Water governance 2.0: A review and second generation research agenda', *Water Resources Management*, 27, 11, 3945–57.

Boisson de Chazournes, L. (2013) *Fresh Water in International Law*, Oxford University Press, Oxford, UK.

Drahos, P., Taylor, V. and van der Heijden, J. (eds) (2016, forthcoming) *Regulation, Institutions and Networks*, Australian National University Press, Acton, Australia.

Holley, C., Gunningham, N. and Shearing, C. (2012) *The New Environmental Governance*, Earthscan, Abingdon, UK.

Mazibuko v. City of Johannesburg (2009) 28 ZACC.

Holling, C. S. (ed.) (1978) *Adaptive Environmental Assessment and Management*, Wiley, Chichester, UK.

McCaffrey, S. (2007) *The International Law of Watercourses*, 2nd edn, Oxford University Press, Oxford, UK.

Moynihan, R. (2014) 'The rising role of regional approaches in international water law: Lessons from the UNECE Water Regime and Himalayan Asia for Strengthening Trans-boundary Water Cooperation', *Review of European Community and International Environmental Law*, 23, 1, 43–58.

Norman, E., Cook, C., and Cohen, A. (eds.) (2015) *Negotiating Water Governance*, Ashgate, UK.

Odum, E. (1953) *Fundamentals of Ecology*, Saunders, Philadelphia, PA.

Rhodes, R. A. W. (2007) 'Understanding governance: ten years on', *Organization Studies*, 28, 8, 1243–64.

Yu, X. (2011) 'Trans-boundary Water Pollution Management: Lessons Learned from River Basin Management in China, Europe and the Netherlands', *Utrecht Law Review*, 7, 1, 188–203.

Index

Please note that page numbers relating to Notes contain the letter ‘n’ followed by note number

For Product Safety Concerns and Information please contact our EU representative GPSR@taylorandfrancis.com
Taylor & Francis Verlag GmbH, Kaufingerstraße 24, 80331 München, Germany

www.ingramcontent.com/pod-product-compliance
Ingram Content Group UK Ltd.
Pitfield, Milton Keynes, MK11 3LW, UK
UKHW020020200926
475702UK00014B/1983

* 9 7 8 1 1 3 8 3 6 4 0 4 2 *